Mental Health Interventions for Climate Emergency Displacement

Jeffrey Kurebwa
Bindura University of Science Education, Zimbabwe

IGI Global
Scientific Publishing
Publishing Tomorrow's Research Today

Vice President of Editorial Melissa Wagner
Director of Acquisitions Mikaela Felty
Director of Book Development Jocelynn Hessler
Production Manager Mike Brehm
Cover Design Jose Rosado

Published in the United States of America by
IGI Global Scientific Publishing
701 East Chocolate Avenue
Hershey, PA, 17033, USA
Tel: 717-533-8845
Fax: 717-533-7115
Website: https://www.igi-global.com E-mail: cust@igi-global.com

Library of Congress Cataloging-in-Publication Data

Names: Kurebwa, Jeffrey, 1982- editor
Title: Mental health interventions for climate emergency displacement / edited by Jeffrey Kurebwa.
Description: Hershey, PA : IGI Global Scientific Publishing, [2026] | Includes bibliographical references and index. | Summary: "This book aims to examine the mental health challenges faced by individuals displaced by climate emergencies"-- Provided by publisher.
Identifiers: LCCN 2025022515 (print) | LCCN 2025022516 (ebook) | ISBN 9798337332819 hardcover | ISBN 9798337332826 paperback | ISBN 9798337332833 ebook
Subjects: LCSH: Environmental refugees--Mental health | Emigration and immigration--Psychological aspects | Climatic changes--Psychological aspects
Classification: LCC RC451.4.R43 M455 2026 (print) | LCC RC451.4.R43 (ebook)
LC record available at https://lccn.loc.gov/2025022515
LC ebook record available at https://lccn.loc.gov/2025022516
British Cataloguing in Publication Data
A Cataloguing in Publication record for this book is available from the British Library.

Table of Contents

Detailed Table of Contents

stigma, and structural injustices. Lastly, it makes legislative recommendations to increase equal access to mental health care and fortify community networks.

This study explores the relationship between depression, anxiety, stress, religious coping, and quality of life among farmers. The study aims to shed light on the unique challenges faced by farmers and the impact of these challenges on their mental health and overall quality of life. A total of 130 farmers from the Malwa district of the Punjab province in India completed a questionnaire that included measures of depression, anxiety, stress, religious coping, and quality of life. The results indicate that depression, anxiety, and stress are significantly related to lower quality of life among farmers. These findings suggest that farmers face unique challenges that can have a significant impact on their mental health and quality of life, and that the use of religious coping strategies may be a helpful resource for improving their well-being.

Climate change is increasingly displacing populations worldwide, giving rise to climate-driven migration with profound mental health consequences. This chapter explores the global approaches to addressing mental health in the context of climate-induced displacement, highlighting the psychological burdens faced by migrants—including trauma, anxiety, depression, and loss of identity. Through case studies and comparative policy analysis, the chapter examines disparities in mental health care access, cultural sensitivity, and global governance frameworks. It advocates for integrated, rights-based, and culturally responsive mental health systems and proposes a comprehensive framework that embeds mental health into climate adaptation and migration policies. The chapter calls for urgent, collaborative action to address this growing global health challenge.

Preface

INTRODUCTION

The escalating climate emergency has emerged not only as a defining environmental challenge of our time but also as a profound public health crisis. While the physical health impacts of climate change have long been recognized, it is becoming increasingly clear that its psychological consequences demand equal attention. Climate-related displacement, in particular, has been linked to heightened risks of depression, post-traumatic stress disorder (PTSD), anxiety, complicated grief, and substance misuse, among other psychiatric conditions. The uncertainty, disruption, and loss that accompany forced displacement exacerbate mental health vulnerabilities by severing social ties, eroding community support systems, and limiting access to vital resources such as food, shelter, and healthcare. These dynamics underscore the urgent need for focused mental health interventions tailored to the experiences of displaced populations.

Vulnerability to climate-induced psychological distress is not evenly distributed. Children, adolescents, the elderly, pregnant women, individuals with chronic illnesses, and marginalized or racialized communities are disproportionately affected. Indigenous populations, in particular, face the compounded burden of ecological grief, cultural disruption, and systemic disempowerment. New terminologies such as *eco-anxiety*, *ecological grief*, and *climate emotions* have emerged in scholarly discourse to capture the complexity of these experiences, reflecting both the universality and cultural specificity of climate-related mental health impacts. Researchers and organizations have also begun identifying adaptive coping strategies, including approaches rooted in community trust, meaning-making, and "active hope," which reinforce resilience while avoiding the pathologization of natural emotional responses.

Despite the growing recognition of climate change's mental health impacts, significant gaps remain in evidence-based interventions and scalable support systems.

Many existing initiatives are fragmented, limited to early-stage pilots, and primarily concentrated in developed countries. Yet, as climate change intensifies, the need for sustainable and context-sensitive interventions grows more pressing, particularly for marginalized and displaced populations. A global consensus has identified the evaluation of clinical and non-clinical psychosocial interventions for appropriateness, feasibility, and scalability as a key research priority. This book contributes to that endeavor by systematically examining emerging strategies and highlighting pathways for building resilience in communities most affected by climate emergencies.

The objectives of this volume are fourfold: first, to examine the unique mental health challenges faced by individuals displaced by climate-related events; second, to advance evidence-based interventions that effectively support displaced populations; third, to explore strategies for fostering resilience through coping mechanisms, community engagement, and social support; and finally, to provide actionable recommendations for policymakers seeking to address the mental health dimensions of climate displacement.

This book is intended to serve as a resource for academics, students, government agencies, non-governmental organizations, and climate experts who are dedicated to advancing mental health support in the face of environmental disruption. By bringing together diverse perspectives and emerging research, we aim to strengthen the collective capacity to address the psychosocial dimensions of climate change and to ensure that mental health is recognized as an integral component of climate resilience and adaptation.

ORGANIZATION OF THE BOOK

Chapter 1: The Interconnection Between Climate Change and Mental Health and Well-Being

The opening chapter provides a narrative review of the multiple pathways through which climate change undermines mental health and well-being. It highlights how climate change operates as a risk amplifier, destabilizing socioeconomic, cultural, and environmental conditions that support psychological stability. With rising global temperatures, water scarcity, and extreme weather events, communities—particularly those already marginalized—face escalating stressors that contribute to depression, anxiety, and distress. Importantly, the chapter underscores how heightened awareness of climate threats and global inaction generates significant psychological strain, while also acknowledging that these same emotional reactions can motivate individuals and communities toward proactive climate engagement.

Chapter 2: The Role of Community Support in Addressing Mental Health Challenges from Climate Displacement

This chapter explores the vital role of community networks in mitigating the mental health impacts of climate displacement. Through a focus on conditions such as PTSD, anxiety, and depression, it emphasizes the importance of social capital—bonding, bridging, and linking—as a cornerstone of resilience and healing. The discussion highlights successful community-led and NGO-supported initiatives that integrate cultural sensitivity and local participation, while also pointing to barriers such as stigma, structural inequities, and limited services. By examining both best practices and persistent obstacles, the chapter offers recommendations for strengthening community support systems and ensuring equitable access to mental health care.

Chapter 3: Exploring the Relationship Between Depression, Anxiety, Stress, Religious Coping, and Quality of Life Among Farmers

The third chapter centers on farmers in Punjab's Malwa district, who face distinctive pressures exacerbated by environmental change. Drawing on survey data from 130 participants, it demonstrates a clear link between depression, anxiety, and stress with diminished quality of life. The chapter importantly highlights religious coping as a protective factor, suggesting that faith-based practices may offer meaningful psychological support to farming communities under duress. By framing these findings in the broader context of rural livelihoods and agricultural stress, the chapter contributes valuable insights into context-specific interventions for mental health and well-being.

Chapter 4: Global Approaches to Mental Health in the Era of Climate-Driven Migration

This chapter takes a global lens to examine the mental health consequences of climate-induced migration, including trauma, identity loss, and profound psychological burdens. Using case studies and comparative policy analysis, it highlights disparities in access to mental health care across regions, underscoring the need for culturally responsive and rights-based approaches. The chapter advocates for integrating psychosocial support into broader climate adaptation and migration policies, offering a comprehensive framework that reimagines global governance. In doing so, it emphasizes the urgency of collaborative, international solutions to address one of the most pressing humanitarian challenges of our time.

Chapter 5: Psychological Consequences of Climate-Induced Statelessness

The fifth chapter focuses on the profound psychological toll of climate-induced statelessness, particularly in coastal and island nations where rising seas erase homes, livelihoods, and even citizenship. It examines how loss of legal identity compounds trauma, intensifying risks of depression, PTSD, and identity crises. By illuminating the lived realities of displaced populations who confront the disappearance of both homeland and future prospects, the chapter situates mental health at the heart of the conversation on legal and humanitarian responses to statelessness.

Chapter 6: Technology-Enabled Psychosocial Interventions for Climate-Affected Populations

This chapter turns to the promise and challenges of digital mental health strategies for communities impacted by climate disasters. With billions living in climate-sensitive regions, scalable solutions are urgently needed. The chapter evaluates interventions such as tele-counseling, mental health apps, and online peer support, while critically addressing barriers related to infrastructure, digital literacy, stigma, and ethics. Through analysis of programs such as India's Tele-MANAS and Australia's MindSpot, it highlights how technology can supplement traditional care models, offering inclusive, culturally sensitive pathways to resilience in climate-affected populations.

Chapter 7: Evaluating Policies and Initiatives and Proposing Sustainable Solutions for the Mental Health of Child Laborers

Child labor, particularly among rag pickers in India, remains a persistent challenge deeply intertwined with poverty and social exclusion. This chapter reviews 122 scholarly sources to assess the mental health implications of child labor and evaluates policy responses to date. Findings reveal significant gaps between intended outcomes and actual impact, with mental health often overlooked as a central concern. The chapter calls for a multi-stakeholder strategy that combines government, civil society, and community engagement to address both the psychological vulnerabilities and the socioeconomic drivers of child labor.

Chapter 8: Climate Change and Women's Health: The Need for Gender-Specific Data and Policy Interventions

Women's health is uniquely shaped by climate risks due to intersecting biological, social, and economic vulnerabilities. This chapter highlights how food and water insecurity, maternal health risks, and limited healthcare access disproportionately impact women, particularly in marginalized communities. A key emphasis is placed on the lack of gender-disaggregated data, which hampers the development of effective, targeted policies. Drawing on case studies and literature analysis, the chapter advocates for integrating gender-sensitive approaches into climate research and policy, ensuring that women's health and resilience are prioritized in adaptation planning.

Chapter 9: Lost Childhoods: Climate Displacement and Child Marriages in the Indian Sundarbans

This chapter addresses the intersection of climate displacement, poverty, and child marriage in the Sundarbans region of India. In the wake of displacement and livelihood loss, families often resort to early marriage as a coping mechanism, particularly for girls. The chapter underscores how such practices not only strip children of their right to childhood but also create lasting mental health burdens for young brides. By framing child marriage as both a gendered humanitarian crisis and a mental health issue, the chapter highlights the urgent need for interventions that safeguard children's futures in climate-vulnerable regions.

CONCLUSION

The chapters in this volume collectively illuminate the urgent and complex intersection between climate change, displacement, and mental health. From the erosion of community networks and livelihoods to the deeply personal struggles of identity, grief, and trauma, the contributions underscore that the impacts of climate change reach far beyond the physical environment—they reverberate within the very core of human well-being. What emerges across these diverse perspectives is a shared recognition that mental health must no longer be treated as a secondary concern in climate discourse. Rather, it is an essential dimension of resilience, adaptation, and justice.

This book also highlights a critical duality: while climate change generates profound psychological burdens, it simultaneously catalyzes innovation, resilience, and collective agency. Whether through technology-enabled interventions, culturally grounded community support, or gender- and child-sensitive policy frameworks, the chapters demonstrate pathways toward more holistic and inclusive responses. They remind us that addressing mental health in the context of climate displacement requires more than clinical solutions—it demands systemic transformation, cross-sector collaboration, and policies that prioritize dignity, equity, and sustainability.

It is our hope that this volume not only advances scholarly understanding but also serves as a resource for policymakers, practitioners, educators, and advocates committed to shaping a future where climate resilience includes psychological well-being. By centering mental health in the conversation on climate adaptation, we can ensure that vulnerable populations are not merely protected from environmental threats but are also empowered to thrive amidst them.

Chapter 1
The Interconnection Between Climate Change and Mental Health and Well-Being

Jeffrey Kurebwa
https://orcid.org/0000-0002-8371-8055
Bindura University of Science Education, Zimbabwe

Thaddeus Mumba Mahela
University of Namibia, Namibia

ABSTRACT

This chapter presents a narrative review of the various pathways through which climate change influences mental health and well-being. Climate change serves as a risk amplifier by undermining the conditions that are essential for maintaining good mental health, which include socioeconomic, cultural, and environmental factors, as well as living and working environments. The disruptive effects of increasing global temperatures and extreme weather phenomena, such as heat waves or water scarcity, exacerbate the existing stressors faced by individuals and communities. This situation has harmful consequences for mental health, particularly for those groups that are already marginalized both within and between nations. The awareness of and experiences related to rising climate threats, combined with the inaction surrounding climate issues, can lead to significant psychological distress; however, intense emotional reactions can also inspire proactive climate engagement.

DOI: 10.4018/979-8-3373-3281-9.ch001

INTRODUCTION

Climate change and mental health are justifiably considered two of the most significant challenges confronting societies worldwide. Nevertheless, there has been a lack of adequate focus on the interplay between these two crises and their shared underlying causes. The assertion that 'the climate crisis is a health crisis' is acknowledged by global health authorities, including the World Health Organization (WHO) (2022). Reports from the Intergovernmental Panel on Climate Change (IPCC) (2022) have also indicated that rising global temperatures will lead to dire repercussions for human health. Although the detrimental impacts of climate change on physical health have been recognized for an extended period, the implications for mental health have been less thoroughly documented, with pertinent reviews and policy briefing reports only emerging more prominently in recent years (Charlson et al., 2021). Mental health and well-being are intricately linked to climate change through various and multifaceted pathways. Therefore, revealing the concealed costs of climate change on mental health and emotional wellbeing, as well as making the hidden advantages of climate action apparent, necessitates a synthesis and acknowledgment of contributions from a range of disciplines. By understanding the shared causes of the interconnected challenges of climate change and mental health, and by developing common solutions that support both individuals and the environment, policymakers, health system leaders, and mental health professionals will be better equipped to establish systems and communities that are resilient to the global challenges of the Anthropocene.

The escalating unmet mental health needs of individuals globally represent a pressing international health issue, even in the absence of the adverse impacts of the climate crisis. In 2010, the worldwide economic burden of mental disorders, encompassing both direct healthcare expenses and indirect costs associated with lost productivity, was estimated at $2.5 trillion (Trautmann et al., 2016), with projections indicating an increase to $16 trillion by 2030 (Patel et al., 2018). Depression ranks as a primary cause of disability on a global scale, and there is a notable rise in the prevalence of emotional disorders and self-harm, particularly among females (McManus et al., 2019). No mental health system worldwide is equipped to address the current demand, resulting in significant 'treatment gaps' between the need for and the availability of mental health services (Frankish et al., 2018). Mental health systems are under considerable strain; globally, 50% of individuals with mental health disorders do not receive treatment, a figure that escalates to nearly 90% in low-resource environments (Rolland et al., 2020). Importantly, it is estimated that 75% of mental health issues begin before the age of 18, and today's youth face the additional challenge of maturing in the context of a climate crisis that they did not create (Hurley et al., 2022). Climate change jeopardizes the aspirations of nations striving to achieve universal healthcare by amplifying both mental and physical health demands while simultaneously disrupting service delivery, as climate-related events threaten health infrastructure and supply chains.

When the effects of climate change on human health were acknowledged, mental health was initially regarded as a secondary issue, with infectious diseases and physical injuries resulting from severe weather events being among the first health consequences to be emphasized. Physical and mental health are deeply interconnected, exhibiting significant co-morbidity among various physical and mental health disorders (Bolton et al., 2015). The COVID-19 pandemic has vividly demonstrated that physical health impacts have both direct and indirect effects on mental health and overall well-being. Despite substantial evidence, the inseparability of mental and physical health is frequently overlooked. In the last decade, the body of evidence illustrating the numerous connections between climate change and mental health has significantly strengthened (Hayes & Poland, 2018). This development has occurred amidst rising temperatures, increasing severity and frequency of extreme weather events, and the resultant damage to both human and natural systems, coupled with inadequate political response and heightened public awareness regarding the present and future ramifications of the climate crisis.

As our comprehension of the links between climate change and mental health, as well as emotional wellbeing, continues to evolve, a significant necessity has emerged for advanced and interdisciplinary 'systems' thinking. This approach should incorporate and synthesize pertinent literature from various fields to grasp the extensive and intricate nature of this complex subject. This chapter contributes to the existing body of literature reviews concerning climate change and mental health by first extending beyond studies that directly address climate change. For instance, it explores relevant literature regarding experiences related to fires and floods, which are not always explicitly associated with climate change. Secondly, it encompasses both academic and grey literature. Furthermore, this chapter acknowledges the substantial growth of climate psychology literature in recent years, particularly regarding the mental health effects stemming from witnessing or being aware of climate change, as well as the connections to coping mechanisms, adaptation, agency, and action. To adequately prepare communities for present and future climate threats, it is essential to understand the psychological barriers and facilitators of climate action, as well as the factors that assist individuals in navigating the crisis while enhancing local capacities to endure and flourish in an evolving climate, which includes psychological adaptation and emotional resilience.

This research posits that mental health and emotional well-being exist on a continuum, allowing individuals to experience varying degrees of mental health at different times. Furthermore, optimal mental health is characterized by a state of well-being where individuals can effectively manage and thrive in all areas of their lives and relationships (World Health Organization, 2005). The research encompasses not only the outcomes associated with suicide and clinically recognized mental health issues (such as new instances of mental disorders, exacerbated symptoms, and increased mortality and morbidity among those with pre-existing mental conditions), but also evaluates the effects on broader population mental health and wellbeing indicators. Additionally, it recognizes that health, including mental health, is not solely an individual phenomenon; rather, it is influenced by a variety of social, environmental, and economic factors (determinants of mental health) (Patel et al., 2018).

Although the climate crisis acts as a risk multiplier, exacerbating pre-existing vulnerabilities and disadvantages (Vergunst & Berry, 2021), it also presents an opportunity multiplier for climate action. The climate emergency necessitates that leaders facilitate a swift societal transition from fossil fuels to low-carbon alternatives, while also investing in adaptation strategies for communities facing the challenges posed by a warming planet. The design and execution of such policies can be most effectively accomplished by comprehending the psychology surrounding climate change, as well as identifying potential win-win scenarios for climate action that can enhance the factors contributing to positive mental health and well-being.

MENTAL HEALTH AND WELLBEING

Despite its growing prevalence, the term 'mental health' remains challenging to define succinctly. Mental health is a comprehensive concept; akin to physical health, every individual possesses mental health that can fluctuate between better and worse states at any given moment. Therefore, good mental health transcends mere absence of illness or disorder, embodying positive wellbeing that fosters flourishing (WHO, 2021). Promoting mental health and wellbeing amid the climate crisis necessitates an understanding of the importance of creating conditions that not only avert illness but also empower individuals to thrive. Furthermore, individuals may exhibit symptoms that can be classified as a mental illness or disorder. Mental disorders are characterized by 'significant disturbances in thinking, emotional regulation, or behaviour' (WHO, 2021), and may be clinically identified when patterns of abnormal psychological and behavioural symptoms become established and hinder functioning. A considerable number of individuals do not (or are unable to) seek appropriate assistance for these symptoms and may lack engagement with support services; however, community prevalence studies acknowledge the substantial burden of disease that mental disorders impose globally (Global Burden of Diseases, 2019). Although mental disorders are typically diagnosed based on the symptoms detailed in the Diagnostic Statistical Manual, they are also recognized for their heterogeneity, overlap, and dependence on biological, experiential, interpersonal, and environmental factors. This study employs the term 'mental disorders' to denote individuals exhibiting symptoms that align with what clinical mental health professionals would classify as a 'mental disorder', while also acknowledging that many individuals do not and cannot access care, thus precluding them from receiving such diagnoses, and that diagnosis may be beneficial for some individuals while not for others.

This study will concentrate on healthy psychological responses to abnormal events, as well as clinically diagnosable mental disorders. It is important to recognize that distress can manifest in both contexts, often stemming from traumatic experiences; however, the latter is typically more linked to chronic distress that hinders an individual's capacity to perform daily tasks. Chronic and debilitating distress, particularly about the climate crisis, can be easily understood, yet it may necessitate support to enhance and safeguard quality of life. Consistent with the findings of Patel et al. (2018), this study acknowledges that mental health and mental disorders exist along a continuum, ranging from mild, temporary distress to chronic, progressive, and severely disabling conditions. While the binary method of diagnosing mental disorders is beneficial for clinical practice, it fails to accurately represent the diversity and complexity of the mental health needs of individuals and populations. The climate crisis does not typically give rise to new forms of mental disorders; rather, it exacerbates existing distress related to the existential and lived threats posed by climate change. In this sense, climate change acts as a risk multiplier, heightening the likelihood of experiencing symptoms that can be classified as mental disorders, intensifying existing symptoms and distress, or both, especially among groups that are already at heightened risk due to structural inequalities. Furthermore, it can diminish the resources available to individuals, communities, and health systems, thereby affecting their ability to cope with and build resilience against stressors and distress.

Mental health and physical health are closely interconnected, as evidenced by the high prevalence of comorbidities (the occurrence of multiple health conditions simultaneously in an individual) (De Hert et al., 2011) and a correlation at the population level between robust physical health and sound mental health. Consequently, the well-documented risks to physical health posed by the climate crisis (such as the increased transmission of infectious diseases) or policy modifications that impact physical health generally also carry implications for mental health. Given that physical and mental health represent two facets of the same reality, "initiatives that focus solely on enhancing physical wellbeing while neglecting mental and social wellbeing are likely to be unsuccessful." (Stewart-Brown, 1998).

CLIMATE CHANGE AND MENTAL HEALTH

Rising global temperatures are anticipated to exacerbate pre-existing climate-related risks while also generating new threats to both natural and human systems. Some of these threats are localized, whereas others are more widespread. As global temperatures continue to rise, these risks are projected to intensify, with many expected to adversely affect human health (IPCC, 2014). The implications of these climate risks extend to mental health and emotional well-being through various mechanisms. These mechanisms range from more 'direct' and overt climate-related experiences (for instance, the first-hand experience of a heatwave or flood) to longer-term direct effects that disrupt the social and economic determinants of mental health, as well as more 'indirect' pathways associated with climate awareness or witnessing climate-related destruction and the inaction of leaders. It is important to recognize that numerous pathways of influence may be regarded as 'indirect'. This perception arises from the diverse range of ways in which climate change exerts 'downstream' effects on mental health that do not stem from a singular event (for example, experiences of sexual violence or conflict during forced migration due to drought, which is itself exacerbated by climate change). This complexity complicates the task of quantifying the true impact of climate change on mental health and emotional well-being. Although the terms direct and indirect impacts are commonly employed, they are defined in various ways throughout the literature, and a clear separation between direct and indirect impacts is often impractical, particularly given the global nature of climate change, which exposes everyone to some degree of the consequences of a warming planet. Likewise, emerging evidence indicates that direct encounters with climate change-related incidents, such as wildfires, amplify the 'indirect' psychological distress associated with recognizing the threats posed by ongoing and future climate change.

DISRUPTION TO ENVIRONMENTAL CONDITIONS

The mental health and well-being of both individuals and communities can deteriorate due to the direct consequences of rising global temperatures, along with the increasing frequency and intensity of extreme weather and climate events. This section examines the mechanisms through which alterations in climatic conditions can negatively impact mental health. Additionally, it illustrates how specific demographic groups (such as those with pre-existing mental health issues) face a heightened risk of experiencing severe and compounding adverse effects during extreme weather or climate events, thereby exacerbating existing inequalities and vulnerabilities. Furthermore, the section emphasizes that the repercussions of exposure to extreme weather and climate conditions can accumulate over time (for instance, through repeated experiences of flooding or wildfires), leading to compounded negative mental health outcomes.

Extreme Temperatures

Evidence from various nations indicates that elevated, extreme, and fluctuating temperatures exacerbate numerous mental health and wellbeing issues, including heightened suicide rates, increased hospital visits and admissions for mental disorders, as well as deteriorated population mental health and emotional wellbeing (Middleton et al., 2021). Nonetheless, the connections between mental health and meteorological factors, such as temperature, are intricate and influenced by contextual and mediating elements (for instance, humidity, seasonality, and local baseline temperatures). The adverse mental health effects associated with rising temperatures have predominantly been documented in Europe, the United States (Hu et al., 2020), and East Asia (Kalkistein et al., 2019), although they have also been noted in Mexico, Brazil, Australia, Kazakhstan, and Russia, with certain studies employing models that utilize data from multiple nations (Gupta & Murray, 1992; Burke et al., 2015). Temperature may influence the manifestation of mental illness symptoms or deteriorate overall mental health and wellbeing through various biological, cognitive, and societal mechanisms. These mechanisms encompass physiological alterations, such as 'heat stress' and changes in blood circulation and central nervous system activity, which can result in cognitive and emotional shifts, as well as societal transformations, including diminished economic productivity and heightened conflict and violence (Buguet, 2007). Elevated or extreme temperatures can also disrupt sleep and diminish its quality, which is recognized to elevate mental health risks (Dixon et al., 2014). Moreover, higher, extreme, or more variable ambient temperatures carry implications for physical health. For instance, rising temperatures are associated with direct physical health consequences, such as exacerbated symptoms or mortality from conditions like diabetes or cardiovascular diseases, as well as indirect effects, including increased incidence and geographical dissemination of infectious and waterborne diseases, such as dengue fever. These physical health repercussions can interact with mental health outcomes, further deteriorating both individual and community mental health, and can also be detrimental to overall well-being.

Suicide

Suicide rates are influenced by external events and conditions, exhibiting a positive correlation with local ambient temperature across daily, weekly, and monthly intervals (Liu et al., 2021). Without adequate adaptation, climate change may elevate the risk of suicide by raising ambient temperatures. A systematic review has provided robust evidence supporting a connection between temperature and the frequency of suicide in 15 out of 17 studies. Meta-analyses indicate that for every 1 °C increase in temperature above a local ambient threshold, there is approximately a 1% rise in suicide incidence (Ding et al., 2016), along with a 2.2% increase in overall mortality associated with suicide or mental disorders (Linkowski et al., 1992). In a scenario characterized by unmitigated climate change, one study estimated an additional 22,000 deaths (with a confidence interval of 9,000-40,000) across the USA and Mexico by the year 2050, which is comparable in effect size to that of an economic recession (Gupta & Murray, 1992). The evidence regarding the relationship between suicide deaths and heatwave events, in particular, remains inconclusive, differing from the general correlation between suicide and ambient temperature previously mentioned. Some studies suggest that the risk of suicide increases with extreme temperatures, such as sudden spikes or heat waves, as well as with high temperature variability. However, a recent global modeling study revealed mixed findings concerning the link between suicide risk and the frequency of heatwave events, proposing that humidity may be a more significant factor influencing this risk (Gao et al., 2019). The precise relationship between temperature and suicide risk is contingent upon local contexts and mediating factors, including baseline temperatures and other meteorological variables such as sunlight hours, humidity, or temporal factors like seasonality (Bando et al., 2017). Elevated levels of relative humidity, which are also influenced by the climate crisis, seem particularly pertinent to suicide risk, with women and young individuals being especially vulnerable.

Mental Health Disorders

Elevated temperatures have been associated with deteriorated physical and mental health outcomes for individuals experiencing symptoms of mental disorders. Research indicates that there are higher mortality rates related to heat and an increase in hospital visits and admissions for mental disorder symptoms during heatwaves (Linkowski et al., 1992). The primary source of evidence for these effects is derived from hospital admission statistics. Globally, higher temperatures have been linked to an increased likelihood of mental health-related emergency department visits, particularly for conditions such as bipolar disorder, schizophrenia, dementia, self-harm, and substance misuse (Lee et al., 2018). The evidence regarding the effects of heat on more prevalent mental disorders, such as anxiety and depression, is less consistent; however, studies have indicated a rise in depressive language among the general population during periods of heat, alongside some evidence suggesting exacerbated anxiety symptoms (Ding et al., 2016). Furthermore, there is a notable and often overlooked increase in mortality risk during heatwaves for individuals exhibiting symptoms that align with mental disorder criteria (Ding et al., 2016), highlighting that this vulnerable population is frequently under-recognized in public health communications concerning the management of heat-related hazards.

Effects on Population Mental Health and Wellbeing

Psychological distress, defined as troubling or upsetting feelings or thoughts, along with negative emotions, has been observed to increase within the general population during periods of elevated temperatures or extreme heat. Research conducted globally, particularly in the Global North, indicates that rising temperatures—sometimes exceeding a locally relevant threshold—are linked to: heightened psychological distress, diminished positive emotions (Baylis et al., 2018), an increase in negative emotions, greater fatigue, a decline in overall wellbeing (Beecher et al., 2016), the use of 'depressive' language in social media communications (Gupta & Murray, 1992), and a rise in self-reported mental health challenges (Ondar & Akay, 2014). Furthermore, humidity may exacerbate the effects of heat, whereas moderately warm or 'comfortably' warm weather, along with increased hours of sunshine, may enhance wellbeing (Austin et al., 2020) and could obscure the effects of temperature if studies fail to adequately account for these variables. The societal, cognitive, and biological alterations induced by heat that can aggravate mental disorder symptoms or elevate suicide risk—such as disruptions in sleep or cognitive changes—may also contribute to the overall decline in mental wellbeing across the broader population.

Temperature Exposure and Adaptation

The ambient temperatures to which an individual is subjected are influenced by their local surroundings, including the availability of air conditioning and green spaces. Dense urban areas tend to create 'heat islands' (Mushangwe et al., 2021). Marginalized groups such as People of Colour, low-income communities, those facing housing insecurity, and incarcerated individuals often lack access to the cooling benefits provided by tree cover, green spaces, and adequate housing (Henderson et al., 2022). Consequently, they are more likely to experience mentally detrimental levels of heat, especially in the absence of investments aimed at heat adaptation in a changing climate. Given that individuals with mental health disorders are disproportionately represented among the homeless, incarcerated populations, and those living in inadequate housing, their heightened vulnerability to heat exacerbates existing challenges that are likely to worsen due to climate change. Although individuals and communities may find ways to adapt to increasing temperatures, particularly through investments in cooler urban environments (such as enhancing tree cover), these initiatives have not yet proven adequate. Burke et al. (2018) reported a lack of evidence for effective adaptation to rising temperatures in the USA thus far, which could mitigate the suicide risk linked to elevated temperatures. Therefore, it is essential to closely monitor the mental health implications of climate adaptation policies, as highlighted by Gupta & Murray (1992).

Extreme Climate Conditions

The IPCC characterizes an extreme weather event as one that is uncommon for a specific location and time of year, whereas an extreme climate event is described as a series of extreme weather occurrences that endure for a certain duration, such as a season (IPCC, 2022). Experiencing an extreme weather or climate event, or coping with its consequences, can be profoundly distressing. Although there is inherent uncertainty in attributing any single extreme weather or climate event to climate change, the likelihood and intensity of such events are expected to rise in a changing climate. Extreme weather and climate occurrences are linked to a higher incidence of post-traumatic stress disorder (PTSD), depression, anxiety, severe psychological stress and distress, substance abuse, and suicidal tendencies (Kar, 2019). These psychological effects have been noted across all age groups, including children, adolescents, adults, and the elderly, as well as among healthcare workers involved in disaster response (McGuire, 2018). The psychological ramifications of climate extremes have not garnered adequate political focus, despite their growing prevalence and significant long-term implications for individuals and communities. While the financial repercussions of floods, wildfires, or hurricanes are often measured in terms of fatalities or property damage, the number of individuals who suffer psychological effects from any disaster can surpass those who are physically harmed by a ratio of 40 to 1 (Rhodes et al., 2010).

While PTSD resulting from extreme weather events may evoke thoughts of globally significant occurrences like Hurricane Katrina, persistent extremes such as recurrent flooding due to unusually high and prolonged rainfall can also result in enduring mental health issues and cumulative psychological effects (Fernandez, 2017). Symptoms are likely to manifest more rapidly and with greater intensity during subsequent exposures, for instance, when individuals face multiple floods, cyclones, droughts, or wildfires; or one disaster followed by another (Center for Disease Control and Prevention, 2017). The psychological and social repercussions of climate-related disasters may take time to emerge. Although some research indicates that the prevalence and severity of psychological disorders peak within the initial 6-12 months following an event like a flood or hurricane (Cruz et al., 2020), other studies have observed an increase over time, with higher rates of PTSD among Hurricane Katrina survivors after 12 months compared to 5-8 months (20.9% versus 14.9%, respectively) (French et al., 2019). Anxiety related to flooding has been documented in individuals even 2.5-5 years post a significant flooding incident (Silveira et al., 2021). Several years following the 2009 Australian 'Black Saturday' bushfires, a small percentage of individuals continued to experience PTSD, depression, or severe psychological distress, with incidence rates increasing for the most affected communities and those who had feared for their lives, lost loved ones, or faced additional stressors (e.g., PTSD rates were 15.6% in highly affected, 7.2% in moderately affected, and 1.0% in low-affected communities) (Wasserstein & La Greca, 1998). A decade later, one-fifth of individuals from severely affected regions were found to have a probable psychological disorder.

Biodiversity

Biodiversity is fundamentally connected to the essential needs for life, such as clean air, safe water, and food (IPCC, 2022). Interaction with biodiverse natural settings serves as a significant environmental factor influencing mental health and overall wellbeing (Wang et al., 2019). Research indicates that time spent in biodiverse natural environments positively affects wellbeing, enhancing mood and happiness, boosting cognitive abilities, increasing vitality, and alleviating stress, anxiety, and depression (Gianfred et al., 2021). Individuals who experience a stronger connection to nature tend to enjoy better health and well-being and are more inclined to engage in conservation efforts (Portner et al., 2021). The advantages are amplified in areas with higher biodiversity. Despite the numerous advantages offered by biodiverse ecosystems, global biodiversity is currently experiencing an unprecedented decline, exacerbated by factors such as land use changes. The loss of biodiversity is closely intertwined with climate change (Hong et al., 2019). Climate change significantly contributes to biodiversity loss, while ecosystems are crucial for sequestering greenhouse gases and aiding climate adaptation, for instance, by providing communities with enhanced protection against extreme weather events. The relationship between biodiversity and the climate crisis presents both a challenge and a potential for improving mental health and emotional well-being. This relationship indicates that climate change poses a risk to the biodiverse ecosystems that are vital for human livelihoods and emotional health. On the other hand, it also suggests that initiatives aimed at reducing greenhouse gas emissions can be complementary to efforts in conserving and restoring biodiversity (Atterdemo et al., 2017).

Air Quality

The combustion of fossil fuels stands as the primary driver of climate change, which concurrently leads to air pollution. Furthermore, climate change can exacerbate air pollution through heightened temperatures, increased humidity, and the occurrence of wildfires. Air pollution, particularly in the form of particulate matter (PM) and nitrogen oxides (NOx), has been associated with an elevated risk of mental health issues, potentially through pathways involving inflammation and neuronal damage (Chen et al., 2022). Research indicates that exposure to various elements of air pollution correlates with increased incidences of depression, anxiety, bipolar disorder, psychosis, schizophrenia, neurodegenerative diseases such as dementia, and a heightened risk of suicide, particularly with short-term exposure (Carey et al., 2018). Those who are subjected to higher levels of air pollution demonstrate a greater likelihood of utilizing mental health services, even when adjusting for factors such as socioeconomic status, age, or ethnicity. Longitudinal studies focusing on mental health service users with psychotic and mood disorders revealed a significant rise in both inpatient days and community mental health service engagement after one year for individuals residing in areas with elevated PM and NOx pollution levels (Tone et al., 2018).

EFFECTS ON SOCIO-ECONOMIC AND POLITICAL CONDITIONS

Climate change has the potential to disrupt and destabilize the social, economic, and political factors that contribute to good mental health and well-being, including stable societies, effective governance, and manageable living costs. While climate change can adversely affect these determinants, they can also shape its course; for instance, nations characterized by trustworthy and transparent governance often experience not only enhanced mental well-being but also reduced carbon emissions.

GLOBAL SOCIAL INEQUALITIES

Social injustice and inequalities represent significant risk factors for deteriorating mental health and the emergence of mental disorders (Rao, 2018). Climate change intensifies pre-existing health disparities, including those related to mental health, as the populations most adversely impacted by climate change—thereby compromising their mental health and overall wellbeing—are often those already marginalized in society. This group includes individuals with prior mental health issues. Moreover, climate change exacerbates inequalities in social and economic opportunities, creating a vicious cycle. Nations and contexts characterized by greater income disparities are linked to elevated levels of greenhouse gas emissions and poorer mental health outcomes (Kelly et al., 2021). More equitable and democratic societies typically exhibit traits such as reduced poverty, enhanced education, and improved healthcare, which contribute positively to climate resilience, a greater willingness to implement climate mitigation strategies, and higher satisfaction of individual needs, resulting in better wellbeing outcomes with lower energy consumption. In contrast, societies that "prioritize economic growth beyond moderate incomes and extractive sectors" tend to experience poorer well-being outcomes alongside increased energy consumption (Steinberger & Roberts, 2010). Climate injustice, which refers to the phenomenon where individuals, communities, and nations that have contributed the least to greenhouse gas emissions suffer the most severe consequences, generates legitimate anger and distress, further elevating mental health risks. Addressing inequalities in resource distribution is essential for enhancing human mental health and well-being. Furthermore, "the amount of energy required for a high global level of human development is decreasing," with opportunities available to significantly boost wellbeing in low-resource environments with only modest increases in energy consumption. Enhancing equity within social and economic frameworks not only benefits overall well-being but also fosters greater participation and trust in governance and climate mitigation efforts, creating an interactive and self-reinforcing process.

VIOLENT CONFLICTS

Climate change has the potential to intensify conflict and violence. This escalation can occur through direct effects, such as rising temperatures that increase instances of interpersonal violence. Conflict serves as both a catalyst and a result of disruptions to the socioeconomic, cultural, and environmental stability of societies, leading to the breakdown of community bonds and social cohesion, a decrease in habitable and arable land, heightened food and water insecurity, pressures on health systems, and forced migration. As climate change generates a multitude of interconnected disruptions and diminishes the capacity of governments to deliver public services, populations facing resource scarcity may become more prone to violence, resulting in heightened intergroup conflict, terrorism, sexual violence, domestic violence, and conflicts or wars both within and between nations (Burke et al., 2015). The mental health repercussions of conflict and violence are well-documented, encompassing post-traumatic stress disorder, depression, anxiety, and substance abuse (Hsiang & Burke, 2014). Nations embroiled in conflict are particularly vulnerable to the impacts of climate change, partly due to geographical factors, but mainly because of their diminished capacity to adapt (Black et al., 2011). Insecurity hampers the ability of individuals, communities, and governments to manage climate shocks, such as those resulting from extreme weather events. The collective and coordinated efforts necessary to transform societies and bolster the climate resilience of local communities can be severely restricted during periods of armed conflict. Furthermore, international support for adaptation initiatives—whether from community organizations like the Red Cross or governmental bodies—may also diminish. These overlapping crises and heightened vulnerabilities can significantly affect mental health and overall well-being. Additionally, conflict can harm the environment and exacerbate climate change, for instance, through the contamination of air, water, and soil due to warfare or through deforestation and burning (Bowles et al., 2015).

FORCED MIGRATION

Climate change is expected to become an increasingly significant factor influencing migration throughout this century, as extreme weather events, rising sea levels, and associated social transformations compel individuals to leave their native regions, resulting in the emergence of 'climate migrants' or 'climate refugees'. In 2021, 23.7 million individuals were internally displaced due to non-conflict-related disasters, with the majority being linked to weather phenomena. A frequently referenced estimate suggests that by 2050, there could be as many as 200 million climate refugees due to climate change and its related consequences, including flooding, rising sea levels, droughts, shortages of food and arable land, and conflicts (Reifels, 2018). Additionally, temporary evacuations or local displacements caused by climate-related disasters are likely to become more common, such as forced relocations or loss of homes following wildfires. As individuals are compelled to alter their livelihoods and lifestyles—either as a preventive measure in response to deteriorating conditions (e.g., drought) or to ensure safety during extreme weather events—those from rural areas may find themselves in impoverished urban settings, confronting new and complex risks (Black et al., 2011). The specific experiences and impacts on migrating individuals are likely to be profoundly influenced by the nature of the migration itself, whether it involves planned relocation versus emergency response, forced versus voluntary movement, or migration within a country versus across borders. The distinction between planned and emergency evacuations can correlate with improved mortality rates and better physical and mental health outcomes for those affected (Gleeson et al., 2020).

Forced displacement or evacuation following climate-related disasters has been associated with mental health issues such as depression, anxiety, and PTSD. The loss of one's home, community, and culture, along with the navigation of potentially dangerous and challenging journeys to safety, results in significant and persistent stress, a diminished sense of identity, and the severance of supportive relationships. First and second generation immigrants face a twofold increase in the risk of psychosis compared to their non-immigrant counterparts (Torres & Casey, 2017). This risk is not uniformly distributed across different host countries and ethnic groups, indicating the significance of social context. Although evidence is scarce regarding the impact of climate migration in comparison to other migration causes, the difficulties encountered by climate refugees are likely to mirror some of the challenges faced by refugees in general, who exhibit elevated rates of mental disorders (Fussel & Lowe, 2014). The mental health implications of climate migration specifically remain under-researched, although there is some evidence indicating adverse effects (Fitspatrick, 2021). For instance, Indigenous Australians compelled to migrate to urban areas due to droughts affecting the pastoralism industry (Koslov et al., 2021) have experienced a decline in mental wellbeing. Individuals displaced by Hurricanes Katrina and Harvey in the United States exhibited heightened psychological distress, stress, and posttraumatic stress when compared to those who remained in their original communities (Neyna et al., 2021). Once again, outcomes are significantly influenced by the level of stress exposure and the support available. Following Hurricane Sandy in the United States, for example, individuals who opted for supported relocation (through a voluntary home buyout program) were less likely to report increased stress than those who decided to rebuild in their original locations with limited support (Campo-Arias & Herazo, 2014).

Displacement or migration induced by climate change may result in a 'risk exchange', where migration enhances certain aspects of physical and mental health while deteriorating others (James et al., 2019). Various factors will affect the choice to remain or relocate, and if migration is effectively planned and supported, some individuals and communities could reap benefits. The degree to which the experience is positive or traumatic—both before, during, and following the move—hinges on the conditions of the situation being abandoned and the characteristics of the destination country or location, including its cultural receptiveness and the adequacy of support systems for migrants. The assistance available for those who remain compared to those who migrate will influence outcomes, whether within a single country or across borders. For instance, the comparative stress of staying versus relocating will be shaped by rising insurance premiums for regions facing heightened climate risks, or the challenges in obtaining aid for post-disaster reconstruction in such areas (Campo-Arias & Herazo, 2014).

Access to suitable resources enhances mental health and wellbeing outcomes. This encompasses assistance in processing traumas related to migration, as well as access to pertinent health, education, employment, and social resources. Social determinants appear to play a particularly significant role in various forced-migration contexts, where 'cultural bereavement' poses mental health threats (Siriwardhana & Stewart, 2013), potentially occurring alongside solastalgia, grief, and a sense of loss related to place in instances of climate migration. Factors such as social exclusion, economic hardship, stigma, and discrimination significantly contribute to the deterioration of mental health among migrants (Palmer & Ward, 2007). A study conducted in the UK revealed that refugees faced heightened levels of stress, anxiety, and depression due to post-migration stressors, including unemployment or underemployment, victimization, inadequate housing, or homelessness (Bonilla-Escobar et al., 2017). These elements adversely affected overall long-term health. Meta-analyses focusing on the mental health of disaster-affected refugees have aimed to pinpoint factors associated with an increased risk of poor mental health outcomes and resilience. Specific risk factors identified for poorer mental health outcomes include residing in institutional settings or rural areas, experiencing limited economic opportunities, being older or female, and possessing higher educational attainment or a more favorable socioeconomic status before the disaster (Fussel & Lowe, 2014). In the context of conflict-induced forced migration, elements such as family and community support, collective identity, and religious affiliation have been correlated with resilience. These investigations underscore the necessity of providing adequate structural, social, and cultural resources for climate migrants as essential mental health interventions.

Access to opportunities for significant cultural and spiritual practices has a profound impact on the mental health of climate migrants. For communities whose cultures are deeply connected to the land, such as Indigenous Communities, preserving cultural traditions post-climate migration can be particularly challenging, thereby intensifying the adverse effects of being forcibly displaced from their homeland. Research involving conflict-displaced individuals from Colombia, Ethiopia, and Bangladesh has highlighted the advantages of upholding cultural traditions and the importance of community-based mental health care support (Vince, 2022).

GOVERNANCE

Governance, whether at local, national, or global levels, has the potential to either foster or hinder the wellbeing of individuals and the environment. Governance characterized by transparency, inclusivity, trustworthiness, and opportunities for participation is associated with enhanced wellbeing among citizens (Weintrobe, 2021). Effective climate action, spanning from local to global initiatives, fundamentally relies on sound governance. Successful adaptation to climate change necessitates the incorporation of local knowledge and the engagement of local communities, which in turn requires government investment. Similarly, the backing for climate mitigation strategies is most robust when there are avenues for meaningful participation in decision-making processes involving all stakeholders affected (across various sectors and societal levels), fostering societal trust, social cohesion, and enhancing climate governance. This dynamic not only relies on but can also enhance the wellbeing of community members. In a similar vein, gender equity contributes to improved health and wellbeing for women, while countries with greater political power for women tend to exhibit lower greenhouse gas emissions (Zeira, 2022). Conversely, climate change and its associated risks can destabilize governments, leading to subsequent disruptions that ultimately exacerbate mental health issues. For instance, 'the convergence of climate risks and conflict further worsens food and economic insecurity and health disparities, limits access to essential services, while weakening the capacity of governments, institutions and societies to provide support. The impact of this overlap is not only wide-ranging; it is also far-reaching. Ripple effects can shape mobility, patterns of transhumance or access to resources on a continental scale.' (Black et al., 2011).

The protection of mental health and wellbeing ultimately relies on governments implementing policies that facilitate a swift transition away from fossil fuels. It is essential that these policies are executed thoughtfully to safeguard individuals who may experience adverse effects, such as those needing to alter their livelihoods. Furthermore, governance can play a crucial role in mitigating the impact of climate change on mental health outcomes. For instance, conflicts may arise when agricultural farmers and cattle herders are compelled by a changing climate to vie for limited resources. Such situations are exacerbated in the absence of robust governance, and the resulting conflicts can further undermine governance structures. By actively enhancing the factors that contribute to good mental health and wellbeing, including investments in reliable and inclusive governance, the adverse effects of climate change on community wellbeing can be alleviated.

DISRUPTION TO CULTURAL CONDITIONS

The values, attitudes, and narratives inherent in a specific culture, as well as those on a global scale, are mutually influential with respect to climate change and the outcomes related to mental health and wellbeing. In this context, we define cultural conditions as the values that a society strives to uphold, which form the foundation of its economies and governance. This encompasses the perspectives individuals within a society hold regarding their relationships, connections, and responsibilities to one another and to the natural world, as well as their perceived roles and purposes in life and society, alongside the narratives that a community shares about itself. Cultures not only shape mental health and wellbeing but also influence how these concepts are perceived, understood, and supported. The extensive and varied impacts of climate change disrupt cultural practices and lifestyles, potentially inflicting intricate damage to identity, purpose, mental health, and well-being. This is particularly relevant for individuals and communities with deep cultural and spiritual ties to their landscapes, notably among many Indigenous populations. For instance, the melting of sea ice in Canada poses a significant threat to Inuit communities, who depend on this ice not only for their livelihoods but also for their essential identity as a people, which is closely linked to their wellbeing. An Inuit elder poignantly remarked: "We are people of the sea ice. And if there's no more sea ice, how do we be people of the sea ice?" (Burroughs & Rindfleisch, 2002).

In contrast, both climate change and deteriorating mental health and wellbeing stem from specific cultural conditions. Psychotherapist Sally Weintrobe has pointed out that 'cultures of uncare' and 'exceptionalism'—the notion that standard limitations, including those of natural systems, do not pertain to a particular individual or culture—are prevalent in numerous countries and contexts. These cultural phenomena adversely affect mental health and wellbeing while simultaneously exacerbating climate change (Moldes & Ku, 2020). Key cultural factors include materialism, hyper-individualism, the remnants of colonialism, and a disconnection between individuals and their communities, as well as between humanity and nature. The emergence of neoliberal economic policies and culture has been linked to various and extensive mental health issues (Dittmar et al., 2014). Materialism and excessive consumption are significant contributors to the climate and ecological crises, as the production, transportation, and disposal of superfluous or disposable goods escalate the use of fossil fuels and other non-renewable resources, leading to increased waste. To remain within planetary limits, wealthier nations must curtail their overconsumption. Although this often aligns with narratives of sacrifice, substantial evidence indicates that materialistic values are inversely related to wellbeing, which can be partially attributed to the individualistic essence of materialism being contrary to collectivist principles. Conversely, diminishing materialistic values can enhance wellbeing.

Cultivating environments that enhance connections between individuals, their communities, and the natural world can alleviate feelings of loneliness and promote overall wellbeing. These cultures and avenues for connection are also essential for effective climate change mitigation and adaptation strategies (Pottner et al., 2022). A disproportionate emphasis on individual-level climate initiatives has faced criticism for diverting attention from the responsibilities of governments, businesses, and other institutions, while simultaneously increasing the chances of individuals feeling overwhelmed and anxious. Likewise, a mental healthcare culture that solely addresses individual-level factors contributing to mental health issues can similarly detract from the focus on broader societal contexts that contribute to distress, as well as communal opportunities and conditions that are more effective in preventing and addressing mental health challenges (Kelman et al., 2021).

Numerous nations and communities that are particularly susceptible to climate change are concurrently grappling with the remnants of colonialism. Indigenous populations in Australia have underscored that "the effects of climate change mirror the experiences of colonization. When the initial settlers [Europeans] invaded

Australia, erroneously and disrespectfully labeling it terra nullius, or no one's land, they brought with them numerous outsiders carrying various diseases. Presently, the anthropogenic effects of climate change are introducing a new form of disease to our nation" (Capaldi et al., 2014). The legacies of colonialism also diminish the capacity of, for instance, small island developing states to adapt to climate change, as their livelihoods may rely on cash crops and fragile infrastructures. Colonialism diverted many societies from practices that were more environmentally protective or from traditional food and agricultural sources that were more resilient to local conditions. Consequently, the impacts on mental health and wellbeing can be intensified by the 'structural violence of colonialism,' which heightens economic difficulties and the inability to satisfy basic needs in the face of a changing climate. 'Climate coloniality' manifests today in the fact that many of the groups and nations most adversely affected by climate change are those that have contributed the least to the issue, while these same nations frequently possess diminished influence in international negotiations, as highlighted at COP26 (307). The cultures of colonization, which alienate individuals from their lands, ecosystems, and traditional sustainable practices in harmony with the non-human world, have both contributed to and perpetuated the climate crisis (Pritchard et al., 2020). Movements for climate justice and decolonization aim to rectify the power and justice disparities established during colonialism and that persist today.

Cultures that prioritize a connection to land and nature offer advantages for both mental health and the mitigation of climate change. The concept of nature connectedness, which refers to the degree to which individuals identify with the natural world, is a fundamental aspect of the identity of numerous Indigenous cultures globally; however, this connection has diminished in highly urbanized and westernized societies. Individuals who possess a stronger bond with nature tend to report greater life satisfaction, experience more positive emotions, and exhibit higher levels of vitality. Research indicates that "feeling a part of nature has been shown to significantly correlate with life satisfaction, vitality, meaningfulness, happiness, mindfulness, and lower cognitive anxiety" (Boykoff & Boykoff, 2007). Humans are inherently predisposed to appreciate and enjoy natural environments, as well as to seek connections with the natural world and other living beings, a phenomenon referred to as the "biophilia hypothesis" (Ford & King, 2015). Those who feel more connected to nature are also more inclined to engage in protective actions for the environment and to endorse climate policies. Cultures that maintain a strong connection to the natural world, including Indigenous communities, are particularly well-positioned to lead local climate initiatives. Involvement in such programs, such as the 'Caring for Country' initiatives in Australia (Vu et al., 2019), yields positive outcomes for mental health and overall well-being.

Narratives and stories constitute an essential component of culture. The media undoubtedly plays a significant role in conveying climate narratives. As the frequency and intensity of extreme weather events rise, the amount of news coverage dedicated to climate change and how it is reported also evolve. Engagement with media coverage affects well-being and mental health; adverse emotional reactions to climate change are associated with increased exposure to information regarding its impacts. Media narratives, in various contexts, intersect with political, socioeconomic, and environmental factors. For instance, an analysis of 37,000 news articles from 45 countries and territories revealed that media narratives in affluent nations were more inclined to highlight domestic policies and scientific evidence of climate change, whereas narratives in less wealthy nations tended to concentrate on international relations and the natural dimensions of climate change (Parry et al., 2022). Another analysis indicated that climate change coverage is generally more prevalent in Global North countries, focusing more on the societal challenges and implications of climate change in the Global South. Global and local cultural narratives interact with regional social values and attitudes, influencing psychological and emotional reactions to climate change, as well as individual and community behaviors (Elgar et al., 2021). For example, as previously noted, extreme heat has been associated with an increase in 'depressive' language within social media posts. Furthermore, media narratives can exert influence on politicians, potentially facilitating or obstructing effective action to mitigate and adapt to climate change, along with the related impacts on individuals' vulnerability to climate issues and anxiety regarding inaction on climate change.

DISRUPTION TO LIVING AND WORKING CONDITIONS

Living and working conditions pertain to the access, availability, and opportunities individuals possess regarding employment, housing, education, nutrition, water, and welfare services. In this section, we present instances of how climate change can interfere with these conditions and the related risk of deteriorating mental health and well-being, which may exacerbate existing disadvantages for certain groups within society.

FOOD AND WATER SECURITY

Climate change is influencing food and water security through alterations in precipitation patterns, rising temperatures, and extreme weather phenomena. It is expected to persist in adversely affecting food security by diminishing crop yields, disrupting food storage and distribution systems, and escalating prices or price volatility. This situation directly impacts mental health and wellbeing through the effects of malnourishment and malnutrition (which can hinder brain development in children), and indirectly through the psychological distress and anxiety stemming from uncertainty regarding individuals' ability to obtain sufficient and suitable food for themselves and their families. Such stress responses can exacerbate symptoms of depression and anxiety. Furthermore, diminished access to traditional foods due to climate change has been reported to negatively affect mental health within Indigenous communities (Mushayi et al., 2020). Water insecurity has both direct and indirect effects on mental health outcomes, with documented increases in psychological stress, despair, depression, and anxiety among individuals facing insecure water supplies. The mechanisms involved may include: stress and insecurity arising from water scarcity and material deprivation, family displacement and loss of connection to people and places, social stigma, experiences of injustice, loss of autonomy, conflict and violence, dehydration affecting the efficacy of psychoactive medications, and concerns regarding health threats and physical ailments, including the spread of diseases and the impacts of heat. Water-related hazards are experienced disproportionately, with communities that already face systemic inequities and vulnerabilities being more severely impacted. This includes individuals who are 'poor, women, children, Indigenous Peoples, and the elderly in all locations, particularly in the Global South.' Mental health repercussions have been documented globally, affecting Indigenous communities in Peru, Australian farmers, and various communities (especially women) in the USA, Bolivia, Nepal, India, and Iran (Gaber et al., 2021).

LIVELIHOODS

Access to a secure and stable means of living is a crucial social and economic factor influencing mental health, where the ability to engage fully in society and realize one's potential is integral to the concept of good mental health. Climate change poses a threat to livelihoods through its effects on various sectors, including agriculture, as previously discussed, or through the detrimental impact of elevated temperatures and polluted air on cognitive function and work capacity. These disruptions to livelihoods can lead to heightened levels of stress and trauma, increased substance abuse, and suicidal thoughts (Bhullar et al., 2022).

DISRUPTION TO INDIVIDUAL PSYCHOLOGY AND INTERACTIONS WITH SOCIAL AND COMMUNITY NETWORKS

Individual psychological reactions to climate change are shaped by the social and community networks in which individuals are situated - for instance, by the extent to which they can access support systems for collective action and the values upheld by the groups they are part of. In contrast, social and community networks may evolve in response to the aspirations and anxieties of individuals, as exemplified by the emergence of social movements like the 'Fridays for Future' youth climate initiative. Given the intricate interconnections among these layers, this section will examine them and their interactions collectively, while also acknowledging how they are influenced by other layers, such as the cultural context of a society and the related media narratives.

PSYCHOLOGICAL RESPONSES TO CLIMATE AWARENESS AND THE IMPACT ON MENTAL HEALTH AND WELLBEING

As the awareness of the gravity of the climate crisis grows, fueled by heightened personal experiences of climate-related extreme events and the corresponding media coverage, the distress associated with climate awareness is also on the rise. This is evidenced by the increasing frequency of terms like 'eco-anxiety' in both academic and popular media discussions. Nevertheless, there is still a notable deficiency in data regarding the prevalence, characteristics, and intensity of emotional and psychological reactions to the climate crisis, as well as their effects on mental health and overall well-being. Although the body of literature on this topic is expanding and evolving rapidly, significant gaps remain in the standardization of metrics and psychological constructs, the impact on various global demographics, the identification of the most effective coping strategies, and the examination of longitudinal changes. This section delves into the variety and prevalence of psychological and emotional responses that have been documented thus far. Intense psychological reactions to the climate crisis are frequently a component of a healthy and adaptive response to the perceived threat, and therefore should not be viewed as pathological (Pihkala, 2022). However, these reactions can transform into persistent sources of stress and, in the absence of adequate support, adversely affect mental health and well-being. To prevent the risk of pathologization, we have preserved a distinction in the subsequent two subsections between the psychological responses themselves and their potential impacts on mental health and well-being.

CONCLUSION

Although mental health has frequently been regarded as the overlooked 'poorer cousin' of physical health in terms of research, awareness, and support, there has been a growing global recognition of the critical significance of mental health and wellbeing for quality of life. This recognition emphasizes its inseparability from physical health and its reliance on a variety of interconnected factors, including individual, social, economic, cultural, political, and environmental conditions (known as mental health determinants). Consequently, the Lancet Commission on Global Mental Health and Sustainable Development has called for the establishment of "environments that foster mental health for everyone" (McManus et al., 2019). The climate crisis constitutes a mental health crisis, jeopardizing advancements in mental health and wellbeing across all contexts. This jeopardy arises from direct encounters with elevated temperatures and extreme weather events, ongoing disruptions to the factors that contribute to good mental health, and the psychological distress associated with confronting an existential threat that is inadequately addressed by those in power. As a risk multiplier, the climate crisis exacerbates existing vulnerabilities to poor mental health and well-being among groups and nations that are already at a disadvantage. Many of the conditions that generate these disadvantages and exacerbate mental health issues—such as ineffective governance, significant social and income disparities, cultures of extractivism, and the pollution of air and ecosystems through fossil fuel use—are the same societal challenges that sustain the climate crisis.

The answers to the climate crisis and its adverse impacts on mental health and wellbeing are available, and the inaction of leaders in implementing these solutions only exacerbates the psychological and moral harm. As noted by Hurley et al. (2022): "It is entirely unjust to expect a generation of young individuals to cultivate greater psychological resilience to confront climate change. However, due to decades of negligence by prior generations, they are being compelled to do so. We have a responsibility to children and young people to prioritize the mitigation of climate change at its origin, while simultaneously investing in evidence-based resources to bolster their mental wellbeing amidst this ongoing crisis" (De Hert et al., 2011). Climate action at every level is contingent upon and intricately linked to psychological reactions to climate change, mental health and wellbeing, and the sense of agency to tackle the issue. Both sound mental health and the necessary actions for a secure climate rely on our relationships with nature and one another, and the transformations that fulfil society's demand for a secure climate can also yield significant benefits for mental health and wellbeing.

The connections between climate change and mental health discussed in this paper are extensive and often reciprocal; comprehending the numerous ways in which our mental health is intertwined with the health of the planet necessitates a holistic or systems approach, along with a synthesis of evidence from various fields (Frtize et al., 2007). This narrative review presents what we consider to be the most thorough overview to date of the evidence pertinent to the emerging yet expanding domain of climate change and mental health. By extending our focus beyond literature that explicitly addresses climate change to incorporate insights from related literature, such as studies on the mental health effects of hurricanes or forced migration, and by including grey literature, we have amalgamated a varied knowledge base to create a more comprehensive understanding of how climate change influences mental health and well-being. Furthermore, this review surpasses the synthesized evidence base of earlier reviews (Patel et al., 2018) by summarizing the climate psychology literature, which has experienced significant growth in recent years, with substantial evidence surfacing since the publication of previous reviews. Future research should focus on clarifying:

(1) The developing understanding of which factors contribute to vulnerability or resilience against the extensive impacts of the climate crisis;
(2) Whether any specific sub-groups of individuals are significant for targeting interventions, and
(3) Which interventions are most effective for various contexts and climate-related stressors?

This review has demonstrated that the psychological, emotional, and social impacts of climate change will intensify as more individuals directly confront its effects, coupled with a growing awareness of the crisis, particularly in light of insufficient remedial measures. This situation underscores the pressing need for world leaders to expedite efforts aimed at reducing greenhouse gas emissions to prevent these disproportionate burdens on society, while also aiding communities in adapting to a warming planet. Actions must be rooted in the requirements of citizens to guarantee the equity and acceptability of the transition to a low-carbon economy, ensuring that resources are allocated where they are most necessary and effective. We have emphasized the various pathways through which climate change can affect mental health and wellbeing; comprehending these pathways and their variations across different cultural contexts is essential for formulating suitable interventions and remedial actions. In conclusion, although the risks that climate change poses to mental health and wellbeing are considerable, the advantages of climate action are equally significant. The challenges presented by climate change offer a critical opportunity to create a more sustainable and healthier future for everyone.

REFERENCES

Austin, E. K., Rich, J. L., Kiem, A. S., Handley, T., Perkins, D., & Kelly, B. J. (2020). Concerns about climate change among rural residents in Australia. *Journal of Rural Studies*, *75*, 98–109. DOI: 10.1016/j.jrurstud.2020.01.010

Bando, D. H., Teng, C. T., & Volpe, F. M. Masi, E de., Pereirame, L.A., & Braga, A.L. (2017). Suicide and meteorological factors in São Paulo, Brazil, 1996-2011: a time series analysis. *Braz J Psychiatry*, 39:220–7. https://doi.org/DOI: 10.1590/1516-4446-2016-2057

Baylis, P., Obradovich, N., Kryvasheyeu, Y., Chen, H., Coviello, L., & Moro, E. (2018). Weather impacts expressed sentiment. *PLoS One*, *13*(4), e0195750. DOI: 10.1371/journal.pone.0195750 PMID: 29694424

Bhullar, N., Davis, M., Kumar, R., Nunn, P., & Rickwood, D. (2022). Climate anxiety does not need a diagnosis of a mental health disorder. *The Lancet. Planetary Health*, *6*(5), e383. DOI: 10.1016/S2542-5196(22)00072-9 PMID: 35550075

Black, R., Bennett, S. R. G., Thomas, S. M., & Beddington, J. R. (2011). Migration as adaptation. *Nature*, *478*(7370), 447–449. DOI: 10.1038/478477a PMID: 22012304

Bolton, J. M., Walld, R., Chateau, D., Finlayson, G., & Sareen, J. (2015). Risk of suicide and suicide attempts associated with physical disorders: A population-based, balancing score-matched analysis. *Psychological Medicine*, *45*(3), 495–504. DOI: 10.1017/S0033291714001639 PMID: 25032807

Bonilla-Escobar, F., Osorio-Cuellar, G., Pacichana-Quinayaz, S. G., Sanchez-Renteria, G., & Fandino-Losada, A. (2017). Do not forget culture when implementing mental health interventions for violence survivors. *Cienc Saude Coletiva*, 22 (9):3053–9. https://doi.org/DOI: 10.1590/1413-81232017229.12982016

Bowles, D. C., Butler, C. D., & Morisetti, N. (2015). Climate change, conflict and health. *Journal of the Royal Society of Medicine*, *108*(10), 390–395. DOI: 10.1177/0141076815603234 PMID: 26432813

Boykoff, M. T., & Boykoff, J. M. (2007). Climate change and journalistic norms: A case-study of US mass-media coverage. *Geoforum*, *38*(6), 1190–1204. DOI: 10.1016/j.geoforum.2007.01.008

Buguet, A. (2007). Sleep under extreme environments: Effects of heat and cold exposure, altitude, hyperbaric pressure and microgravity in space. *Journal of the Neurological Sciences*, *262*(1–2), 145–152. DOI: 10.1016/j.jns.2007.06.040 PMID: 17706676

Burke, M., Hsiang, S. M., & Miguel, E. (2015). Global non-linear effect of temperature on economic production. *Nature, 527*(7577), 235–239. DOI: 10.1038/nature15725 PMID: 26503051

Burke, M., Hsiang, S. M., & Miguel, E. (2015). Climate and Conflict. *Annual Review of Economics, 7*(1), 577–617. DOI: 10.1146/annurev-economics-080614-115430

Burroughs, J. E., & Rindfleisch, A. (2002). Materialism and Well-Being: A Conflicting Values Perspective. *The Journal of Consumer Research, 29*(3), 348–370. DOI: 10.1086/344429

Campo-Arias, A., & Herazo, E. (2014). Stigma and mental health in victims of Colombia's internal armed conflict in a situation of forced displacement. *Allman, editor. Estigma Salud Ment En Pers Vict Confl Armado Interno Colomb En Situacion Desplazamiento Forzado*, 43(4):212–7.

Carey, I. M., Anderson, H. R., Atkinson, R. W., Beevers, S. D., Cook, D. G., & Strachan, D. P. (2018). Are noise and air pollution related to the incidence of dementia? A cohort study in London, England. *BMJ Open, 8*(9), e022404. DOI: 10.1136/bmjopen-2018-022404 PMID: 30206085

Charlson, F., Ali, S., Benmarhnia, T., Pearl, M., Massazza, A., & Augustinavicius, J. (2021). Climate Change and Mental Health: A Scoping Review. *International Journal of Environmental Research and Public Health, 18*(9), 4486. PMID: 33922573

Chen, S., Oliva, P., & Zhang, P. (2022). *Air Pollution and Mental Health: Evidence from China*. National Bureau of Economic Research (NBER Working Papers). Report No.: w24686. http://www.nber.org

De Hert, M., Correll, C. U., Bobes, J., Cetkovich-Bakmas, M., Cohen, D., & Asai, I. (2011). Physical illness in patients with severe mental disorders. I. Prevalence, impact of medications and disparities in health care. *World Psychiatry; Official Journal of the World Psychiatric Association (WPA), 10*(1), 52–77. DOI: 10.1002/j.2051-5545.2011.tb00014.x PMID: 21379357

Ding, N., Berry, H. L., & Bennett, C. M. (2016). The importance of humidity in the relationship between heat and population mental health: Evidence from Australia. *PLoS One, 11*(10), e0164190. DOI: 10.1371/journal.pone.0164190 PMID: 27727320

Dittmar, H., Bond, R., Hurst, M., & Kasser, T. (2014). The relationship between materialism and personal well-being: A meta-analysis. - PsycNET. *Journal of Personality and Social Psychology, 107*(5), 879–924. DOI: 10.1037/a0037409 PMID: 25347131

Dixon, P. G., Sinyor, M., Schaffer, A., Levitt, A., Haney, C. R., & Ellis, K. N. (2014). Association of weekly suicide rates with temperature anomalies in two different climate types. *International Journal of Environmental Research and Public Health, 11*(11), 11627–11644. DOI: 10.3390/ijerph111111627 PMID: 25402561

Elgar, F. J., Pickett, W., Pförtner, T. K., Gariépy, G., Gordon, D., & Georgiades, K. (2021). Relative food insecurity, mental health and wellbeing in 160 countries. *Social Science & Medicine, 268*, 113556. DOI: 10.1016/j.socscimed.2020.113556 PMID: 33293171

Ford, J. D., & King, D. (2015). Coverage and framing of climate change adaptation in the media: A review of influential North American newspapers during 1993–2013. *Environmental Science & Policy, 48*, 137–146. DOI: 10.1016/j.envsci.2014.12.003

Frankish, H., Boyce, N., & Horton, R. (2018). Mental health for all: A global goal. *Lancet, 392*(10157), 1493–1494. DOI: 10.1016/S0140-6736(18)32271-2 PMID: 30314858

Gaber, N., Silva, A., Lewis-Patrick, M., Kutil, E., Taylor, D., & Bouier, R. (2021). Water insecurity and psychosocial distress: Case study of the Detroit water shutoffs. *J Public Health, 43*(4), 839–845. DOI: 10.1093/pubmed/fdaa157 PMID: 32930795

Gao, J., Cheng, Q., Duan, J., Xu, Z., Bai, L., & Zhang, Y. (2019). Ambient temperature, sunlight duration, and suicide: A systematic review and meta-analysis. *The Science of the Total Environment, 646*, 1021–1029. DOI: 10.1016/j.scitotenv.2018.07.098 PMID: 30235587

Gleeson, C., Frost, R., Sherwood, L., Shevlin, M., Hyland, P., & Halpin, R. (2020). Post-migration factors and mental health outcomes in asylum-seeking and refugee populations: A systematic review. *European Journal of Psychotraumatology, 11*(1), 1793567. DOI: 10.1080/20008198.2020.1793567 PMID: 33312450

Global Burden of Diseases. (2019). Mental Disorders Collaborators. Global, regional, and national burden of 12 mental disorders in 204 countries and territories, 1990–2019: A systematic analysis for the Global Burden of Disease Study 2019. *The Lancet. Psychiatry, 9*(2), 137–150.

Gupta, S., & Murray, R. M. (1992). The Relationship of Environmental Temperature to the Incidence and Outcome of Schizophrenia. *The British Journal of Psychiatry, 160*(6), 788–792. DOI: 10.1192/bjp.160.6.788 PMID: 1617362

Hayes, K., & Poland, B. (2018). Addressing Mental Health in a Changing Climate: Incorporating Mental Health Indicators into Climate Change and Health Vulnerability and Adaptation Assessments. *International Journal of Environmental Research and Public Health*, *15*(9), 1806. DOI: 10.3390/ijerph15091806 PMID: 30131478

Henderson, S. B., McLean, K. E., Lee, M. J., & Kosatsky, T. (2022). Analysis of community deaths during the catastrophic 2021 heat dome. *Environmental Epidemiology (Philadelphia, Pa.)*, *6*(1), e189. DOI: 10.1097/EE9.0000000000000189 PMID: 35169667

Hsiang, S. M., & Burke, M. (2014). Climate, conflict, and social stability: What does the evidence say? *Climatic Change*, *123*(1), 39–55. DOI: 10.1007/s10584-013-0868-3

Hu, J., Wen, Y., Duan, Y., Yan, S., Liao, Y., & Pan, H. (2020). The impact of extreme heat and heat waves on emergency ambulance dispatches due to external cause in Shenzhen, China. *Environ Pollut*, *261*, 114156. DOI: 10.1016/j.envpol.2020.114156 PMID: 32092626

Hurley, E. A., Dalglish, S. L., & Sacks, E. (2022). Supporting young people with climate anxiety: Mitigation, adaptation, and resilience. *The Lancet. Planetary Health*, *6*(3), e190. DOI: 10.1016/S2542-5196(22)00015-8 PMID: 35278385

Intergovernmental Panel on Climate Change. (2014). [*Summary for Policymakers.* https://www.ipcc.ch]. *Climatic Change*, *2014*, •••.

James, P., Iyer, A., & Webb, T. L. (2019). The impact of post-migration stressors on refugees' emotional distress and health: A longitudinal analysis. *European Journal of Social Psychology*, *49*(7), 1359–1367. DOI: 10.1002/ejsp.2589

Kelman, I., Ayeb-Karlsson, S., Rose-Clarke, K., Prost, A., Ronneberg, E., & Wheeler, N. (2021). A review of mental health and wellbeing under climate change in small Island developing states (SIDS). *Environ Res Lett ERL Web Site*, *16*(3), 033007. DOI: 10.1088/1748-9326/abe57d PMID: 34149865

Lee, S., Lee, H., Myung, W., Kim, E. J., & Kim, H. (2018). Mental disease-related emergency admissions attributable to hot temperatures. *The Science of the Total Environment*, *616*, 688–694. DOI: 10.1016/j.scitotenv.2017.10.260 PMID: 29126638

Linkowski, P., Martin, F., & De Maertelaer, V. (1992). Effect of some climatic factors on violent and non-violent suicides in Belgium. *Journal of Affective Disorders*, *25*(3), 161–166. DOI: 10.1016/0165-0327(92)90001-m PMID: 1527270

Liu, J., Varghese, B., Hansen, A., Xiang, J., Zhang, Y., & Dear, K. (2021). Is there an association between hot weather and poor mental health outcomes? A systematic review and meta-analysis. *Environment International*, •••, 153. PMID: 33799230

McManus, S., Gunnell, D., Cooper, C., Bebbington, P. E., Howard, L. M., & Brugha, T. (2019). Prevalence of non-suicidal self-harm and service contact in England, 2000–14: Repeated cross-sectional surveys of the general population. *The Lancet. Psychiatry*, *6*(7), 573–581. DOI: 10.1016/S2215-0366(19)30188-9 PMID: 31175059

Middleton, J., Cunsolo, A., Pollock, N., Jones-Bitton, A., Wood, M., & Shiwak, I. (2021). Temperature and place associations with Inuit mental health in the context of climate change. *Environmental Research*, •••, 198. PMID: 33857460

Moldes, O., & Ku, L. (2020). Materialistic cues make us miserable: A meta-analysis of the experimental evidence for the effects of materialism on individual and societal well-being. *Psychology and Marketing*, *37*(10), 1396–1419. DOI: 10.1002/mar.21387

Palmer, D., & Ward, K. (2007). 'Lost': Listening to the voices and mental health needs of forced migrants in London. *Medicine, Conflict, and Survival*, *23*(3), 198–212. DOI: 10.1080/13623690701417345 PMID: 17822063

Parry, S., McCarthy, S. R., & Clark, J. (2022). Young people's engagement with climate change issues through digital media–a content analysis. *Child and Adolescent Mental Health*, *27*(1), 30–38. DOI: 10.1111/camh.12532 PMID: 34904349

Patel, V., Saxena, S., Lund, C., Thornicroft, G., Baingana, F., & Bolton, P. (2018). The Lancet Commission on global mental health and sustainable development. *Lancet*, *392*(10157), 1553–1598. PMID: 30314863

Pihkala, P. (2022). Toward a Taxonomy of Climate Emotions. *Front Clim*, 3(3:738154).

Pritchard, A., Richardson, M., Sheffield, D., & McEwan, K. (2020). The relationship between nature connectedness and eudaimonic well-being: A meta-analysis. *Journal of Happiness Studies*, *21*(3), 1145–1167. DOI: 10.1007/s10902-019-00118-6

Reifels, L. (2018). *Reducing Mental Health Risks in Emergencies: MHPSS through a Disaster Risk Reduction Lens*. MHPSS Network Webinar: MHPSS & DRR.

Siriwardhana, C., & Stewart, R. (2013). Forced migration and mental health: Prolonged internal displacement, return migration and resilience. *International Health*, *5*(1), 19–23. DOI: 10.1093/inthealth/ihs014 PMID: 24029841

Stewart-Brown, S. (1998). Emotional wellbeing and its relation to health. *BMJ (Clinical Research Ed.)*, *317*(7173), 1608–1609. DOI: 10.1136/bmj.317.7173.1608 PMID: 9848897

Trautmann, S., Rehm, J., & Wittchen, H. U. (2016). The economic costs of mental disorders: Do our societies react appropriately to the burden of mental disorders? *EMBO Reports*, *17*(9), 1245–1260. PMID: 27491723

Vergunst, F., & Berry, H. L. (2021). Climate Change and Children's Mental Health: A Developmental Perspective. *Clinical Psychological Science*, •••, 21677026211040788. PMID: 35846172

Vince, G. (2022). How scientists are coping with 'ecological grief'. *The Observer*. https://www.theguardian.com

Vu, H. T., Liu, Y., & Tran, D. V. (2019). Nationalizing a global phenomenon: A study of how the press in 45 countries and territories portrays climate change. *Global Environmental Change*, *58*, 101942. DOI: 10.1016/j.gloenvcha.2019.101942 PMID: 32288343

Weintrobe, S. (2021). *Psychological roots of the climate crisis: neoliberal exceptionalism and the culture of uncare*. Bloomsbury Publishing.

WHO. (2021). *Mental health: strengthening our response*. https://www.who.int

WHO. (2021). *Mental disorders*. https://www.who.int

World Health Organization (WHO). (2005). *Promoting mental health: concepts, emerging evidence, practice: a report of the World Health Organization*. Department of Mental Health and Substance Abuse in collaboration with the Victorian Health Promotion Foundation and the University of Melbourne. World Health Organization.

Zeira, A. (2022). Mental Health Challenges Related to Neoliberal Capitalism in the United States. *Community Mental Health Journal*, *58*(2), 205–212. DOI: 10.1007/s10597-021-00840-7 PMID: 34032963

ADDITIONAL READING

Charlson, F., Ali, S., Benmarhnia, T., Pearl, M., Massazza, A., Augustinavicius, J., & Scott, J. G. (2021). Climate change and mental health: A scoping review. *International Journal of Environmental Research and Public Health*, *18*(9), 4486. PMID: 33922573

Cianconi, P., Betrò, S., & Janiri, L. (2020). The impact of climate change on mental health: A systematic descriptive review. *Frontiers in Psychiatry*, *11*, 74. PMID: 32210846

Cuomo, C. J. (2011). Climate change, vulnerability, and responsibility. *Hypatia*, *26*(4), 690–714. https://www.jstor.org/stable/41328876

Ebi, K. L., & Hess, J. J. (2020). Health risks due to climate change: Inequity in causes and consequences. *Health Affairs, 39*(12), 2056–2062. PMID: 33284705

Hayes, K., Blashki, G., Wiseman, J., Burke, S., & Reifels, L. (2018). Climate change and mental health: Risks, impacts and priority actions. *International Journal of Mental Health Systems, 12*(28), 1–12. DOI: 10.1186/s13033-018-0210-6 PMID: 29881451

Jelks, N. O., Jennings, V., & Rigolon, A. (2021). Green gentrification and health: A scoping review. *International Journal of Environmental Research and Public Health, 18*(3), 1–23. DOI: 10.3390/ijerph18030907 PMID: 33494268

Kumar, N. (2018). Urban climate-health strategies. *Harvard Public Health Review (Cambridge, Mass.), 17*, 1–6. https://www.jstor.org/stable/10.2307/48515214

Perez, V. W., & Egan, J. (2016). Knowledge and concern for sea-level rise in an urban environmental justice community. *Sociological Forum, 31*(S1), 885–907. https://www.jstor.org/stable/24878752

World Health Organization. (2013). *Building back better: sustainable mental health care after emergencies: Overview*. www.who.int

KEY TERMS AND DEFINITIONS

Climate grief/ Ecological Grief/ Eco-Grief: the grief felt about experienced or anticipated ecological losses, including the loss of species, ecosystems, and meaningful landscapes due to acute or chronic environmental change.

Eco-anxiety/Climate Anxiety: the chronic fear of environmental doom.

Eco-distress/Climate Distress: a way of describing the wide range of thoughts and emotions people may experience when they hear bad news about our planet and the environment.

Solastalgia: The distress experienced when your home environment is changing in ways perceived to be profoundly negative, either by events related to climate change but also scenarios such as the establishment of a mine or a volcanic eruption. Changes that attack our sense of home, even without dislocation, can create a deep homesickness.

Chapter 2 The Role of Community Support in Addressing Mental Health Challenges From Climate Displacement: Impacts of Community Networks

Primrose Sumasumo
https://orcid.org/0009-0008-6057-4117
Scientific and Industrial Research and Development Centre, Zimbabwe

ABSTRACT

This chapter examines the role that community support plays to address the mental health issues that people who have been displaced by climate change experience. The chapter looks into how psychological distress such as anxiety, depression, and post-traumatic stress disorder, is increased by climate displacement, especially for vulnerable groups including women, children, and Indigenous communities. The chapter emphasizes the role that social capital bonding, bridging, and linking plays in building resilience and helping the speeding up of the healing process. It examines successful community led and NGO partnered mental health programs that incorporate local involvement and cultural sensitivity. The chapter also covers obstacles to getting community support, such as a lack of services, cultural difficulties, stigma, and structural injustices. Lastly, it makes legislative recommendations to increase equal access to mental health care and fortify community networks.

DOI: 10.4018/979-8-3373-3281-9.ch002

INTRODUCTION

Climate change is now associated with an increase in global displacement and a rise in mental health problems, which places a double burden on vulnerable populations around the world. Stronger extreme weather events, such as hurricanes, floods, and droughts, are displacing millions of people every year, distressing livelihoods, and hindering social networks, according to the Intergovernmental Panel on Climate Change (IPCC) (Arias et al., 2021). For example, nearly 20 million people were internally displaced by climate-related risks in 2023 alone; this number is expected to increase as global warming increase (IIED, 2022). Displacement causes severe psychological and physical harm; research shows that impacted groups have higher rates of anxiety, depression, and post-traumatic stress disorder (PTSD) (WHO, 2021).

Systemic injustices increase the effects of displacement on mental health, due to financial inequalities and cultural displacement. Marginalized groups including women, children, and Indigenous communities face increased vulnerability. The intergenerational trauma of climatic disasters is highlighted by the fact that 29% of children in Indonesia who were exposed to frequent tidal floods expressed mild depression and 93% indicated moderate anxiety (Ellis, 2022). Similar to this, migration brought on by drought in Kenya's Turkana region increased economic desperation and gender-based violence, with women turning to survival sex work (Ellis, 2022). These results highlight the ways that climate displacement affects social determinants of health, such as safety, economic stability, and community cohesiveness, in addition to surroundings.

Support from the community becomes a vital barrier against the decline of mental health in this situation. Through the reconstruction of collective identity and agency, community support defined as official and informal networks that offer emotional, social, and practical resources promotes resilience. For instance, by combining ancient customs with contemporary planning, participatory relocation initiatives in Indonesia helped flood-displaced communities regain hope and social cohesiveness (Burrows et al., 2021).On the other hand, diminished social connections like those seen in the Lake Chad Basin of Niger increase the risk of mental illness, with relocated families reporting substance misuse and domestic violence due to a lack of resources (Ellis, 2022).

This chapter looks at how community-driven tactics, like peer networks, culturally sensitive therapies, and group coping techniques, can lessen trauma and aid in healing. Additionally, it criticizes the stigma, language hurdles, and disjointed health systems that disproportionately impact displaced people as obstacles to effective support (UN, 2023). This analysis attempts to promote fair, culturally aware frameworks for mental health care in a time of climate-driven relocation by combining data from frontline areas.

LITERATURE REVIEW

The Role of Social Capital in Recovery and Resilience

A foundational component of community support is social capital which offers a framework for comprehending the ways in which interconnection and shared resources affect mental health. Social capital is defined as networks, norms, and trust which enables the community to act together effectively by pursuing shared objectives According to (Putnam, 2000), social capital is the intangible resources that communities can use to boost resilience and ease the process of recovering from disaster. It is defined as the networks, norms, and trust that allow members to act together more effectively to accomplish common objectives. It fosters collaboration, support amongst individuals, and a sense of shared responsibility. It is not just the sum of the individual ties; rather, it is the collective value produced from these connections. Communities with strong social capital are more resilient to traumas the groups would feel less powerless and loneliness. For instance migrants from African nations with significant climate vulnerability, for example, reported larger correlations between social capital loss and emotional problems than migrants from less vulnerable locations (Di Giorgi et al., 2020). Similarly, refugees who had greater cognitive social capital (like trust) and structural social capital (like community involvement) demonstrated better mental health outcomes, which were mediated by a stronger feeling of coherence (van Sint Fiet et al., 2022). In addition, a study from Putnam's claim that social capital encourages cooperation and shared accountability, empowering groups to overcome hardship together (Putnam, 2000). Social capital can be divided into three categories bonding, bridging, and linking capital to better understand its complex nature (Szreter & Woolcock, 2004).

Bonding Capital

Relationships within homogeneous groups, such as family, close friends, and neighbours, are referred to as bonding capital. These intimate bonds serve as an important source of emotional support, a feeling of community, and a protective barrier against loneliness and stress. Since people depend on familiar ties to help them deal with the trauma of loss and uncertainty, bonding capital can be especially important for preserving psychological well-being in the setting of climate induced migration (Szreter & Woolcock, 2004). Populations maintaining family and neighbourly relationships reported better psychosocial well-being as they have emotional support and a sense of belonging. For example, displaced families in the Lake Chad Basin of Niger who lacked bonding capital reported higher rates of domestic violence and substance misuse (UN, 2023). Studies conducted after a disaster in New Orleans also show that bonding capital, which promotes mutual aid and trust, lessens PTSD symptoms. The Hurricane Katrina survivors who relied on family networks for emotional support recovered more quickly compared to those who were socially isolated in New Orleans (Rhodes et al., 2010). In refugee camps where interventions like bonding capital which includes peer-led support groups, have helped rebuild collective identity and lessen stigma, interventions that increase bonding (Villalonga-Olives et al., 2023).

Bridging Capital

Bridging capital refers to the connections between different groups or communities by providing access to a variety of resources and viewpoints, bridging capital helps displaced people integrate into their new communities. These wider networks make it easier to access a variety of materials, data, and viewpoints, which encourages creativity and problem-solving. In order to help displaced populations navigate new institutions, find work, and assimilate into new communities, bridging capital can be quite helpful. Bridging capital can also lessen prejudice and increase social cohesion by creating ties across social divisions. Relocation projects in Indonesia used bridging capital to provide access to housing, healthcare, and livelihood opportunities by connecting flood-displaced communities with NGOs and local governments (Burrows et al., 2021). However, bridge attempts are frequently hampered by discrimination in host communities. For example, systematic exclusion of climate migrants in Europe worsened depression and anxiety (Ellis, 2022). By encouraging empathy between the host and displaced populations, multilevel interventions like cultural exchange workshops can combat stigma (Villalonga-Olives et al., 2023).

Linking Capital

Linking Capital Involves connections to individuals or institutions in positions of power or authority. Institutional access and advocacy addressing structural injustices requires linking money, which ties marginalized populations to power institutions. Due to cultural insensitivity, bureaucratic obstacles, and language problems, displaced populations frequently do not have access to mental health care (UN, 2023). Funding for psychosocial support has been successfully obtained by initiatives that prepare community leaders to push for legislative reforms, such as Kenya's drought-affected women's coalitions (IIED, 2022). Communities can receive necessary services, advocate for their needs, and have an impact on policy decisions because to these vertical links. For disadvantaged and marginalized groups, who might not have the means or clout to take care of their own needs, linking capital is especially crucial. Similar to this, women in Kenya's Turkana region lost their communal safety due to drought-induced migration, which resulted in gender-based violence and survival sex work (IIED, 2022). People become isolated and psychologically susceptible as a result of these disturbances, which also destroy trust and collective identity.

Therefore, the vital role that social capital plays in mental health has been emphasized more and more in recent research. Higher levels of social capital, for example, were linked to decreased rates of anxiety and depression in older people, according to research by (Bai et al., 2020) This implies that social connection-building therapies which could be helpful in fostering mental health in older adults. In a similar study (Villalonga-Olives et al., 2022), shows that social capital mediates the association between refugees' mental health outcomes and their exposure to traumatic experiences. According to their research, robust social networks can mitigate the negative impacts of trauma by offering a feeling of security, encouragement, and self-determination. These studies highlight the potential of social capital as a protective factor in the face of adversity, emphasizing the need for interventions that strengthen community connections and promote social inclusion.

THE ROLE OF RESILIENCE BUILDING

Resilience is defined by the American Psychological Association as the capacity to bounce back swiftly from adversity requires community support. But resilience is not only a personal trait, it is linked to the social and environmental setting in which people live (Posklinsky & Nouman, 2025). Building resilience requires community support, which offers a variety of tools and systems that encourage constructive adaptation in the face of hardship. These techniques include exchanging experiences, facilitating access to necessary resources, and establishing a collective identity. Community support contributes to resilience through several mechanisms:

Collective Identity

The development of a shared identity is one of the main ways that community support increases resilience. Mutual understanding and a sense of belonging are created when people share their ideas and experiences, which promotes support and solidarity. Through shared experiences, a strong collective identity fosters a sense of belonging and offers emotional support and resilience. By strengthening common beliefs, customs, and stories, this bond prevents isolation and aids people in preserving their sense of direction and continuity in the face of change. This is supported by research, that connects group identification to better mental health, such as increased self-esteem and decreased rates of anxiety and sadness (Ellis, 2022).

For displaced populations who could have lost their homes, means of subsistence, and social networks this shared identity might be especially crucial (Farny & Dentoni, 2025). Additionally, collective identity gives displaced people the ability to fight for their rights. Strong group identification promotes collaboration, raises awareness, and influences community-beneficial policy changes. Research indicates that social identity promotes resilience and improves psychological well-being (Farny & Dentoni, 2025). However, obstacles like language and cultural dispersion make it difficult to create and preserve this identity. Programs for community assistance must take care of these, establishing welcoming environments that respect many identities and promote a sense of community. The literature highlights the value of belonging to several groups and suggests that having a variety of social relationships might improve one's well-being. For displaced people and communities to heal and rehabilitate, collective identity is therefore essential (Ellis, 2022).

Shared Experiences

Additionally, groups of people who have gone through comparable traumatic experiences might offer special kinds of assistance that promote emotional processing and recovery. The idea of shared experiences is crucial to the mental health and general well-being of displaced communities and is fundamentally linked to collective identity. Even though the circumstances of each individual can vary greatly, displacement is by definition a shared experience (USCRI, 2024). Those who have been forcibly dislocated can develop a sense of solidarity, understanding, and collective resilience by acknowledging and validating these shared experiences.

From the trauma of leaving one's home and losing loved ones to the difficulties of adjusting to new surroundings and the prejudice encountered as refugees or internally displaced people, shared experiences cover a broad spectrum of situations and feelings (Leap & Thompson, 2018). By recognizing these shared experiences, the community may develop empathy and support. People can find safe places to share their stories, express their feelings, and get support and validation from others who have experienced similar things through support groups, healing circles, and community-based rituals. According to (Leap & Thompson, 2018) this process of group healing can assist people in understanding their experiences, lowering emotions of guilt and shame, and creating more flexible coping mechanisms. Community-based support groups, storytelling projects, and group therapy are all powerful means of establishing forums for the recognition and discussion of common experiences, which promotes healing and a feeling of group strength.

Access to Resources

Lastly, the mental health and general well-being of displaced people are significantly influenced by their access to resources. Losing homes, livelihoods, and social support systems as a result of displacement frequently makes people and families vulnerable and in need of aid. For displaced people to have fundamental needs met and their lives stabilized, they must have enough access to resources including food, shelter, medical treatment, education, and legal assistance wellbeing (Ejiohuo et al., 2024). In order to support mental health and well-being, these resources are critical for providing for basic requirements and lowering stress. Inadequate availability of these services might worsen pre-existing mental health issues and lead to the emergence of new ones. For example, persistent stress, anxiety, and depression can be caused on by food shortages and substandard housing. Inadequate access to medical treatment can lead to untreated mental and physical health issues, which further impacts wellbeing (Ejiohuo et al., 2024). Furthermore, the failure to obtain education and employment might lower self-esteem and limit prospects for social integration. Addressing these resource shortfalls requires a multifaceted approach that includes governments, humanitarian organizations, and local people. This includes providing immediate help, pushing for policy reforms to promote fair resource access, and encouraging displaced people to engage in life-changing decision-making processes.

The significance of context in comprehending resilience has been emphasized by recent studies. (Ungar, 2006) highlights that social support and community resources play a role in determining resilience, which is not just an individual characteristic. In order to encourage constructive adaptation in the face of adversity, his work emphasizes the necessity of interventions that address variables at the individual and community levels. Similarly, community-based interventions that foster social cohesiveness and collective efficacy can improve resilience among communities affected by disasters, according to a study by (Patel et al., 2011)These results imply that interventions that foster social inclusion, fortify community networks, and enable communities to act collectively can be especially successful in fostering resilience. By fostering a sense of belonging, providing access to essential resources, and promoting collective healing, community support can play a vital role in helping individuals and communities recover from the mental health impacts of climate-induced displacement.

METHODOLOGY

In this chapter, the data on mental health and psychosocial support (MHPSS) for displaced people is compiled from a comprehensive analysis of peer-reviewed literature. The literature search, using terms like "displacement," "refugees," "mental health," "psychosocial support," and "cultural competence," the literature search used a number of databases, including PubMed and Google Scholar. Studies that addressed the views of displaced people and program providers were given priority under the inclusion criteria, with an emphasis on empirical data about the efficacy of interventions, care barriers, and culturally appropriate methods.

In order to uncover important elements influencing mental health outcomes, such as stigma, mistrust, and community support mechanisms, data extraction involved thematic synthesis. Analysis of intervention approaches such as community savings groups, Self-Help Plus (SH+), and cooperative frameworks including local governments and non-governmental organizations were also included in the methodology. Variability in study designs and circumstances, as well as difficulties extrapolating results to a variety of displaced groups, are among the limitations.

Data extraction entailed theme synthesis to identify significant elements influencing mental health outcomes, such as stigma, mistrust, and community support systems. The research also included a review of intervention methods such as Self-Help Plus (SH+), community savings clubs, and collaborative frameworks involving NGOs and local governments. Limitations include study design and context diversity, as well as difficulties in generalizing findings among varied displaced populations.

DISCUSSION

Intervention Programs: Leveraging Community Support

In order to effectively address the complex mental health needs of communities displaced by conflict and climate change, community assistance is essential. The enlarged examples that follow show how community-led projects and NGO collaborations use social capital, cultural significance, and participatory methods to build resilience, lessen psychological suffering, and encourage long-term healing.

Community-Led Mental Health Initiatives

The program is organized on community centres, which provide displaced people with safe havens. The initiative relies heavily on community leaders and local groups to mobilize resources and identify the community's unique mental health needs. Peer support groups, counselling, and workshops on culturally appropriate mental health are just a few of the services the organization provides. Community Led Mental Health Initiative for displaced groups who frequently experience trauma, loss, and uncertainty, community led mental health programs place a strong emphasis on empowerment, cultural sensitivity, and local ownership. Community centres have emerged as crucial locations for displaced people to obtain psychosocial care ingrained in local social and cultural reality in the context of Jordan's Syrian refugee crisis (Panichkriangkrai et al., 2022).

A variety of services are offered by these institutions, such as individual counselling, psychoeducation courses, peer support groups, and culturally appropriate healing techniques like music, storytelling, and customary rituals. These programs' participatory structure guarantees that mental health services are co-created with community people rather than enforced, which increases acceptability and confidence. To close the gaps between displaced communities and official health systems, local leaders and volunteers receive training on how to recognize mental health issues, run group sessions, and make referrals (Panichkriangkrai et al., 2022). Through the normalization of discussions about mental health inside well-known social networks, this grassroots method also combats stigma, a major obstacle to care in many cultures. In addition, to emphasizing resilience building via shared experiences and group bonding, these workshops frequently integrate traditional healing methods (Larrieta et al., 2022)Studies have demonstrated that these community-led mental health programs can considerably lessen participants' symptoms of anxiety, sadness, and PTSD

The effectiveness of these community-led programs is supported by empirical data. According to (Peconga & Hogh Thogersen, 2020), Syrian refugees who took part in these sessions reported notable improvements in their PTSD, anxiety, and depression symptoms. In addition, participants reported experiencing a revitalized sense of empowerment and belonging, both of which are critical for psychological healing. These results are consistent with other studies showing that shared experiences and social connections are protective factors against mental health decline in displacement environment (Tol et al., 2011). Furthermore, community-led initiatives that address the socioeconomic determinants of mental health frequently include resilience-building elements including dispute mediation, livelihood assistance, and skill development. These programs assist displaced people in rebuilding their life beyond short-term psychological alleviation by encouraging agency and self-efficacy, hence increasing holistic well-being.

NGO Partnerships for Community Support

Program Example 2: Community Support Partnership with Non-profits Another successful strategy for providing displaced populations with comprehensive mental health and psychosocial support (MHPSS) is the collaboration of NGOs with local communities. Through its community-based mental health initiatives carried out in various settings, such as Bangladesh, Jordan, and Uganda, the International

Rescue Committee (IRC) serves as an example of this strategy (IIED, 2022). By educating local employees and community members in trauma-informed care, culturally sensitive counselling, and community involvement techniques, the IRC's methodology emphasizes capacity building. In environments with limited resources, this decentralized strategy increases the number of mental health professionals and guarantees that services are suitable for the local culture and context. In order to help communities effectively respond to mental health crises, training also covers stress management strategies, psychosocial first aid, and referral networks.

The IRC promotes social events, community discussions, and awareness campaigns to improve social cohesiveness and lessen stigma in addition to clinical interventions. Through these events, displaced people can re-establish social connections, exchange stories, and work together to overcome obstacles (EU, 2022). For instance, it has been demonstrated that community discussions in refugee settlements improve knowledge of mental health concerns and promote actions related to seeking help. These community projects run by NGOs have a well-established impact. According to, (Bolton et al., 2007), community-based mental health interventions significantly decreased the symptoms of anxiety and depression in Ugandan refugees when compared to control groups. By integrating services within reputable community organizations, these programs' collaborative and participatory nature promotes long-lasting gains in mental health.

Additionally, NGO collaborations improve access to specialized care when required by establishing connections between displaced communities and official health institutions. In order to overcome systemic obstacles to mental health treatment like financing gaps, discrimination, and language access, they also support policy changes.

BARRIERS TO COMMUNITY SUPPORT

The success of community support in meeting the mental health needs of displaced populations may be hampered by a number of obstacles, despite the substantial potential advantages. In addition to more general structural problems with resource limitations and social inequality, these obstacles frequently result from the disruptive character of relocation itself. It is essential to comprehend these obstacles in order to create focused plans to lessen their effects and guarantee that community assistance reaches the people who need it most.

Social Isolation

Displacement frequently upends pre-existing social networks, which makes impacted people feel more socially isolated. People who are compelled to leave their homes and towns may become isolated and lonely as a result of losing touch with their neighbours, family, and friends. Losing social ties can have a significant effect on mental health by raising the risk of anxiety, depression, and other mental health issues. Feelings of loneliness and isolation can be made worse by the trauma of relocation and the strain of acclimating to a new setting. Social isolation has been shown to substantially impact mental health outcomes, especially for refugees and displaced people who may already be dealing with stressors connected to trauma (Forray et al., 2024).

After being displaced, some communities are more susceptible to social isolation. For instance, older people may find it challenging to make new friends and adjust to new surroundings. Due to social stigma or physical restrictions, people with disabilities may encounter additional obstacles to social involvement. Discrimination and exclusion may be experienced by members of marginalized communities, further separating them from the general population. To meet their unique needs and encourage social inclusion, these vulnerable groups need focused interventions. According to studies, refugees with disabilities frequently face additional difficulties, such as stigma and restricted access to community resources, which can make them feel even more alone (Forray et al., 2024). A number of strategies can be used to help displaced populations overcome social isolation and foster a sense of community:

Community-Building Events

It can be quite successful to plan social gatherings and activities that unite people and promote a feeling of community. These activities can include everything from volunteer opportunities and community gardening initiatives to sporting competitions and cultural festivals. The secret is to give people the chance to engage with one another, exchange stories, and form bonds. It has been demonstrated that community participation programs promote social cohesiveness and refugee mental health outcomes (Versey et al., 2023).

In-depth, well-planned community-building activities can serve as effective healing and integration catalysts. These gatherings can be designed to meet the unique psychological requirements of displaced communities in addition to providing a venue for social contact. For instance, guided meditation sessions can encourage relaxation and stress reduction, while therapeutic arts and crafts workshops can offer a nonverbal means of processing trauma. People can develop resilience and make sense of their experiences by including aspects of narrative therapy and collaborative storytelling (Versey et al., 2023). It is crucial to use culturally sensitive methods involving local cultural brokers and community leaders in the planning process guarantees that events are inclusive of many customs and beliefs, relevant, and respectful.

Additionally, these gatherings can provide a forum for sharing important information about resources like career prospects, legal assistance, and mental health services, enabling people to adjust to their new surroundings more confidently. Community gatherings have been shown to greatly lessen feelings of loneliness and foster social networks, both of which are essential for mental health (Riza et al., 2020). Community building activities can be extremely important in reducing the long-term psychological effects of relocation and advancing sustainable integration by creating a feeling of shared identity and support.

Outreach Initiatives

In order to promote resilience and enhance mental health outcomes, outreach workers' emotional support and help with coping strategy development are essential. According to research, these outreach programs especially those that include consistent communication and emotional support help participants feel more connected and have less symptoms of anxiety and depression (Phang et al., 2023). Additionally, programs that incorporate peer support components help people feel like they belong, which is important for those who are lonely. These outreach initiatives help isolated people's overall quality of life by attending to their practical and emotional needs and making it easier for them to receive necessary resources like housing aid and mental health treatment (Jiménez et al., 2021). These initiatives use a range of tactics, such as home visits to evaluate living situations and foster trust, frequent phone calls to offer emotional support and track wellbeing, and the use of community navigators to link people with available resources and support services. Community navigators are essential to good communication and comprehending the particular difficulties experienced by isolated people because they frequently share cultural origins with the people they assist (Jiménez et al., 2021).

Technology-Based Solutions

For those who have trouble obtaining in-person services, using technology to link them to virtual communities and online support groups can be especially beneficial. Online platforms give people a convenient and safe place to talk about their experiences, get peer support, and find out about mental health resources. It is becoming more widely acknowledged that digital tools are a useful tool for connecting displaced people, particularly when social alienation is occurring.

Studies have indicated that digital health services, such as online support groups, can greatly increase people's activation and participation in health management. By giving users access to essential information and resources, these platforms empower users in addition to facilitating communication (LaMonica et al., 2021). According to a scoping study, for example, digital health services are used in a variety of settings and enable people to interact with peers and healthcare providers in an efficient manner. This is important for those who are unable to access traditional healthcare (Rosenlund et al., 2023)

Furthermore, the COVID-19 epidemic has sped up the implementation of digital health solutions, proving their usefulness in preserving social ties and offering assistance during times of loneliness. Research shows that members of online support groups experienced better mental health outcomes, such as less loneliness and more social support (Leonardsen et al., 2020). This move to digital solutions emphasizes how crucial it is to incorporate technology into mental health treatment, especially for disadvantaged groups that might find it difficult to receive in-person assistance. In summary, technology-based solutions are essential for establishing community, helping people connect with the support networks they need, and improving mental health outcomes particularly for those who are displaced or lonely.

Resource Constraints

Financial Limitations

Many communities struggle financially, making it difficult for them to offer displaced populations proper mental health care and support services. Resources are usually allocated to more urgent requirements like food, shelter, and healthcare, leaving little money for mental health efforts. The availability of mental health specialists, support groups, and other crucial services required to address the mental health needs of refugees and displaced people is significantly hampered by this lack of funding.

Inadequate funding for mental health services is a major obstacle to providing these groups with appropriate community care, according to recent studies. For example, in low- and middle-income countries (LMICs) impacted by humanitarian disasters, financial and human resource limitations are significant barriers to scaling up mental health and psychosocial support (MHPSS) interventions, according to a comprehensive study (Troup et al., 2021). According to the review, a lot of mental health services have trouble finding long-term funding, which restricts their ability to effectively serve the needs of displaced people.

Financial constraints also frequently result in a shortage of qualified mental health practitioners and inadequate infrastructure to support mental health efforts, according to a study on the mental health of refugees in LMICs. Many people do not obtain the care they require as a result, creating a substantial treatment gap (Dickson et al., 2024). The lack of resources has an impact on both the accessibility of services and the standard of care given, since mental health practitioners may be overworked and unable to provide their patients with the support they need.

Furthermore, even though mental health is vital to the general wellbeing of displaced populations, money for other urgent humanitarian needs may take precedence over mental health programs. In order to guarantee that displaced people, receive comprehensive care, a World Health Organization (WHO) report emphasizes the importance of incorporating mental health into larger health and humanitarian responses and calls for greater investment in mental health services (Forray et al., 2024). Providing mental health services to displaced persons is severely hampered by financial constraints. Improving access to care and guaranteeing that mental health needs are sufficiently satisfied in humanitarian circumstances depend on addressing these barriers

Logistical Challenges

Issues like poor infrastructure and transportation can make it very difficult for displaced people to get community support. It can be difficult for many displaced people to get mental health services or go to community events since they live in isolated places with few transportation choices. Inadequate public transit systems make this problem worse by separating people from necessary services and support systems. According to research, displaced people' vulnerabilities are exacerbated by logistical obstacles, such as transportation problems, which have a substantial impact on their capacity to obtain essential services (Forray, Oltean et al. 2024).

Infrastructural Deficits

The capacity to plan community-based events is further restricted by inadequate infrastructure, such as the lack of community centers and meeting rooms. Support groups, workshops, and other community events that encourage social interaction and mental health are challenging to organize without suitable spaces. One of the biggest obstacles to building social support networks among displaced persons has been found to be the absence of physical locations for community interaction. According to studies, communal spaces are necessary for promoting interpersonal relationships and interactions, both of which are critical for mental health rehabilitation (Bajbouj et al., 2022).

Furthermore, the lack of facilities specifically designed to provide mental health services might result in a dependence on improvised solutions, which are sometimes insufficient to deliver all-encompassing care. Access to mental health support for displaced people may become even more difficult as a result of fragmented resources that are unable to address their holistic requirements (Forray, Oltean et al., 2024). To summarize, displaced populations' access to mental health care is severely hampered by logistical difficulties and infrastructural deficiencies. Resolving these obstacles is crucial to enhancing these people's general wellbeing and making sure they get the assistance they require.

Access to community support may also be hampered by logistical issues like poor infrastructure and transportation. It might be challenging for displaced people to get mental health services or go to community events since they may reside in isolated places with few transit choices. The capacity to plan community-based events may also be hampered by inadequate facilities, such as community centers and meeting rooms. According to research, logistical obstacles have a major influence on displaced populations' capacity to obtain essential services, which exacerbates their vulnerability.

Stigma and Mistrust

Stigma and mistrust are deeply embedded barriers that greatly hinder mental health care for displaced people. Mental health stigma persists in many displaced communities, formed by cultural and gendered attitudes that view psychological pain as a sign of weakness or moral failing (Dickson et al., 2024). Individuals are discouraged from revealing their issues or seeking help because they fear discrimination, social rejection, or ostracism from their community. Such views encourage a culture of secrecy and denial, leading many people to hide their symptoms or avoid mental health treatment entirely.

These issues are made worse by mistrust of health systems. Trust in official institutions and service providers is undermined by the trauma, persecution, or unfavourable interactions that many displaced people have had with authorities in their home countries or while migrating (Sukiasyan, 2024). For example, Syrian refugees, have shown reluctance to seek mental health care because they fear it may lower their social status or cause family shame (Kiliç et al., 2024). Women's autonomy to seek assistance may also be limited by gender conventions, especially in patriarchal cultures.

In addition, Engaging with mental health services can be hindered by worries about confidentiality, insensitive or culturally inappropriate care, and the possibility that getting therapy could compromise their legal status or asylum applications (EU, 2022). Sometimes displaced persons find it difficult to express their needs or feel understood due to a lack of culturally competent providers and translators, which exacerbates their feelings of mistrust and alienation. The cumulative effect of stigma and mistrust has results in reduced use of mental health and psychosocial support programs, continued psychological suffering, and growing social isolation. Addressing these challenges requires culturally sensitive, community-based approaches that create trust, improve mental health literacy, and actively engage displaced communities in the design and implementation of care (EU, 2022).

Communication and Language

For displaced people, language barriers are a major barrier, particularly for refugees and migrants who frequently speak minority or indigenous languages that the host nation's providers are not familiar with. Therapeutic rapport, treatment planning, and proper diagnosis are all severely hampered by the lack of trained translators or bilingual mental health specialists. Inappropriate care or a misdiagnosis may arise from miscommunications regarding symptoms, cultural manifestations of distress, and treatment expectations (Ellis, 2022).

Delays or fragmented care may result from inconsistent funding or poorly organized interpretation services, even when they are available. For instance, a dearth of Arabic-speaking therapists and insufficient interpretation support are two major reasons why Syrian refugees in Turkey commonly report having trouble receiving mental health treatment (Kiliç et al., 2024).

Lack of Culturally Competent Care

A lack of culturally competent care exacerbates displaced populations' mental health difficulties. Many healthcare practitioners lack proper training in cultural humility, trauma-informed care, or the unique needs of refugees, resulting in care that can be insensitive, irrelevant, or even dangerous (Lau & Rodgers, 2021). This divide may create re-traumatization, mistrust, and disengagement from mental health treatment, because displaced individuals may feel misunderstood or disregarded (WHO, 2021).

Culturally competent care goes beyond language translation; providers must understand their patients' cultural backgrounds, migration histories, and the unique stressors of displacement. Understanding patients' cultural origins, utilizing culturally appropriate evaluation instruments, and modifying interventions to conform to culture norms and values are all components of providing culturally competent care.

For instance, incorporating family involvement, community rituals, or traditional healing techniques can improve outcomes and acceptance (Larrieta et al., 2022).

Using culturally relevant screening instruments, such as the Refugee Health Screener-15 (RHS-15), and adapting therapies to reflect cultural norms are critical stages (Lau & Rodgers, 2021). For example, enlisting family members, community leaders, or using traditional healing traditions might improve the acceptability and effectiveness of programs for refugee clients (Theodosopoulos et al., 2024). Research has demonstrated that family-centered models, bicultural facilitators, and community partnerships boost trust, lower obstacles to care, and enhance mental health outcomes.

Organizational flexibility, humility, and a dedication to continuous learning are also essential. Successful programs frequently collaborate with refugee communities to co-design services, hire cultural brokers or ethnic paraprofessionals, and adjust routines to reflect community rhythms (Lau & Rodgers, 2021). Finally, guaranteeing refugee participation in planning, implementing, and assessing mental health care is crucial for developing true cultural competency and building trust (Lau & Rodgers, 2021).

Fragmentation and Prioritization

Mental health and psychosocial support (MHPSS) services are often provided as temporary or parallel initiatives in humanitarian catastrophes rather than being integrated into the larger health system. Care continuity problems and inefficiencies result from this fragmentation. Despite its crucial relevance, emergency response operations frequently place a higher priority on housing needs and physical health, making mental health a secondary consideration (USCRI, 2024). Furthermore, funding for mental health initiatives is usually short-term and restricted, relying on donor priorities that can change quickly. Further limiting availability, this financial uncertainty makes it difficult to train local providers and build sustainable services (Otorkpa et al., 2024).

Barriers related to geography and logistics Geographic accessibility is a significant barrier, even in cases where mental health care are available. People who have been displaced frequently reside in isolated or unsafe places, such as unofficial urban settlements or rural camps, that are distant from medical facilities. Movement may be restricted by inadequate or dangerous transportation infrastructure and security concerns, especially for women and vulnerable population (WHO, 2021). The closest mental health clinics, for instance, are hours away, necessitating expensive and unreliable transportation, according to Syrian refugees in Lebanon and Jordan (Otorkpa et al., 2024). These practical difficulties limit follow-up care and deter regular attendance, which lowers the efficacy of interventions.

Effect on Results for Mental Health Disparities in mental health among displaced groups are made worse by the lack of supports. Anxiety, depression, PTSD, and other problems frequently deteriorate without access to proper care, resulting in long-term impairment and a lower quality of life. Secondary issues like substance misuse, social isolation, and family dissolution are also more likely to occur when early intervention is lacking (Dumke, 2024). Additionally, many displaced people rely entirely on unofficial support networks due to the lack of community-based mental health services, which may not be enough to handle severe or complicated mental health requirements. This disparity emphasizes how vital it is to increase service accessibility via creative, scalable, and culturally aware models.

STRATEGIES FOR ENHANCEMENT

Collaboration Models

Fostering collaboration among NGOs, local governments, and community organizations results in a more comprehensive and coordinated support system for displaced people. NGOs provide specialized knowledge in mental health therapies and direct service delivery, whereas local governments provide infrastructure, policy support, and access to public health networks. Community organizations provide vital insights into the cultural context and life experiences of displaced populations, ensuring initiatives are relevant and trustworthy (EU, 2022).

Multi-sectoral partnerships are valuable, as evidenced by successful models like those supported by the UNHCR and the WHO's Group of Friends of Health for Refugees and Host Communities. These partnerships create community-based protection initiatives like safe spaces and support groups, increase the ability of local staff, and include mental health and psychosocial support (MHPSS) into national health systems (EU, 2022). For instance, Malta's strategy integrates governmental organizations, non-governmental organizations, and community centers to provide comprehensive mental health services, language instruction, and social inclusion initiatives to stateless and forcibly displaced individuals (WHO, 2021).

Projects like STRENGTHS across Europe and the Middle East demonstrate how academics, NGOs, and local partners can collaborate to scale up evidence-based, community-led mental health interventions while adapting them to varied regulatory and cultural contexts (EU, 2022). These multi-stakeholder approaches guarantee that resources, expertise, and local knowledge are used to address both clinical and social determinants of mental health, ultimately enhancing access, lowering stigma, and strengthening resilience among displaced populations.

Training Programs

Capacity-building efforts to improve mental health care for displaced people should focus the creation of comprehensive training programs for community leaders and volunteers. These programs are critical for providing individuals with the skills needed to identify and successfully treat mental health concerns in their communities. Training must go beyond basic mental health literacy to include active listening and communication skills, trauma-informed care principles, cultural competency, and understanding of professional referral mechanisms. Give migrants the tools they need to change policy. Through seminars and mentorship, the Empowering Voices initiative teaches underserved communities how to lobby for mental health services, engage legislators, and obtain financing.

Effective training programs should include mental health literacy components, which will help community leaders understand prevalent psychological conditions and symptoms. In addition, using active listening and empathic communication strategies will improve their ability to provide assistance (Reis et al., 2022) . Trauma-informed care concepts are crucial as they emphasize providing safe and supportive environments for people who have been through trauma. Cultural competence is also essential for ensuring that community members interact respectfully and efficiently with varied groups. Furthermore, training should provide knowledge of local mental health services and referral channels, allowing community leaders to connect individuals to appropriate care when necessary (Reis et al., 2022).

A range of delivery techniques should be used to optimize these training programs' effectiveness and reach. While accessible online modules enable participants to learn at their own speed, in-person workshops can promote experiential learning and community involvement. Mentorship programs that match community people with professional mental health experts might improve learning by providing helpful advice and assistance (Singh et al., 2023). To make sure these training programs meet community needs, they must be continuously evaluated. Monitoring community mental health outcomes can assist determine how training affects general mental health and well-being, and feedback surveys can shed light on the training's applicability and effectiveness.

Peer Support Networks

Peer support networks are becoming more widely acknowledged as important tools for improving mental health and wellbeing, especially for vulnerable groups like those who have experienced trauma or displacement. These networks use the strength of common experiences to help people who are going through similar things feel more connected, less alone, and more supportive of one another. Peer support networks have the potential to improve mental health outcomes by fostering safe and encouraging environments that foster cultural competency, trauma-informed care, and mental health literacy.

One of the primary advantages of peer support networks is their ability to increase mental health literacy. Participants obtain essential knowledge about mental health disorders, coping strategies, and relevant resources by sharing experiences and learning from one another. This improved understanding can empower people to seek treatment when necessary and to encourage others in their community who are struggling. According to research, peer support can improve mental health outcomes by making people feel more understood and validated in their experiences (Repper & Carter, 2011). Furthermore, peer support can assist to lessen the stigma associated with mental illness by encouraging people to seek care without fear of being judged or discriminated against (Corrigan, 2004).

For peer support networks to be effective, cultural competence and trauma-informed care principles must be incorporated. Understanding the effects of trauma on people and building resilience via supportive relationships are key components of trauma-informed peer support (EU, 2022). This method aids in making sure that network connections foster healing and recovery rather than unintentionally re-traumatizing people. Peer networks have promoted mutual aid and lessened stigma in refugee camps. Self-Help Plus (SH+) programs, for instance, improved psychosocial outcomes by teaching community members to facilitate group sessions (Karyotaki et al., 2023).

Additionally, cultural competence is essential because it guarantees that peer support programs are considerate of participants' varied backgrounds and needs (Sue et al., 1992). Peer support networks can better meet the complex needs of people from diverse cultural backgrounds and foster more effective support by educating peer supporters in trauma-informed practices and cultural competency.

Collaborative Funding

In order to overcome resource limitations, it is crucial to obtaining sufficient funds is crucial for offering displaced populations long-term mental health care and community support initiatives. This calls for a multipronged strategy that aggressively seeks funding from a range of sources, such as philanthropic organizations, governments, and non-governmental organizations (NGOs).

The careful planning and submission of grant bids is one important tactic. The target population's needs, the program's precise goals, the evidence-based interventions that will be used, and the expected results should all be clearly stated in these well designed proposals. They must also exhibit a thorough comprehension of the goals and policies of the financing organization.

Aside from filing grants, planning fundraising activities can be a successful strategy for increasing awareness and obtaining funding (Dickson et al., 2024). These might be anything from little neighbourhood get-togethers to grand galas or charity performances. Maximizing participation and fundraising potential can be achieved by including celebrities, local companies, and community leaders.

It's also critical to advocate for more government funding for mental health. Lobbying government representatives, educating the public on the mental health requirements of displaced people, and promoting legislative reforms that give mental health first priority are all part of this. Arguments for further government funding may be made by proving the long-term advantages of investing in mental health and the cost-effectiveness of mental health treatments. For example, by emphasizing the trauma associated with drought, women's coalitions in Kenya were able to successfully lobby for financing for psychosocial treatment (IIED, 2022).

Initiatives for collaborative funding mark a significant paradigm shift in the financing and provision of mental health care for displaced people. By deliberately combining resources, networks, and knowledge, this method generates a synergistic environment that surpasses the limitations of individual businesses operating in silos and achieves a bigger collective impact (USCRI, 2024). The basic principle of collaborative finance is the pooling of funds from various sources, including governments, non-governmental organizations, philanthropic foundations, and even private sector groups. This varied funding base creates a more robust and resilient financial foundation by lowering the risks associated with relying solely on one source of funding. Furthermore, it enables the funding of larger, more ambitious projects that would be beyond the scope of any one organization.

Collaborative funding programs make use of each partner organization's distinct capabilities and expertise in addition to financial contributions. One NGO may have extensive experience in trauma-informed care, for instance, while another may have a solid reputation for community mobilization. By combining these complementary skill sets, collaborative projects can deliver more comprehensive and culturally sensitive services that are tailored to the specific needs of the displaced population (Bogic et al., 2015). Additionally, a more integrated and coordinated approach to the provision of mental health services is promoted by collaborative funding. Organizations can prevent duplication of effort, expedite referral processes, and guarantee that people receive a continuum of care that meets their various needs by cooperating. For displaced communities, who frequently deal with a complicated web of issues like trauma, displacement, poverty, and social isolation, this coordinated approach is especially important (Bogic et al., 2015).

A Steering Committee

A steering committee or advisory board of members from each participating institution is frequently established as part of successful joint funding efforts. This committee is in charge of managing the project, making sure it stays in line with its goals, and keeping an eye on its advancement. To preserve confidence and transparency, partners must collaborate and communicate on a regular basis. (Eloul et al., 2013) Promoting laws that give mental health and community support top priority is an essential first step in enacting long-lasting, structural change that helps displaced people. This lobbying goes beyond merely increasing awareness; it calls for a calculated and persistent effort to influence policymakers and the legislative environment in a way that supports mental health and wellbeing. For example the refugees in Community Savings Groups in the Lake Chad Basin united to provide relief to members in need, promoting camaraderie and lowering internal strife (Villalonga-Olives et al., 2022). Policymakers can lessen the negative effects of climate displacement on mental health while promoting resilient, just societies by tackling the loss of social capital through community-driven, culturally grounded approaches.

Lobbying national, regional, and local government representatives is a crucial component of policy advocacy. This entails establishing rapport with elected officials and their staff, supplying them with fact-based data regarding the mental health need of displaced people, and promoting particular legislative reforms that would meet those requirements.

A variety of topics might be the subject of lobbying activities, including boosting financing for mental health services, facilitating greater access to mental health care, and making sure that mental health is incorporated into larger development and humanitarian initiatives. Creating a conducive political climate for policy change also requires increasing public understanding of mental health issues. Numerous strategies, such as public forums, media campaigns, neighbourhood gatherings, and social media campaigns, can be used to accomplish this. It is feasible to lessen stigma and foster public support for policy change by educating the public on the frequency of mental health issues among displaced populations, the effects of trauma and displacement on mental health, and the efficacy of mental health therapies.

Advocacy is Pushing for Modifications to Laws and Rules

This could entail attempting to pass new legislation that safeguards displaced people's rights to mental health care, revising current legislation to make sure it takes mental health needs into account, and opposing rules that impose obstacles to mental health care. Advocates might, for instance, try to amend immigration laws that deny displaced people access to mental health care or contest laws that discriminate against people with mental illnesses (EU, 2022).

Research continuously shows that promoting policies can be a very successful way to boost financing and support for mental health services for displaced people. Advocates can guarantee that resources are distributed in a way that best serves the needs of the most vulnerable and foster a more supportive atmosphere for mental health care by influencing laws and policies. Forming alliances with other groups and interested parties, creating succinct and straightforward policy proposals, and successfully presenting those recommendations to the public and decision-makers are all crucial to maximizing the impact of policy advocacy initiatives (EU, 2022).

Community-Based Participatory Research (CBPR)

Community-Based Participatory Research (CBPR) is a collaborative strategy that actively engages community people at all stages of intervention development, ensuring that programs are community-driven and culturally relevant. CBPR strategies have been found to improve the relevance, acceptability, and effectiveness of mental health interventions for refugees by taking into account their lived experiences and cultural views (Oppenheim et al., 2019). For example, the Positive Minds for Refugees (PMR) initiative employed CBPR to collect input from refugees via focus groups, which resulted in the modification of intervention content to better address cultural norms, social integration, and self-care requirements (Siddiq et al., 2023). Reviews and implementation studies frequently note that CBPR develops trust, boosts engagement, and enhances outcomes by making interventions more sensitive to the unique needs of displaced people.

Additionally, by fostering inclusive collaborations between academics, service providers, and community residents, CBPR encourages capacity building and shared decision-making. This strategy is particularly helpful in the field of refugee mental health, where community ownership and cultural concerns are crucial for implementation and sustainability (Filler et al., 2021).

CONCLUSION

Displaced populations like women, children, and Indigenous people are disproportionately affected by the severe and complex mental health issues associated with climate displacement. System injustices, such as social exclusion, cultural displacement, and economic inequality, have compounding consequences that make psychological suffering worse and make rehabilitation more difficult. Case studies from Indonesia, Kenya, and Niger demonstrate how climate induced displacement affects important social determinants of health like safety, economic stability, and community cohesiveness in addition to physical surroundings. One of the most important protective factors in reducing these negative mental health consequences is community support. In its bonding, bridging, and linking forms, social capital offers displaced people access to resources, emotional support, and avenues for institutional engagement and advocacy. However, broken social networks, cultural hurdles, and structural gaps in service delivery make it difficult to maintain social capital after displacement.

Good intervention programs show that by building resilience, restoring social connections, and expanding access to care, culturally aware, community-led strategies, and NGO collaborations can greatly enhance mental health outcomes. However, obstacles such as restricted access to services, linguistic and cultural disparities, stigma, and prejudice still stand in the way of providing fair mental health care. Policymakers, practitioners, and community members can collaborate to build more resilient and just societies that prioritize the mental health of displaced populations, giving them the chance to recover, start over, and make a significant contribution to their new communities. This can be achieved by combining these strategies: sustainable funding, proactive policy advocacy, and empowering community initiatives. The ultimate objective is to promote a society in which everyone, irrespective of their status as displaced, has access to the care and support they require to thrive and where mental health is acknowledged as a fundamental human right.

REFERENCES

Arias, P., Bellouin, N., Coppola, E., Jones, R., Krinner, G., Marotzke, J., Naik, V., Palmer, M., Plattner, G.-K., & Rogelj, J. (2021). [the physical science basis. Contribution of Working Group I to the Sixth Assessment Report of the Intergovernmental Panel on Climate Change; technical summary.]. *Climatic Change*, •••, 2021.

Bai, Z., Xu, Z., Xu, X., Qin, X., Hu, W., & Hu, Z. (2020). Association between social capital and depression among older people: Evidence from Anhui Province, China. *BMC Public Health*, *20*, 1–11. PMID: 33066764

Bajbouj, M., Ta, T. M. T., Hassan, G., & Hahn, E. (2022). Editorial: The Nine Grand Challenges in Global Mental Health [Editorial]. *Frontiers in Psychiatry, Volume 12 - 2021*. https://doi.org/DOI: 10.3389/fpsyt.2021.822299

Bogic, M., Njoku, A., & Priebe, S. (2015). Long-term mental health of war-refugees: A systematic literature review. *BMC International Health and Human Rights*, *15*(1), 29. DOI: 10.1186/s12914-015-0064-9 PMID: 26510473

Bolton, P., Bass, J., Betancourt, T., Speelman, L., Onyango, G., Clougherty, K. F., Neugebauer, R., Murray, L., & Verdeli, H. (2007). Interventions for depression symptoms among adolescent survivors of war and displacement in northern Uganda: A randomized controlled trial. *Journal of the American Medical Association*, *298*(5), 519–527. PMID: 17666672

Burrows, K., Pelupessy, D. C., Khoshnood, K., & Bell, M. L. (2021). Environmental Displacement and Mental Well-Being in Banjarnegara, Indonesia. *Environmental Health Perspectives*, *129*(11), 117002. https://doi.org/doi:10.1289/EHP9391 PMID: 34747632

Corrigan, P. (2004). How stigma interferes with mental health care. *The American Psychologist*, *59*(7), 614. PMID: 15491256

Di Giorgi, E., Michielin, P., & Michielin, D. (2020). Perception of climate change, loss of social capital and mental health in two groups of migrants from African countries. *Annali dell'Istituto Superiore di Sanita*, *56*(2), 150–156. PMID: 32567564

Dickson, K., Ko, S. Y. J., Nguyen, C., Minchenko, D., & Bangpan, M. (2024). Mental health and psychosocial support programmes for displaced populations in low-and middle-income countries (LMICs): A systematic review of process, perspectives and experiences. *Cambridge Prisms: Global Mental Health*, *11*, e62. PMID: 38774885

Dumke, L. (2024). *Mental health care for refugees in Germany: Needs and barriers*. Universitätsbibliothek Bielefeld.

Ejiohuo, O., Onyeaka, H., Unegbu, K. C., Chikezie, O. G., Odeyemi, O. A., Lawal, A., & Odeyemi, O. A. (2024). Nourishing the Mind: How Food Security Influences Mental Wellbeing. *Nutrients*, *16*(4). Advance online publication. DOI: 10.3390/nu16040501 PMID: 38398825

Ellis, D. (2022). *Social Identification and Mental Health*. https://www.news-medical.net/health/Social-Identification-and-Mental-Health.aspx

Eloul, L., Quosh, C., Ajlani, R., Avetisyan, N., Barakat, M., Barakat, L., Ikram, M. W., Shammas, L., & Diekkamp, V. (2013). Inter-agency coordination of mental health and psychosocial support for refugees and people displaced in Syria. *Intervention (Amstelveen, Netherlands)*, *11*(3), 340–348.

EU. E. C. (2022). *Fostering responsive mental health systems in the Syrian refugee crisis*. https://cordis.europa.eu/project/id/733337

Farny, S., & Dentoni, D. (2025). Social identity and place-based dynamics in community resilience building for natural disasters: An integrative framework. *Ecology and Society*, *30*(2).

Filler, T., Benipal, P. K., Torabi, N., & Minhas, R. S. (2021). A chair at the table: A scoping review of the participation of refugees in community-based participatory research in healthcare. *Globalization and Health*, *17*(1), 103. DOI: 10.1186/s12992-021-00756-7 PMID: 34488810

Forray, A. I., Oltean, O., Hanft-Robert, S., Madzamba, R., Liem, A., Schouten, B., Anthonissen, C., Swartz, L., Chereche , R. M., Higgen, S., Hall, B. J., & Mösko, M. (2024). Uncovering multi-level mental healthcare barriers for migrants: A qualitative analysis across China, Germany, Netherlands, Romania, and South Africa. *BMC Public Health*, *24*(1), 1593. DOI: 10.1186/s12889-024-19046-z PMID: 38877460

IIED. I. I. f. E. a. D. (2022). *People fleeing impacts of climate change suffering physical and mental health crises*. https://www.iied.org/people-fleeing-impacts-climate-change-suffering-physical-mental-health-crises

Jiménez, F. N., Brazier, J. F., Davoodi, N. M., Florence, L. C., Thomas, K. S., & Gadbois, E. A. (2021). A Technology Training Program to Alleviate Social Isolation and Loneliness Among Homebound Older Adults: A Community Case Study [Community Case Study]. *Frontiers in Public Health, Volume 9 - 2021*. https://doi.org/DOI: 10.3389/fpubh.2021.750609

Karyotaki, E., Sijbrandij, M., Purgato, M., Acarturk, C., Lakin, D., Bailey, D., Peckham, E., Uygun, E., Tedeschi, F., Wancata, J., Augustinavicius, J., Carswell, K., Välimäki, M., van Ommeren, M., Koesters, M., Popa, M., Leku, M. R., Anttila, M., Churchill, R., & Barbui, C. (2023). Self-Help Plus for refugees and asylum seekers: An individual participant data meta-analysis. *BMJ Mental Health*, *26*(1). Advance online publication. DOI: 10.1136/bmjment-2023-300672 PMID: 37524517

Kiliç, C., Kaya, E., Karadağ, Ö., & Üner, S. (2024). Barriers To Accessing Mental Health Services Among Syrian Refugees: A Mixed-Method Study. [Suriye Kökenli Sığınmacıların Ruh Sağlığı Hizmetlerine Ulaşmasını Engelleyen Etmenler: Karma Yöntemli Bir Çalışma.]. *Türk Psikiyatri Dergisi = Turkish Journal of Psychiatry*, *35*(2), 87–94. DOI: 10.5080/u27044 PMID: 38842150

LaMonica, H. M., Davenport, T. A., Ottavio, A., Rowe, S. C., Cross, S. P., Iorfino, F., Jackson, T. A., Easton, M. A., Melsness, J., & Hickie, I. B. (2021). Optimising the integration of technology-enabled solutions to enhance primary mental health care: A service mapping study. *BMC Health Services Research*, *21*(1), 68. DOI: 10.1186/s12913-021-06069-0 PMID: 33451328

Larrieta, J., Esponda, G. M., Gandhi, Y., Simpson, N., Biriotti, M., Kydd, A., Eaton, J., & Ryan, G. K. (2022). Supporting community-based mental health initiatives: Insights from a multi-country programme and recommendations for funders. *BMJ Global Health*, *7*(5), e008906. PMID: 35537763

Lau, L. S., & Rodgers, G. (2021). Cultural Competence in Refugee Service Settings: A Scoping Review. *Health Equity*, *5*(1), 124–134. DOI: 10.1089/heq.2020.0094 PMID: 33778315

Leap, B., & Thompson, D. (2018). Social solidarity, collective identity, resilient communities: Two case studies from the rural US and Uruguay. *Social Sciences*, *7*(12), 250.

Leonardsen, A.-C. L., Hardeland, C., Helgesen, A. K., & Grøndahl, V. A. (2020). Patient experiences with technology enabled care across healthcare settings- a systematic review. *BMC Health Services Research*, *20*(1), 779. DOI: 10.1186/s12913-020-05633-4 PMID: 32838784

Oppenheim, C. E., Axelrod, K., Menyongai, J., Chukwuezi, B., Tam, A., Henderson, D. C., & Borba, C. P. C. (2019). The HEAAL Project: Applying Community-Based Participatory Research (CBPR) Methodology in a Health and Mental Health Needs Assessment With an African Immigrant and Refugee Faith Community in Lowell, Massachusetts. *Journal of Public Health Management and Practice*, *25*(1), E1–e6. DOI: 10.1097/phh.0000000000000707 PMID: 30507809

Otorkpa, C., Otorkpa, O. J., Olaniyan, O. E., & Adebola, O. A. (2024). Mental health issues of children and young people displaced by conflict: A scoping review. *PLOS Mental Health*, *1*(6), e0000076.

Panichkriangkrai, W., Detkong, T., Kruahong, N., Thamarangsi, T., & Tangcharoensathien, V. (2022). Mental health preparedness and response during the COVID-19 pandemic: From global to national implementation. *Global Mental Health (Cambridge, England)*, *9*, 111–114. PMID: 36606233

Patel, V., Chowdhary, N., Rahman, A., & Verdeli, H. (2011). Improving access to psychological treatments: Lessons from developing countries. *Behaviour Research and Therapy*, *49*(9), 523–528. PMID: 21788012

Peconga, E. K., & Hogh Thogersen, M. (2020). Post-traumatic stress disorder, depression, and anxiety in adult Syrian refugees: What do we know? *Scandinavian Journal of Public Health*, *48*(7), 677–687. DOI: 10.1177/1403494819882137 PMID: 31814514

Phang, J. K., Kwan, Y. H., Yoon, S., Goh, H., Yee, W. Q., Tan, C. S., & Low, L. L. (2023). Digital intergenerational program to reduce loneliness and social isolation among older adults: Realist review. *JMIR Aging*, *6*(1), e39848. PMID: 36598801

Posklinsky, I., & Nouman, H. (2025). Community Building in Crisis: Interventions With Displaced Communities in Temporary Housing Shelters. *Journal of Community Psychology*, *53*(1), e23170. DOI: 10.1002/jcop.23170 PMID: 39711305

Putnam, R. D. (2000). *Bowling alone: The collapse and revival of American community*. Simon and schuster.

Reis, A. C., Saheb, R., Moyo, T., Smith, C., & Sperandei, S. (2022). The impact of mental health literacy training programs on the mental health literacy of university students: A systematic review. *Prevention Science*, *23*(4), 648–662. PMID: 34272642

Repper, J., & Carter, T. (2011). A review of the literature on peer support in mental health services. *Journal of Mental Health (Abingdon, England)*, *20*(4), 392–411. PMID: 21770786

Rhodes, J., Chan, C., Paxson, C., Rouse, C. E., Waters, M., & Fussell, E. (2010). The impact of hurricane Katrina on the mental and physical health of low-income parents in New Orleans. *The American Journal of Orthopsychiatry*, *80*(2), 237–247. DOI: 10.1111/j.1939-0025.2010.01027.x PMID: 20553517

Riza, E., Kalkman, S., Coritsidis, A., Koubardas, S., Vassiliu, S., Lazarou, D., Karnaki, P., Zota, D., Kantzanou, M., & Psaltopoulou, T. (2020). Community-based healthcare for migrants and refugees: a scoping literature review of best practices. Healthcare, Rosenlund, M., Kinnunen, U.-M., & Saranto, K. (2023). The use of digital health services among patients and citizens living at home: scoping review. *Journal of Medical Internet Research, 25*, e44711.

Siddiq, H., Elhaija, A., & Wells, K. (2023). An Integrative Review of Community-Based Mental Health Interventions Among Resettled Refugees from Muslim-Majority Countries. *Community Mental Health Journal, 59*(1), 160–174. DOI: 10.1007/s10597-022-00994-y PMID: 35751790

Singh, I., Kumar, S., & Panchal, J. (2023). Employee Training and Development Enhancing Employee Performance-A Study. *ResearchGate. Retrieved October, 14*, 2024.

Sue, D. W., Arredondo, P., & McDavis, R. J. (1992). Multicultural counseling competencies and standards: A call to the profession. *Journal of Counseling and Development, 70*(4), 477–486.

Sukiasyan, S. (2024). The Mental Health of Refugees and Forcibly Displaced People: A Narrative Review. *Consortium Psychiatricum, 5*(4), 78–92. DOI: 10.17816/cp15552 PMID: 39980620

Szreter, S., & Woolcock, M. (2004). Health by association? Social capital, social theory, and the political economy of public health. *International Journal of Epidemiology, 33*(4), 650–667. PMID: 15282219

Theodosopoulos, L., Fradelos, E. C., Panagiotou, A., Dreliozi, A., & Tzavella, F. (2024). Delivering Culturally Competent Care to Migrants by Healthcare Personnel: A Crucial Aspect of Delivering Culturally Sensitive Care. *Social Sciences, 13*(10), 530. https://www.mdpi.com/2076-0760/13/10/530

Tol, W. A., Barbui, C., Galappatti, A., Silove, D., Betancourt, T. S., Souza, R., Golaz, A., & Van Ommeren, M. (2011). Mental health and psychosocial support in humanitarian settings: Linking practice and research. *Lancet, 378*(9802), 1581–1591. PMID: 22008428

Troup, J., Fuhr, D. C., Woodward, A., Sondorp, E., & Roberts, B. (2021). Barriers and facilitators for scaling up mental health and psychosocial support interventions in low- and middle-income countries for populations affected by humanitarian crises: A systematic review. *International Journal of Mental Health Systems, 15*(1), 5. DOI: 10.1186/s13033-020-00431-1 PMID: 33413526

UN. U. N. (2023). *Review of mental health and well-being policies and practices in United Nations system organizations.* https://docs.un.org/en/JIU/REP/2023/4

Ungar, M. (2006). Resilience across Cultures. *British Journal of Social Work, 38*(2), 218–235. DOI: 10.1093/bjsw/bcl343

USCRI. (2024). U. S. [*Mental Health Awareness Month: Barriers and Access to Mental Health Care.* https://refugees.org/mental-health-awareness-month-barriers-and-access-to-mental-health-care/]. *CR (East Lansing, Mich.), I,* •••.

van Sint Fiet, A., de la Rie, S., van der Aa, N., Bloemen, E., & Wind, T. (2022). The relevance of social capital and sense of coherence for mental health of refugees. *SSM - Population Health, 20,* 101267. PMID: 36281249

Versey, H. S., Murad, S., McPhee, I., & Schwarz, W. (2023). A Bridging-Community (ABC) Project: A Community Building Social Participation Intervention Among Resettled Refugees in Boston. *Journal of International Migration and Integration, 24*(1), 441–465. DOI: 10.1007/s12134-021-00908-z

Villalonga-Olives, E., Wind, T., Armand, A., Yirefu, M., Smith, R., & Aldrich, D. P. (2022). Social-capital-based mental health interventions for refugees: A systematic review. *Villalonga-Olives, E., Wind, TR, Armand, A., Yirefu, M., Smith, R., and Aldrich, DP (2022). Social capital based mental health interventions for refugees: A systematic review.* [Record #43 is using a reference type undefined in this output style.]. *Social Science & Medicine.*

WHO. (2021). *Mental health and forced displacement.* https://www.who.int/news-room/fact-sheets/detail/mental-health-and-forced-displacement

KEY TERMS AND DEFINITIONS

CBPR: community people are included in every stage of the research process through the use of community-based participatory research, or CBPR.

Displacement: When individuals are compelled to leave their usual homes because of conflict, violence, or natural catastrophes.

MHPSS: A complete strategy that includes community-based psychosocial interventions as well as clinical mental health care.

Mutual Aid: is the voluntary sharing of resources and assistance among members of a community.

Peer Networks: are associations of people with similar backgrounds who help one another both practically and emotionally.

PTSD: A mental illness which may develop after being exposed to traumatic experiences.

Resilience: The ability to bounce back from hardship, impacted by personal characteristics as well as social and environmental elements.

Stigma: Discrimination against people with mental health disorders as a result of negative attitudes and beliefs.

Trauma: informed care a method that acknowledges the effects of trauma and incorporates this knowledge into the provision of healthcare is known as.

Chapter 3

Exploring the Relationship Between Depression, Anxiety, Stress, Religious Coping, and Quality of Life Among Farmers

Navneet Kaur
Conestoga College Institute of Technology and Advanced Learning, Canada

Chandani Rani
Amity University, Mohali, India

Mubashir Gull
https://orcid.org/0000-0003-2907-7119
Gandhi Institute of Technology and Management, India

ABSTRACT

This study explores the relationship between depression, anxiety, stress, religious coping, and quality of life among farmers. The study aims to shed light on the unique challenges faced by farmers and the impact of these challenges on their mental health and overall quality of life. A total of 130 farmers from the Malwa district of the Punjab province in India completed a questionnaire that included measures of depression, anxiety, stress, religious coping, and quality of life. The results indicate that depression, anxiety, and stress are significantly related to lower quality of life

DOI: 10.4018/979-8-3373-3281-9.ch003

among farmers. These findings suggest that farmers face unique challenges that can have a significant impact on their mental health and quality of life, and that the use of religious coping strategies may be a helpful resource for improving their well-being.

INTRODUCTION

Punjab is a state in India's northwest. It represents one of the smallest states in the nation, covering only 50,362 sq. km, or 1.5% of India's total surface area, yet supporting 2.5% of the country's people. The rotation of wheat and rice dominates the state's farming pattern (Grover et al., 2017).Amid the height of the green revolution, Punjab had excelled in farming. Its gross domestic product (GDP) from agriculture rose at a rate of 5.7% per year from 1971–1972 and 1985–1986; this was more than twice as fast as the 2.31% growth rate attained at the all–India level during the same time period. Punjab's rich history agricultural growth decreased to 3% per year between 1985–1986 and 2004–2005 as a result of the green revolution, which began to show warning signs after 1985–1985. However, when Punjab's agricultural GDP growth fell to just 1.6% per year from 2005–2006 to 2016–2017, less than half the 3.6% growth of farmland GDP throughout all of India during the same period, the state's farming faced its major issues (Gulati et al., 2021).

Punjab has seen an unprecedented rise in suicides among farmers since 2015. Large loans with high interest rates are taken out by farmers in the state, some of which can be used for personal spending. According to government statistics, a high degree of debt, especially to institutions, is the cause of 51% of suicides among farmers in Punjab (Saini et al., 2022). Another 40-year-old farmer is accused of committing himself by ingesting poison in Punjab in April 2022 after being unhappy over the season's low wheat yield. Due to this, Punjab has had 14 suicides of farmers, 11 of which were from the Malwa region in April (Jagga, 2022). On the one hand, increased input expenditures like rent for land, diesel, and fertilizers, as well as shortfalls in two successive crop seasons, appear to have taken a heavy toll on these farmers. Men are often told to "suck it up" and "take it like a man" when dealing with life's pressures. Toughness and stoicism are expected; emotional vulnerability is seen as a blemish on their manhood. These expectations may discourage men from seeking help when they encounter mental health adversities and instead drive them to destructive coping strategies (Seager et al., 2014).

LITERATURE REVIEW

Farm families have a distinctive connection and structure known as "the agri-family system"(Bennett & Bennett, 1982), which may be a cause of stress. Familial and economic duties are intertwined in farming families, making it such that personal as well as family demands conflict with business needs. This is the main differentiation among agricultural households and non-farming households. The close working relationships between members of the family, often spanning generations, the necessity of making significant financial decisions every day, and reliance on erratic, unpredictable variables like the weather and markets for survival are all traits that are typical of farming households and may add stress to life (Rosenblatt & Anderson, 1982).

About 550 million of the 570 million farms worldwide are family-run, meaning they are owned and operated exclusively by the farmer or the farmer's immediate family (Lowder et al., 2016). Family farming is still an occupation that spans generations (Maciuba et al., 2013) integrating social resources, family values, and financial security to promote farmers' well-being. When a male farmer succeeds in his farming, it has been associated with restoring the family's respect and dignity; when he fails, however, it is associated with social pressure and guilt, which impacts his fundamental sense of self (Bryant & Garnham, 2015), and frequently resulting in heightened financial instability, social exclusion, stigma, and self-criticism.

The agricultural economy for farmers is in trouble due to the growing cost of production and the declining pricing of agricultural products. Farmers are exposed to uncontrollable and unpredictable conditions while working in agriculture. These conditions include unpredictable weather patterns, unpredictable animals and machinery, role conflict, hectic schedules, extended periods of physically taxing and repetitive labour, and social exclusion (Fraser et al., 2005). Acute and ongoing stresses that are too much for a person to handle can increase psychological distress, cause issues with mental wellness, and even lead to suicide (Santos et al., 2021). Given that there have been reports of poor mental health in farmers around the globe, it is possible that structural aspects of agriculture contribute to their psychological well-being. Numerous mental health outcomes, such as depression, anxiety, and psychological morbidity, are all more common in farmers than in non-farmers (Daghagh Yazd et al., 2019).

It is well known that the agricultural sector is stressful (Booth & Lloyd, 2000). Commodity costs, time constraints, and environmental factors are common stresses among farmers (Austin et al., 2018a; Furey et al., 2016). According to a review of the literature, several occupational factors, such as "financial pressures"(Kallioniemi et al., 2016), "poor weather," and farmers' incapacity to prepare for such events, are likely to cause stress among individuals working in the agricultural sector (Daghagh

Yazd et al., 2019), "farm workload," which includes extended hours of work and heavier workloads during periods of high demand (Lunner Kolstrup et al., 2013), the accessibility of "additional workforce"(Brennan et al., 2020), the occurrence of poor animal welfare (Daghagh Yazd et al., 2019), and adherence to governmental regulations (Alpass et al., 2004; Booth & Lloyd, 2000; Daghagh Yazd et al., 2019). According to Daghagh Yazd et al. (2019), younger farmers have greater levels of stress overall, while the specific occupational stresses that they deal with may differ by age cohort. The quality of life is negatively impacted by farm economics and debt levels, which have been highlighted as major stresses affecting farmers (Heo et al., 2020).Farmers in the USA believe that local banks and other lenders are less eager to extend loans or credit than they formerly were (Jones et al., 2018). Farmers spend a lot of time alone, alone in challenging, risky, and isolated situations, with little opportunity for recuperation or socialising (Truchot & Andela, 2018). Many farmers experience mental health issues and low psychological wellness as a result of this secluded way of life (Roy et al., 2013).

Farmers are impacted by climate change, which causes events like floods and droughts. According to studies, droughts cause worry and anxiety among farmers as well as financial difficulties, reduced crops, and higher labour requirements (Austin et al., 2018b; Fennell et al., 2016; Odabasi & Hartarska, 2021). Farmers' quality of life may be impacted by worrying about the weather, bad weather, and agricultural loss caused by the weather (Jones et al., 2018; Kearney et al., 2014a). Farmers had higher rates and prevalences of depression compared to non-farmers, particularly among male farmers who also exhibited higher levels of anxiety (Sanne, 2004; Sanne et al., 2003).

Due to the challenging and stressful nature of farming, suicide rates among farmers are far higher than those in the general population. The economic and occupational risks, remote location, and rural culture associated with farming create significant risk factors for suicide, especially high levels of chronic stress, a reported risk factor for suicide (Polanco-Roman et al., 2016). In Kerala, Karnataka, and Andhra Pradesh, the media initially brought attention to farmer suicides. However, news of farmer suicides in Punjab soon surfaced, which disturbed and was highly unexpected in such an affluent area. Farmers who commit suicide are typically the family's most productive male employee and primary income (Das, 2011). Since the middle of the 1980s, farmer suicides in Punjab have been a major public concern. 2,116 suicides have been documented by the state government since 1986, although this number may only represent the "The tip of the iceberg" since numerous additional incidents may have gone unreported or overlooked (Gill & Singh, 2006).In addition to reducing the amount of labour used in households, mechanisation of major crop harvests and extensive use of biological technology have significantly increased production costs. Reduced returns on the production of food grains were undoubtedly a result of rising costs, stagnating technology, and a near-freeze in the minimum support rates of wheat and rice that made the already unfavourable conditions of trade even worse. The causes of the rise in debt among Punjabi landowners were the narrowing of the margins between profits and production expenses, growing weather uncertainty, and a reliance on loans with higher interest rates from unregulated lenders (Gill & Singh, 2006). This has made matters worse to the point where farmers from the region Punjab have resorted to killing themselves (Taki et al., 2021).

There are a variety of factors that contribute to suicides, with debt playing a significant role. The other elements were financial hardship, agricultural failure, drunkenness, marriage and domestic strife, drug addiction, and ineffective loan use (Gill & Singh, 2006). When examining risk factors for farmer suicide, agricultural values and culture are crucial aspects to be taken into account. Farmers frequently feel a strong sense of duty to their families to carry on their labour, exhibiting tremendous pride in their profession, and have an intimate connection with their frequently ancestral farms. The possibility of agricultural failure, which would mean disappointing family members and frequently losing the labour of previous generations, is very severe (Purc-Stephenson et al., 2023).

Agriculture is a means by which religion and nature interact. Organised religion is a socially acceptable entity for addressing mental wellness issues in rural regions. At both the macro and local levels, religion has an impact on socioeconomic results. Poor farmers take more risks than wealthy farmers do, and this risk-taking behaviour can be attributed to the farmers' intense religiosity (Kahsay et al., 2022) .When farmers face macroeconomic deterioration, religion improves their quality of life (Meyer & Lobao, 2003).

The employment of faith-based behaviours and beliefs to cope with stress is known as religious coping. This is demonstrated by Punjab's farmers' recitation of the Gurbani (*Shri Guru Granth Sahib*, 2000); during times of distress, farmers frequently recite hymns to invoke divine assistance. Belief in Hukam (Divine Will); by attributing agricultural success or failure to God's will, many farmers lessen their worries about circumstances that are beyond their control.

"Think of the Immaculate Naam; whatsoever He desires will come to pass" (SGGS, Ang 109). Farmers in Punjab often attribute crop success or failure to Hukam (divine will), which assists farmers to accept hardships from a spiritual perspective.

"God's Name fills the storehouses, and the farmer's labour yields the produce." (SGGS, Ang 1091). Farmers feel that while their hard work is crucial, success is ultimately determined by divine support, which helps them deal with failures in crops and financial strain.

"The water, lotus, and fish are all he Himself; everything has been created by the Creator." (SGGS, Ang 1121). Farmers who believe in divine protection are better able to cope with the stress of weather variations, debt, and unstable markets.

These passages demonstrate how Sikh spirituality and farming are closely related. Prayer, Hukam (divine will), seva (community support), and Santokh (contentment) are examples of religious coping strategies that assist Punjab's farmers in overcoming stressors including debt, erratic weather, and market swings. Faith provides resilience against adversity by acting as an emotional and psychological anchor.

People are considered to be in a state of wellbeing when they possess "the psychological, social, and physical resources they need to meet a particular psychological, social, and/or physical challenge" (Dodge et al., 2012). A see-saw with wellbeing at the centre serves as an illustration. The physical, psychological, and social resources of each individual are on one side of the seesaw. On the opposite side are the difficulties that the person encounters. Recent research demonstrated a relationship between farmer quality of life and farm size, revenue from farming, meaningful labour, and working conditions(Röös et al., 2019). Income and well-being are related. Accordingly, it comes to light that the disparity between household income and stated ambitions has a significant impact on reported subjective wellbeing (Knight & Gunatilaka, 2012). Income and profitability are related to life satisfaction, according to studies looking beyond the analysis of satisfaction levels to examine the interdependences between various QoL indicators (Mzoughi, 2014). However, aiming to increase income levels frequently requires working harder or raising herd sizes while also becoming older (Molnar, 1985). Farmers work long hours compared to other vocations, often with seasonal peaks and variances (Wojewódzka-Wiewiórska et al., 2020).Dairy farmers may even have a heavier burden(G. B. Hansen et al., 2020), which reduces their quality of life by limiting their free time and preventing them from getting enough sleep (Wojewódzka-Wiewiórska et al., 2020).

The literature on sociology of agriculture acknowledges that a rural community's quality of life is impacted by the agricultural sector's structure. A portion of this literature particularly connects the economic and social structures of agriculture to challenges in a community. Often referred to as the Goldschmidt hypothesis, it asserts that communities with industrial farms owned by absentees are less developed socially and economically than those with a high percentage of family farms. Research has shown for decades that agriculture is a vital component of rural development(Green, 1985). Beginning in the 1940s with Walter Goldschmidt's seminal study of two Californian agricultural communities in 1978(Goldschmidt, 1978), this type of research was subsequently confirmed by the Small Farm Viability Project. Family-run farms dominated one neighbourhood, whereas large-scale farms that employed a lot of wage workers typified the other. Residents of the industrial farming community had lower levels of income and educational attainment, indicating poor socioeconomic situations. Additionally, the property owners of these industrial farms controlled the community's political and economic structure; they were not local residents and showed little concern for the standard and variety of public services. On the other hand, Goldschmidt found that the neighbourhood, which was primarily made up of family farms, had superior socioeconomic circumstances. He came to the conclusion that an economic structure with a larger middle class and more evenly distributed incomes was produced by a higher concentration of family farmers. Additionally, because the family farmers shared a common community and were actively involved in local government and its services, public services were more extensive and of higher quality.

According to Israel et al. (1994), the Cumulative Stress Model (CSM) explains how long-term stressors like poverty, prejudice, violence, and social disadvantage can affect people like farmers. According to Cumulative Stress Model, farmers may experience a "stress pile-up" effect, in which a number of stressors build up and interact to have a cascade of detrimental effects. Internal and external stressors are combined into a single model by the CSM. When evaluating a person's stress experience, CSM takes into account both external and internal elements, such as a lack of social support, a sense of helplessness, and underlying physical and mental health issues. These factors were divided into five main categories by Israel et al. (1994): (a) the psycho-environmental conditions that cause stress; (b) how the individual and community perceive that stress; (c) the short-term reactions to the stressor; (d) the long-term effects of the stress on health and mental health; and (e) factors that reduce or mitigate stress, such as a stable socioeconomic status, a supportive social network, and the ability to make decisions on one's own.(Israel et al., 1994). The buildup of pressures in farming can result in suicide and self-harm (Bjornestad et al., 2021).

The Stress Process Model describes how stress develops, what factors can mitigate it, and how mediators connect stress sources to symptoms (Pearlin et al., 1981). Both stressful events and ongoing life pressures brought on by shifts in self-perception can be stressors. Coping mechanisms and social support are examples of stress mediators. This model depicts crucial elements of the stress response in farmers. Every family deal with stress on a daily basis in some way. But some households are more capable of managing than others. The original Family Stress Theory: ABCX model, which initially created in 1949 by (Reuben, 1949) and McCubbin and Patterson (1983) later modified into Double ABCX, offers insight into family stress and coping over time (Weber, 2011).The American farm crisis of the 1980s provides a pertinent illustration of this notion, showing that while some farm families were able to weather the storm and carry on farming, others were not. Research has demonstrated that farm families who have resilient beliefs about their status when faced with farm downturns seek out and increase their assets as well as coping mechanisms (Caldwell et al., 2004).

According to the ABCX model, the stressor (A) communicates with the family's resources (B), which then connects with the family's perception of the event (C), to result in X. (the outcome or crisis) (McCubbin & Patterson, 1983; Weber 2011). As the initial crisis is stretched to Double ABCX, it causes Aa (a build-up of stressors), which connects with Bb (existing and new resources), and Cc (perception of the stresses and pile up), potentially leading to despair. This affects the family's ability to cope and function in the long run (i.e., Xx), which may result in non-adaption, adaptation, or mal-adaption (McCubbin & Patterson, 1983) . The model explains why a farmer is most at risk of maladaptation amid periods of severe stress if they do not take action to ask for assistance or to seek out social and coping resources. The numerous pressures that affect the farmer, farm family, and farming operation are particularly significant since they also have a detrimental influence on other farmers, the local farm economy, rural community services, and financing institutions and businesses that serve the agricultural industry. Due to the Midwesterners' tendency towards individualism and privacy, the farm family is likely to try to become more independent and work harder as the community's numerous stressors mount. This is because they believe that there aren't many, if any, resources outside the family that they can rely on (*Rural Mental Health Overview - Rural Health Information Hub*, n.d.)

Family Stress Theory

Double ABCX was utilised in earlier studies that were published around the time of the 1980s farm crisis to help us understand why farmers experience stress and how to best support them. Weigel & Weigel (1990) investigated the relationship between resources and stressors as they related to pleasure in multigenerational farm families. The authors emphasise the value of educational materials and initiatives that concentrate on stress reduction, strengthening families, and improving inter-generational relationships.

According to Davis-Brown & Salamon (1987), they created an instrument to help professionals provide therapy to farm families who have lost or are in the process of losing their farm. Programs for families affected by the agricultural issue have primarily focused on offering financial counselling, frequently omitting the trauma experienced by a family facing the loss of a farming operation, according to the authors (p. 373). To put it another way, specialists must comprehend that the agricultural activity is a significant element of the farmers' character in order to properly assist those who are farming.

A framework for comprehending risk and protective factors for farmers' mental health and wellness is offered by the Conservation of Resources (COR) theory. This idea has been used to examine how farmers deal with drought and whether they plan to quit farming (Peel et al., 2016). The theory states that people work to minimise resource loss and to acquire and hold onto various kinds of resources

(Hobfoll, 1988). When people face the possibility of losing money, actually lose money, or fail to recover money after investing, psychological or physical stress results (Hobfoll, 1988). Burnout, depression, or heart disease are serious long-term consequences, whereas frustration, rage, or terror are examples of short-term effects (Hobfoll, 2001). The COR theory states that resources can be divided into four categories: human traits, energy, objects, and condition (Hobfoll, 1988). Social confidence, occupational skills, leadership, self-efficacy, self-esteem, optimism, and hope are examples of characteristics of an individual/human trait. Knowledge, time, and money are examples of energy resources. Physical assets are known as object resources, and for farmers, these can include their equipment, homes, cars, and land. A person's social standing and health, including their work, occupation, community standing, and relationship status, are referred to as condition resources. Farmers' mental health and well-being are impacted by the same problems that endanger their resources.

The study's primary objective is to assess the levels of depression, anxiety, stress, and religious coping among farmers as well as their association with quality of life among farmers.

METHODOLOGY

Participants and Procedure

A purposive sample of 130 farmers from the Malwa district of the Punjab province, with ages ranging from 22 to 60 and a mean age of 41.88 years, were selected for this study. Data was collected in February month of the year 2023.The researcher introduced himself to the respondents before outlining the goal of the study and established friendly relationship with them. The researcher provided a brief introduction to the study's objectives before administering a scheduled questionnaire. Confidentiality was assured to the participants, and their information was solely intended for research purposes only. Responders were provided with paper and pencil questionnaires and instructed to carefully read/listen and follow the questionnaire's directions, and it took each respondent approximately 30 to 40 minutes to complete all of the questions.

Measures Used

Hospital Anxiety and Depression Scale (HADS)

Anxiety ("restless, anxious moods, and thoughts") and depression ("state of loss of interest and diminished pleasure response") in the previous week are two separate subscales on this 14-item measure (Zigmond & Snaith, 1983). According to Zigmond and Snaith (1983), both subscales include seven items each, with responses graded on a 4-point Likert scale (0–3), resulting in a score range of 0–21. The anxiety and depression subscales were assessed independently in accordance with the scale's recommendations by adding the items, and "caseness" was then categorised as Normal (0–7), Mild (8–10), Moderate (11–15), and Severe (> 16) (Zigmond & Snaith, 1983). For anxiety and depression, a number of authors have developed "possible" (cut-point score of 8) and "probable" (cut-point score of 11) case categories (Bartram et al., 2009). The HADS has undergone considerable validation(Bjelland et al., 2002) and has been applied to research on farmers in Norway (Torske et al., 2016), France (Guillien et al., 2017), and England (Booth & Lloyd, 2000). The scale's internal reliability was 0.91, which is excellent.

Farm / Ranch Stress Inventory

A 28-question self-report survey quantifying stresses particular to farmers and ranchers is called the Farm/ Ranch Stress Inventory. On a scale of 1 to 4, respondents rank each stressor as (1) No Stress, (2) A Little Stressful, (3) Moderately Stressful, and (4) Very Stressful (Kearney et al., 2014b). The scale's internal reliability was .924, which is excellent.

Religious Coping

The Brief Religious Coping (RCOPE) was used to measure religious coping. It is a 14-item survey that Pargament designed for elderly US residents. The tool divides religious coping into 7 positive and 7 negative categories, such as "looked for a stronger connection with God" and "questioned God's love for me." According to a four point Likert scale (1 = not at all, 2 = rarely, 3= sometimes and 4 = a great deal), each feature is scored (Pargament et al., 2000). The scale's internal reliability was .767, which is good.

WHOQOL

The World Health Organisation Quality of Life (WHOQOL)-BREF was used to assess quality of life (QoL). The WHOQOL-BREF is an international QOL tool that assesses each person's perceptions in relation to their culture, value systems, and unique objectives, standards, and concerns. It is cross-culturally comparable. It comprises of 26 measures that evaluate the four main aspects of quality of life: environment (9 items), social interactions (3 items), psychological health (6 items), and physical health (8 items). A person's view of their quality of life in each distinct category is reflected in the four domain ratings. The domain scores are scaled in a positive direction; the available ranges for the domains of physical health, psychological health, social interactions, and environment are 8–40, 6–30, 3–15, and 9–45. Every item is scored using a Likert scale with a range of 1 to 5. A technique used frequently to convert domain scores into a scale of 0–100. The scale's internal reliability was 0.964, which is excellent. (World Health Organization, 2004)

Response Accuracy Scale

To measure the accuracy of participants' responses, the Response Accuracy Scale, a single item, was utilized. This customised scale consisted of five points, ranging from "Didn't read the questions in the survey at all" to "Read all questions in the survey at all." Participants were asked to indicate the extent to which they felt they answered the questionnaire appropriately. Only those who selected "read most of the survey" or "read all survey questions" were included in the research (Monaghan et al., 2020).

Data Analysis

Appropriate statistics like Spearman's correlational analysis were employed to examine the association between variables by using the programme Statistical Product and Service Solutions (SPSS 20 Version).

Ethical Issues

The study has complied with ethical standards in order to protect the rights of the participants. To safeguard their privacy, the personal information of participants was not disclosed to anyone. The researchers committed to using the information collected only for the study's intended purpose and in a responsible and ethical manner.

DISCUSSION OF FINDINGS

The results of the study reveal a significant link between farm size and various factors affecting farmers' well-being, including stress, health, depression, anxiety, years of experience in agriculture. Specifically, when examining the relationship between farm stress and size, the data indicated a negative association ($r = -.459$; $p < .01$), suggesting that large farms are associated with lower levels of stress among farmers.

In contrast, when investigating the relationship between farmers' health and farm size. The findings showed a positive association ($r = .556$; $p < .01$), indicating that as farms increase in size, so do the health levels of their owners. This study also identified a comparable correlation between farm size and depression ($r = -.378$; $p < .01$), anxiety ($r = -.554$; $p < .01$), years served in agriculture ($r = -.187$; $p < .05$), and alcohol use/any other behaviour ($r = -.297$; $p < .01$) and working hours with farm size ($r = -.198$; $p < .05$) Table 1.

Table 1. Exploring the correlation link among the various predictor variables with criterion variable

	Qol & GH	Physical Heath	Psychological	Social Relationship	Environmental	Farmers QOL
Depression	-.73**	-.69**	-.73**	-.59**	-.75**	-.77**
Anxiety	-.79**	-.80**	-.80**	-.66**	-.75**	-.82**
Positive Rel. Coping	-0.15	-0.13	-0.13	-0.09	-.19*	-0.16
Negative Rel. Coping	-.26**	-.29**	-.29**	-.27**	-.28**	-.31**
Farm Related Factors	-.59**	-.63**	-.62**	-.61**	-.67**	-.68**
Social Factors	-.63**	-.70**	-.72**	-.62**	-.74**	-.76**
Financial Factors	-.70**	-.70**	-.72**	-.63**	-.80**	-.78**
Farmers Stress	-.71**	-.75**	-.76**	-.69**	-.81**	-.82**
Years in Agriculture	-.42**	-.43**	-.39**	-.34**	-.42**	-.44**
Alcohol Use/Any other	-0.08	-0.07	-0.08	-0.04	-0.02	-0.05
Working hours	-0.16	-.24**	-.25**	-.28**	-.22*	-.25**
Farm size	.57**	.55**	.56**	.42**	.51**	.56**
Irrigation/Water issue	.35**	.29**	.32**	.31**	.49**	.39**

Figure 1. Exploring the correlation among the variables

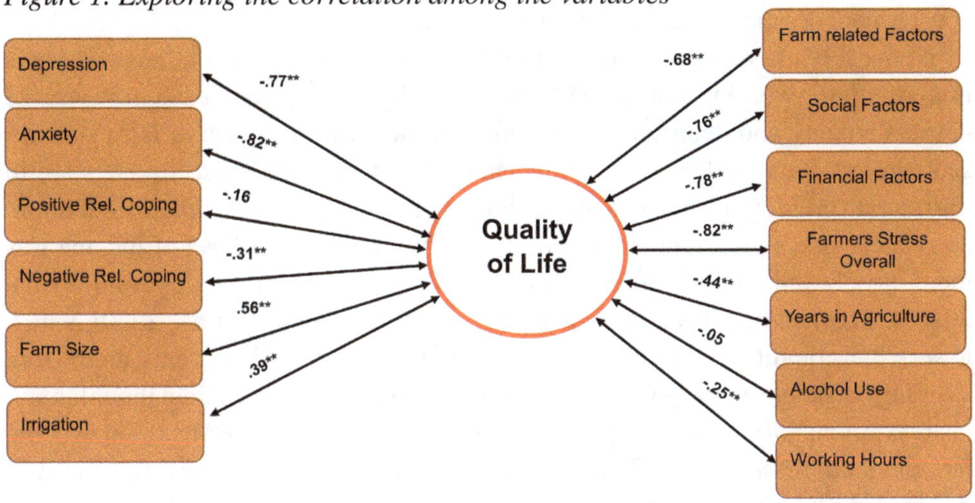

The results of the study illustrate a significant relationship between farmers quality of life and several factors including depression, anxiety, negative religious coping, and farmer stress. Specifically, the results showed negative correlation between Qol and depression ($r = -.77$; $p < .01$), suggesting that an increase in depression is associated with a decrease in QoL. Similarly, there was a negative correlation between QoL and anxiety ($r = -.82$; $p < .01$), negative religious coping ($r = -.31$; $p < .01$), QoL and farmer stress ($r = -.82$; $p < .01$).

In addition, the study found that there is a positive correlation between farmer stress and years in agriculture indicating that and increase in years spent in agriculture is associated with an increase in stress levels ($r = .26$; $p < .01$). Furthermore, there was a positive correlation between farm stress and working hours ($r = .18$; $p < .05$), while there was a negative correlation between farm stress and farm size ($r = -.46$; $p < .01$), and irrigation/water issue ($r = -.44$; $p < .01$) Table 1.

The main purpose of the study was to explore the relationship between stress, anxiety, and depression with the QoL of farmers. Additionally, the study investigated how stress levels were associated with farmers' health, alcohol consumption, farm size, years of experiences in agriculture, and working hours. First, the study's findings revealed an association between farmer stress and farm size, indicating that a reduction in farm size increases farmer stress. In other words, a farm's stressor that produces strain is a lack of money (McGregor et al., 1995). Quality of life has been linked to factors including farm size, revenue from farming, meaningful job, and working conditions for farmers (Röös et al., 2019). The health of farmers, their use of alcohol or other drugs, and the size of their farms are all positively correlated. As farm size expands, so do farmers' health and drug and alcohol usage. According to previous research, farmers have alcohol-related difficulties more frequently than non-farmers do (Jarman et al., 2007).

Noticeably, there is a negative relationship between depression, anxiety, years in agriculture, working hours and farm size, which means decrease in farm size upshots anxiety and depressive symptoms together with increase in farm size upshots in less time for leisure activities and more workload. These findings are in mirror with previous researches. According to a Norwegian study, where agriculture accounted for a significant amount of household income, the level of income was more closely linked to financial stress and poor mental health (Logstein, 2016a, 2016b). Farmers may even carry a greater workload (Hansen et al., 2020), which lowers their quality of life by restricting their free time and prevents them from sleeping sufficiently (Wojewódzka-Wiewiórska et al., 2020). Suicidal thoughts and distress among farmers were significantly more strongly predicted by the size of the farmer's land (Kandlur et al., 2022). Farmers psychological resilience is significantly influenced by economic factors, including their income from both agriculture and non-agriculture, the size and ownership of their land, agricultural expenses, and the costs of insurance and medical care (Tahernejad et al., 2024).

The study's findings also revealed a negative correlation between depression, anxiety, stress, and negative religious coping and Quality of Life. These results agree with those of earlier studies. According to recent research conducted in the Midwest, 2/3 of farmers there stated that they suffered from anxiety-related conditions, and over fifty percent of them had depression (Rudolphi et al., 2020). Farmers in the countryside report lower mental, physical, and overall wellness than their non-farming contemporaries (Brew et al., 2016; Hounsome et al., 2012; Kallioniemi et al., 2016). Another research by Torske et al.,(2016) comparing Norwegian farmers to their siblings and persons in other professions likewise revealed that farmers experienced feelings of depression and anxiety more frequently than those in other occupational categories or their non-farming siblings. This shows how an agricultural job may have an impact on mental health. Farmers' mental health can be impacted by financial issues, worrying about the weather (Kearney et al., 2014a), harsh weather, and crop loss caused by the weather (Jones et al., 2018). According to Heo et al., (2020), important stresses affecting farmers' quality of life are farm finances and debt levels. Farmers spend a lot of time alone, isolation is challenging, risky, and isolated situations, with little opportunity for restoration or socialisation (Truchot & Andela, 2018). Many farmers experience mental health issues and low psychological wellness as a result of this isolated lifestyle (Roy et al., 2013). Similar studies have shown that, when farmers are confronted with macroeconomic deterioration, religion improves their quality of life (Meyer & Lobao, 2003).

Third, the study's findings revealed that stress is positively correlated with a farmer's years spent in agriculture and working hours. Previous studies have shown that farmers may experience stress associated to their work due to a lack of leisure time, a heavy workload, time constraints, and long hours of physically taxing and repetitive work (Fraser et al., 2005b). Additionally, the size of a farmer's land and water issues have a detrimental impact on stress. A larger farm generates more money and keeps the farmer out of stressful financial situations, which reduces stress. Financial strain was identified as the primary stressor among (Kallioniemi et al., 2016; McGregor et al., 1995) participants in earlier investigations. This finding is in line with earlier research, which indicated that the farmer's perception of his or her level of preparedness and ability to deal with climate variations through the use of modified cultivation strategies, the selection of an appropriate crop, and potential access to irrigation (Han et al., 2022) greatly influenced the level of concern.

Another study provide evidence that farmers explained how living in their place of employment provided them with continual reminders of their burden, which they claimed added to the sense of endless, overwhelming work (Thompson et al., 2023). Thus, Farming presents a special combination of interconnected factors that can be detrimental to mental health, including social isolation, extended working hours, and unpredictable revenue because of uncontrollable factors like the weather (B. G. Hansen et al., 2020).

Mental Health Strategies Which can Assist Famers to Improve Their Mental Health

Mental Wellness Awareness and Crisis Training

Farmers with mental health disorders are frequently finally treated by non-mental health professionals like general practitioners because of the scarcity of mental health care providers in rural locations (Lowder et al., 2021). Another significant obstacle to obtaining treatment is the absence of healthcare professionals who are familiar with issues impacting the agricultural community and who understand how to live a farming lifestyle (Roy et al., 2019).

Assistance From Peer Intervention

Peer assistance interventions should be provided to the farmers. Australian farmers who took part in informal support groups, such as having conversation at a nearby mechanic shop during the work day, described experiencing enormous relief from the "healing power that comes from being allowed to express themselves to someone regarding their difficulties" (Perceval et al., 2018).They thought the informal support groups may work well to prevent rural men from death by suicide.

Service Network Expansion

Interventions should promote communication and cooperation between parties offering assistance to distressed farmers, such as agricultural advocacy groups, healthcare professionals, and rural advisers.

Agroecological Learning

By educating farmers on chemical-input-free, ecologically sound polyculture, agroecological education efforts should be implemented with the goal of healing both farmer mental health and local ecology.

Political Advocacy

Farmers should get involved in pro-farmer political action groups that support compassionate farm policies since doing so may help farmers deal with stress and mental health issues (Roy et al., 2019).

Communities can support farmers in improving their mental health by

- Remote and rural communities can guarantee access to in-person, online, and phone mental health resources. Increasing awareness of these resources by launching informational campaigns, such as regarding suicide prevention among farmers that aim to lessen stigma and boost requests for and offers of assistance.
- Preserve solid social connections by keeping friendships and positive relationships with loved ones as a top priority and by taking part in social events and peer support programs.
- By encouraging farmers to participate in a pastime, communities can foster social connections among farmers.
- A national policy should include services and supports specifically designed to address the mental health of farmers (Jones-Bitton, n.d.).
- By being aware of subtle indicators that signal mental health issues, such as body language and tone of voice, healthcare workers can spot individuals with mental health issues, especially men. Improved training is required to guarantee that these experts can identify depression (Paraschakis et al., 2016).

The results of the present study provide evidence to back up the Conservation of Resources theory (Hobfoll, 1988) because farmers' mental health is impacted by the different risks they confront to their resources. The emotional turmoil that might arise when people perceive a loss or threat to their resources may be reflected in the increase in stress from Time 1 to Time 2 (Hobfoll, 1988). In this scenario, price rises for fertiliser, a crucial crop resource, may have challenged farmers' financial and energy resources. Harvest also presents a variety of challenges to farmers' material, financial, and labour resources, particularly time, money, and equipment, as well as a lack of social support options. As per the Conservation of Resources theories, personal characteristics pertain to an individual's worldview, as people with unfavourable dispositions may find it difficult to cope with moments of severe stress (Peel et al., 2016) and working long hours and handling the rigorous duties of farming can be difficult, and it can strain family ties (Riethmuller et al., 2023). In addition to condition resources like relationships and social standing, the protective elements of social support and belonging also represent individual resources like optimism, hope, and self-worth. It is especially crucial to support farmers' mental health by lowering the threat to these conditions and personal resources during times of high stress.

The Goldschmidt hypothesis, which explains that the size and organisation of agricultural operations affect the social and economic well-being of rural communities, is also supported by the study's findings. This hypothesis states that, in comparison to smaller, less capital-intensive farms, larger, more industrialised farms are linked to greater economic resources, better services, and an improved quality of life. According to this study, farmers who oversee larger farms typically report higher incomes and better general well-being. By implying that greater scale and related financial stability could act as buffers against psychological distress, this supports the Goldschmidt viewpoint. The intricate relationship between farm structure, financial gains, and mental health outcomes is highlighted by the fact that farmers who experienced higher levels of stress, anxiety, and depression also reported lower quality of life.

The Cumulative Stress Model (Israel et al., 1994) provides a valuable framework for understanding the psychological burden experienced by farmers. This model conceptualizes stress as arising from both environmental conditions (e.g., economic pressures, isolation, environmental uncertainty) and individual-level factors (e.g., powerlessness, lack of support). The "stress pile-up" effect described by CSM explains how chronic exposure to multiple, interacting stressors can lead to long-term mental health challenges, such as anxiety and depression, which this study identifies as key contributors to a diminished quality of life among farmers. Furthermore, the model highlights the importance of protective factors—such as social support, financial security, and autonomy—which aligns with our finding that increases in farm size are associated with increased well-being. Integrating the CSM into this analysis underscores the need for comprehensive, multi-level interventions to address both the root causes and the psychological consequences of farm-related stress.The results of this study highlight the importance of investigating and addressing these issues urgently. The insight gained from this study can assist policy makers, healthcare providers, and public health professionals in making informed decisions, providing policy recommendations, allowing resources, and offer mental health services and training. It is therefore essential for healthcare providers to identify the specific health needs of individuals working in agriculture and determine whether mental health challenges are unique to the agricultural industry or prevalent throughout rural regions. Prices should be empowered by the government of the crops grown by farmers. This can be possible by guaranteeing equitable prices for all commodities and transforming market structures for openness and direct access, farmers can reach their full potential.

Regardless of the strengths of the present study, there are limitations to be acknowledged. This cross-sectional study only offers a "snap-shot" of the mental health of the participating farmers from Punjab since random sampling was not used, which limits the ability to infer causality. Sample size is one of the most crucial survey-criterion. There is no easy way to determine how large a sample should be because doing so requires a grasp of a variety of concepts, such as statistical power and impact size (Prajapati et al., 2010). Despite the statistically significant connections, our sample size was insufficient to provide sufficient power for comprehensive study. Furthermore, only male farmers were included in the sample data. Other variables like masculinity, perceived social support, resilience, or usage of coping methods have not been examined in relation to farmers. Although the study demonstrates a favourable correlation between large farms, higher income, and improved well-being. However, present study mostly concentrates on large farms and mental health at the individual level, failing to adequately account for the wider community-level effects that the Goldschmidt hypothesis highlights, such as shifts in social cohesiveness, public services, or regional job trends. Causal interpretation is limited by the data's cross-sectional character. It's unclear if having a large farm directly improves mental health, or it is more likely that superior-resourced farmers are just more inclined to keep up larger operations, even though larger farms seem to be linked to improved well-being. Lastly, as Goldschmidt first cautioned, this study did not consider the social trade-offs of large-scale farming, such as greater automation or land consolidation, which could have detrimental long-term effects on rural community life.

CONCLUSION

Bases on our research, farming is a challenging and demanding profession that is linked to increased levels of stress, anxiety, and unhappiness. The future of farming remains uncertain, and this uncertainty may impact the mental health of farmers, according to current agricultural policy. The prevalence of depression among farmers, especially among men, is concerning. To address this issue, preventing strategies such as self-help groups, coping skills training, mental health education programmes, and practical support (with e.g., financial assistance, retraining for individuals who choose to quit farming) are necessary. Farmers face a higher risk of psychological distress due to factors such as unpredictable economic outlook, independent cultures, and limited access to mental health care.

Help-seeking behaviour is negatively impacted by perceived stigma concerning mental wellbeing therapies and services in rural locations, but not in urban areas. Access to conventional mental healthcare is further constrained by the widespread stigma associated with mental illness and strongly ingrained cultural ideals of fortitude, perseverance, and independence In addition to the psychological risk factors already mentioned, which are a typical aspect of farming, new research has focused on the potential cognitive effects of organophosphate exposure. Many people in the farming community believe that exposure to organophosphates caused a variety of these symptoms, which causes great grief to those who experience them because they frequently feel that their symptoms are not being treated seriously. It is necessary to conduct more study on how exposure to organophosphates affects psychological discomfort, mental health, and suicide risk. It may be essential to teach farmers the value of their connections with their family, spouse, and friends in order to reduce stress and ultimately improve wellbeing (Deegan & Dunne, 2022). Future studies also need to take into account other factors like farmers in Punjab's use of coping mechanisms, perceived social support, and masculinity,

As highlighted by the Goldschmidt hypothesis, future research should take a more comprehensive approach to investigate how farm size and structure affect not only individual well-being but also larger community dynamics. Establishing causal links between shifts in farm scale, income stability, and mental health outcomes over time may be made easier with the aid of longitudinal research. Further understanding of how farmers view the trade-offs between financial success and social or emotional stressors may also be possible by utilising qualitative methods. Future studies should also examine how community-level factors, like social support networks, healthcare, and educational opportunities, may moderate the correlation between farm structure and quality of life. The societal effects of farm consolidation could be better understood through comparative research between regions with different agricultural models (such as smallholder versus industrial farming systems), which could also help guide policy initiatives that support rural communities' mental and economic well-being.

REFERENCES

Alpass, F., Flett, R., Humphries, S., Massey, C., Morriss, S., & Long, N. (2004). Stress in dairy farming and the adoption of new technology. *International Journal of Stress Management, 11*(3), 270–281. DOI: 10.1037/1072-5245.11.3.270

Austin, E. K., Handley, T., Kiem, A. S., Rich, J. L., Lewin, T. J., Askland, H. H., Askarimarnani, S. S., Perkins, D. A., & Kelly, B. J. (2018a). Drought-related stress among farmers: Findings from the Australian Rural Mental Health Study. *The Medical Journal of Australia, 209*(4), 159–165. DOI: 10.5694/mja17.01200 PMID: 30041594

Austin, E. K., Handley, T., Kiem, A. S., Rich, J. L., Lewin, T. J., Askland, H. H., Askarimarnani, S. S., Perkins, D. A., & Kelly, B. J. (2018b). Drought-related stress among farmers: Findings from the Australian rural mental health study. *The Medical Journal of Australia, 209*(4), 159–165. DOI: 10.5694/mja17.01200 PMID: 30041594

Bartram, D. J., Yadegarfar, G., & Baldwin, D. S. (2009). A cross-sectional study of mental health and well-being and their associations in the UK veterinary profession. *Social Psychiatry and Psychiatric Epidemiology, 44*(12), 1075–1085. DOI: 10.1007/s00127-009-0030-8 PMID: 19294320

Bennett, J., & Bennett, R. J. (1982). *Of Time and the Enterprise: North American Family Farm Management in a Context of Resource Marginality*. University of Minnesota Press.

Bjelland, I., Dahl, A. A., Haug, T. T., & Neckelmann, D. (2002). The validity of the Hospital Anxiety and Depression Scale. An updated literature review. *Journal of Psychosomatic Research, 52*(2), 69–77. DOI: 10.1016/s0022-3999(01)00296-3 PMID: 11832252

Bjornestad, A., Cuthbertson, C., & Hendricks, J. (2021). An analysis of suicide risk factors among farmers in the midwestern United States. *International Journal of Environmental Research and Public Health, 18*(7), 3563. DOI: 10.3390/ijerph18073563 PMID: 33808131

Booth, N. J., & Lloyd, K. (2000). Stress in farmers. *The International Journal of Social Psychiatry, 46*(1), 67–73. DOI: 10.1177/002076400004600108 PMID: 14529080

Brennan, M., Hennessy, T., & Dillon, E. (2020). Towards a better measurement of the social sustainability of Irish agriculture. *International Journal of Sustainable Development, 23*(3–4), 263–287. DOI: 10.1504/IJSD.2020.115229

Brew, B., Inder, K., Allen, J., Thomas, M., & Kelly, B. (2016). The health and wellbeing of Australian farmers: A longitudinal cohort study. *BMC Public Health*, *16*, 988. DOI: 10.1186/s12889-016-3664-y PMID: 27634298

Bryant, L., & Garnham, B. (2015). The fallen hero: Masculinity, shame and farmer suicide in Australia. *Gender, Place and Culture*, *22*(1), 67–82. DOI: 10.1080/0966369X.2013.855628

Caldwell, T. M., Jorm, A. F., & Dear, K. B. G. (2004). Suicide and mental health in rural, remote and metropolitan areas in Australia. *The Medical Journal of Australia*, *181*(S7), S10–S14. DOI: 10.5694/j.1326-5377.2004.tb06348.x PMID: 15462636

Daghagh Yazd, S., Wheeler, S. A., & Zuo, A. (2019). Key risk factors affecting farmers' mental health: A systematic review. *International Journal of Environmental Research and Public Health*, *16*(23), 4849. DOI: 10.3390/ijerph16234849 PMID: 31810320

Das, A. (2011). Farmers' suicide in India: Implications for public mental health. *The International Journal of Social Psychiatry*, *57*(1), 21–29. DOI: 10.1177/0020764009103645 PMID: 21252353

Davis-Brown, K., & Salamon, S. (1987). Farm families in crisis: An application of stress theory to farm family research. *Family Relations*, *36*(4), 368. DOI: 10.2307/584485

Deegan, A., & Dunne, S. (2022). An investigation into the relationship between social support, stress, and psychological well-being in farmers. *Journal of Community Psychology*, *50*(7), 3054–3069. DOI: 10.1002/jcop.22814 PMID: 35132638

Dodge, R., Daly, A., Huyton, J., & Sanders, L. (2012). The challenge of defining wellbeing. *International Journal of Wellbeing*, *2*(3), 222–235. DOI: 10.5502/ijw.v2i3.4

Fennell, K. M., Jarrett, C. E., Kettler, L. J., Dollman, J., & Turnbull, D. A. (2016). "Watching the bank balance build up then blow away and the rain clouds do the same": A thematic analysis of South Australian farmers' sources of stress during drought. *Journal of Rural Studies*, *46*, 102–110. DOI: 10.1016/j.jrurstud.2016.05.005

Fraser, C. E., Smith, K. B., Judd, F., Humphreys, J. S., Fragar, L. J., & Henderson, A. (2005). Farming and mental health problems and mental illness. *The International Journal of Social Psychiatry*, *51*(4), 340–349. DOI: 10.1177/0020764005060844 PMID: 16400909

Furey, E. M., O'Hora, D., McNamara, J., Kinsella, S., & Noone, C. (2016). The roles of financial threat, social support, work stress, and mental distress in dairy farmers' expectations of injury. *Frontiers in Public Health*, *4*, 126. DOI: 10.3389/fpubh.2016.00126 PMID: 27446893

Gill, A., & Singh, L. (2006). Farmers' suicides and response of public policy: Evidence, diagnosis and alternatives from Punjab. *Economic and Political Weekly*, *41*(26), 2762–2768.

Goldschmidt, W. (1978). *As you sow: Three studies in the social consequences of agribusiness. Monclair: Allanheld*. Osmun and Company.

Green, G. P. (1985). Large-scale farming and the quality of life in rural communities: Further specification of the Goldschmidt hypothesis. *Rural Sociology*, *50*(2), 262–274.

Grover, D., Singh, J. M., Kaur, A., & Kumar, S. (2017). *State Agricultural Profile-Punjab* [Technical report]. Punajb Agricultural University. https://rgdoi.net/10.13140/RG.2.2.29375.87203

Guillien, A., Laurent, L., Soumagne, T., Puyraveau, M., Laplante, J.-J., Andujar, P., Annesi-Maesano, I., Roche, N., Degano, B., & Dalphin, J.-C. (2017). Anxiety and depression among dairy farmers: The impact of COPD. *International Journal of Chronic Obstructive Pulmonary Disease*, *13*, 1–9. DOI: 10.2147/COPD.S143883 PMID: 29296078

Gulati, A., Roy, R., & Hussain, S. (2021). Performance of Agriculture in Punjab. In Gulati, A., Roy, R., & Saini, S. (Eds.), *Revitalizing Indian Agriculture and Boosting Farmer Incomes* (pp. 77–112). Springer Nature., DOI: 10.1007/978-981-15-9335-2_4

Han, G., Schoolman, E. D., Arbuckle, J. G., & Morton, L. W. (2022). Weather, values, capacity and concern: Toward a social-cognitive model of Specialty crop farmers' perceptions of climate change risk. *Environment and Behavior*, *54*(2), 327–362. DOI: 10.1177/00139165211026607

Hansen, B. G., Bugge, C. T., & Skibrek, P. K. (2020). Automatic milking systems and farmer wellbeing–exploring the effects of automation and digitalization in dairy farming. *Journal of Rural Studies*, *80*, 469–480. DOI: 10.1016/j.jrurstud.2020.10.028

Hansen, G. B., Tandberg Bugge, C., & Skibrek, P. K. (2020). *Automatic milking systems and farmer wellbeing–exploring the effects of automation and digitalization in dairy farming. 80*, 469–480. https://doi.org/DOI: 10.1016/j.jrurstud.2020.10.028

Heo, W., Lee, J. M., & Park, N. (2020). Financial-related psychological factors affect life satisfaction of farmers. *Journal of Rural Studies*, *80*, 185–194. DOI: 10.1016/j.jrurstud.2020.08.053

Hobfoll, S. E. (1988). *The Ecology of Stress*. Taylor & Francis.

Hobfoll, S. E. (2001). The influence of culture, community, and the nested-self in the stress process: Advancing Conservation of Resources theory. *Applied Psychology*, *50*(3), 337–370. DOI: 10.1111/1464-0597.00062

Hounsome, B., Edwards, R. T., Hounsome, N., & Edwards-Jones, G. (2012). Psychological morbidity of farmers and non-farming population: Results from a UK survey. *Community Mental Health Journal*, *48*(4), 503–510. DOI: 10.1007/s10597-011-9415-8 PMID: 21556782

Israel, B. A., Checkoway, B., Schulz, A., & Zimmerman, M. (1994). Health education and community empowerment: Conceptualizing and measuring perceptions of individual, organizational, and community control. *Health Education Quarterly*, *21*(2), 149–170. DOI: 10.1177/109019819402100203 PMID: 8021145

Jagga, R. (2022, April 24). *Depressed over low wheat yield, farmer dies by suicide; 14th such death in Punjab in April*. https://indianexpress.com/article/chandigarh/depressed-over-low-wheat-yield-farmer-dies-by-suicide-14th-such-death-in-punjab-in-april-7884497/

Jarman, D., Naimi, T., Pickard, S., Daley, W., & De, A. (2007). *Binge drinking and occupation, North Dakota, 2004–2005*. *4*(4). https://stacks.cdc.gov/view/cdc/4124

Jones, M. S., Reed, D. B., & Hunt, M. L. (2018). Suicide: An unrecognized epidemic among farmers. *Workplace Health & Safety*, *66*(9), 464. DOI: 10.1177/2165079918784055 PMID: 30015597

Kahsay, G. A., Kassie, W. A., Medhin, H., & Hansen, L. G. (2022). Are religious farmers more risk taking? Empirical evidence from Ethiopia. *Agricultural Economics*, *53*(4), 617–632. DOI: 10.1111/agec.12697

Kallioniemi, M. K., Simola, A., Kaseva, J., & Kymäläinen, H.-R. (2016). Stress and burnout among Finnish dairy farmers. *Journal of Agromedicine*, *21*(3), 259–268. DOI: 10.1080/1059924X.2016.1178611 PMID: 27081893

Kandlur, R., Sardana, S., & Richardson-Vejlgaard, R. (2022). The Agrarian distress: Factors explaining the will to live among rural and distressed family farmers. *Psychiatry Research Communications*, *2*(1), 100019. DOI: 10.1016/j.psycom.2021.100019

Kearney, G. D., Rafferty, A. P., Hendricks, L. R., Allen, D. L., & Tutor-Marcom, R. (2014a). A cross-sectional study of stressors among farmers in Eastern North Carolina. *North Carolina Medical Journal*, *75*(6), 384–392. DOI: 10.18043/ncm.75.6.384 PMID: 25402689

Kearney, G. D., Rafferty, A. P., Hendricks, L. R., Allen, D. L., & Tutor-Marcom, R. (2014b). A Cross-sectional study of stressors among farmers in Eastern North Carolina. *North Carolina Medical Journal*, *75*(6), 384–392. DOI: 10.18043/ncm.75.6.384 PMID: 25402689

Knight, J., & Gunatilaka, R. (2012). Aspirations, adaptation and subjective well-being of rural–urban migrants in China. In *Adaptation, Poverty and Development* (Clark, D.A., pp. 91–110). Palgrave Macmillan UK. https://doi.org/DOI: 10.1057/9781137002778_4

Logstein, B. (2016a). Farm-related concerns and mental health status among Norwegian farmers. *Journal of Agromedicine*, *21*(4), 316–326. DOI: 10.1080/1059924X.2016.1211055 PMID: 27420178

Logstein, B. (2016b). Predictors of mental complaints among Norwegian male farmers. *Occupational Medicine*, *66*(4), 332–337. DOI: 10.1093/occmed/kqw019 PMID: 26944593

Lowder, S. K., Sánchez, M. V., & Bertini, R. (2021). Which farms feed the world and has farmland become more concentrated? *World Development*, *142*, 105455. DOI: 10.1016/j.worlddev.2021.105455

Lowder, S. K., Skoet, J., & Raney, T. (2016). The Number, size, and distribution of farms, smallholder afrms, and family farms worldwide. *World Development*, *87*, 16–29. DOI: 10.1016/j.worlddev.2015.10.041

Lunner Kolstrup, C., Kallioniemi, M., Lundqvist, P., Kymäläinen, H.-R., Stallones, L., & Brumby, S. (2013). International perspectives on psychosocial working conditions, mental health, and stress of dairy farm operators. *Journal of Agromedicine*, *18*(3), 244–255. DOI: 10.1080/1059924X.2013.796903 PMID: 23844791

Maciuba, S. A., Westneat, S. C., & Reed, D. B. (2013). Active coping, personal satisfaction, and attachment to land in older African-American farmers. *Issues in Mental Health Nursing*, *34*(5), 335–343. DOI: 10.3109/01612840.2012.753560 PMID: 23663020

McCubbin, H. I., & Patterson, J. M. (1983). The Family Stress Process. *Marriage & Family Review*, *6*(1–2), 7–37. DOI: 10.1300/J002v06n01_02

McGregor, M., Willock, J., & Deary, I. (1995). Farmer stress. *Journal of Farm Management*, *9*, 57.

Meyer, K., & Lobao, L. (2003). Economic hardship, religion and mental health during the midwestern farm crisis. *Journal of Rural Studies*, *19*(2), 139–155. DOI: 10.1016/S0743-0167(02)00069-4

Molnar, J. J. (1985). Determinants of subjective well-being among farm operators: Characteristics of the individual and the firm. *Rural Sociology*, *50*(1), 141–162.

Monaghan, C., Bizumic, B., Williams, T., & Sellbom, M. (2020). Two-dimensional Machiavellianism: Conceptualization, theory, and measurement of the views and tactics dimensions. *Psychological Assessment*, *32*, 277–293. DOI: 10.1037/pas0000784 PMID: 31750680

Mzoughi, N. (2014). Do organic farmers feel happier than conventional ones? An exploratory analysis. *Ecological Economics*, *103*, 38–43. DOI: 10.1016/j.ecolecon.2014.04.015

Odabasi, S., & Hartarska, V. (2021). Farmer suicides: Effects of socio-economic, climate, and mental health factors. *The Journal of Mental Health Policy and Economics*, *24*(2), 61–71. PMID: 34151778

Paraschakis, A., Michopoulos, I., Christodoulou, C., Koutsaftis, F., & Douzenis, A. (2016). Psychiatric medication intake in suicide victims: Gender disparities and implications for suicide prevention. *Journal of Forensic Sciences*, *61*(6), 1660–1663. DOI: 10.1111/1556-4029.13195 PMID: 27643812

Pargament, K. I., Koenig, H. G., & Perez, L. M. (2000). The many methods of religious coping: Development and initial validation of the RCOPE. *Journal of Clinical Psychology*, *56*(4), 519–543. DOI: 10.1002/(sici)1097-4679(200004)56:4<519:aid-jclp6>3.0.co;2-1 PMID: 10775045

Pearlin, L. I., Lieberman, M. A., Menaghan, E. G., & Mullan, J. T. (1981). The stress process. *Journal of Health and Social Behavior*, *22*(4), 337–356. PMID: 7320473

Peel, D., Berry, H. L., & Schirmer, J. (2016). Farm exit intention and wellbeing: A study of Australian farmers. *Journal of Rural Studies*, *47*, 41–51. DOI: 10.1016/j.jrurstud.2016.07.006

Perceval, M., Ross, V., Kõlves, K., Reddy, P., & Leo, D. (2018). *Social factors and Australian farmer suicide: A qualitative study. 18*(1), 1367. https://doi.org/DOI: 10.1186/s12889-018-6287-7

Polanco-Roman, L., Gomez, J., Miranda, R., & Jeglic, E. (2016). Stress-related symptoms and suicidal ideation: The roles of rumination and depressive symptoms vary by gender. *Cognitive Therapy and Research*, *40*(5), 606–616. DOI: 10.1007/s10608-016-9782-0 PMID: 27695146

Prajapati, B., Dunne, M., & Armstrong, R. (2010, July 16). *Sample size estimation and statistical power analyses.* https://www.semanticscholar.org/paper/Sample-size-estimation-and-statistical-power-Prajapati-Dunne/8b940fa7c816608266e71832e021f4cd036ded76

Purc-Stephenson, R., Doctor, J., & Keehn, J. E. (2023). Understanding the factors contributing to farmer suicide: A meta-synthesis of qualitative research. *Rural and Remote Health, 23*(3), 8189. DOI: 10.22605/RRH8189 PMID: 37633833

Reuben, H. (1949). *Families under stress: Adjustment to the crises of war separation and reunion.* Harper.

Riethmuller, M. L., Dzidic, P. L., McEvoy, P. M., & Newnham, E. A. (2023). Change, connection and community: A qualitative exploration of farmers' mental health. *Journal of Rural Studies, 97*, 591–600. DOI: 10.1016/j.jrurstud.2023.01.018

Röös, E., Fischer, K., Tidåker, P., & Nordström Källström, H. (2019). How well is farmers' social situation captured by sustainability assessment tools? A Swedish case study. *International Journal of Sustainable Development and World Ecology, 26*(3), 268–281. DOI: 10.1080/13504509.2018.1560371

Rosenblatt, P. C., & Anderson, R. M. (1982). Interaction in Farm Families: Tension and Stress 1. In *The Family In Rural Society*. Routledge.

Roy, P., Duplessis-Brochu, É., & Tremblay, G. (2019). Responses to adversity faced by farming men: A gender-transformative analysis. *International Journal of Child, Youth & Family Studies : IJCYFS, 10*(1), 49–69. DOI: 10.18357/ijcyfs101201918806

Roy, P., Tremblay, G., Oliffe, J. L., Jbilou, J., & Robertson, S. (2013). Male farmers with mental health disorders: A scoping review. *The Australian Journal of Rural Health, 21*(1), 3–7. DOI: 10.1111/ajr.12008 PMID: 23384130

Rudolphi, J. M., Berg, R. L., & Parsaik, A. (2020). Depression, anxiety and stress among young farmers and ranchers: A pilot study. *Community Mental Health Journal, 56*(1), 126–134. DOI: 10.1007/s10597-019-00480-y PMID: 31583619

Rural Mental Health Overview—Rural Health Information Hub. (n.d.). Retrieved March 24, 2023, from https://www.ruralhealthinfo.org/topics/mental-health

Saini, S., Siraj, H., & Khatri, P. (2022, June 30). *A Fresh Spike in Farmer Suicides in Punjab* |. https://www.theindiaforum.in/article/spike-farmer-suicides-punjab

Sanne, B. (2004). Farmers are at risk for anxiety and depression: The Hordaland health study. *Occupational Medicine, 54*(2), 92–100. DOI: 10.1093/occmed/kqh007 PMID: 15020727

Sanne, B., Mykletun, A., Dahl, A. A., Moen, B. E., & Tell, G. S. (2003). Occupational differences in levels of anxiety and depression: The Hordaland Health Study. *Journal of Occupational and Environmental Medicine*, *45*(6), 628–638. DOI: 10.1097/01.jom.0000069239.06498.2f PMID: 12802216

Santos, E. G. de O., Queiroz, P. R., Nunes, A. D. da S., Vedana, K. G. G., & Barbosa, I. R. (2021). Factors associated with suicidal behavior in farmers: A systematic review. *International Journal of Environmental Research and Public Health*, *18*(12), 6522. DOI: 10.3390/ijerph18126522 PMID: 34204344

Seager, M., Sullivan, L., & barry, J. (2014). Gender-related schemas and suicidality: Validation of the male and female gender scripts questionnaire. *New Male Studies, 3*(3), 32–54. https://discovery.ucl.ac.uk/id/eprint/1472580/9/Gender_related_pdf.pdf

Shri Guru Granth Sahib. (2000Singh, G., Trans.). World Sikh Center Inc.

Tahernejad, A., Sohrabizadeh, S., & Mashhadi, A. (2024). Exploring factors affecting psychological resilience of farmers living in drought-affected regions in Iran: A qualitative study. *Frontiers in Psychology*, *15*, 1418361. DOI: 10.3389/fpsyg.2024.1418361 PMID: 39286558

Taki, F., Witte, M., Jaijee, I. S., Jessel, S., Mauer, E., Lurie, J., & Kaur, G. (2021). Suicide in rural Punjab, India: Implications for ensuring farmer rights. *Journal of Global Health Reports*, *5*, e2021037. DOI: 10.29392/001c.22236

Thompson, R., Hagen, B. N. M., Lumley, M. N., Winder, C. B., Gohar, B., & Jones-Bitton, A. (2023). "An incredible amount of stress before you even put a shovel in the ground": A mixed methods analysis of farming stressors in Canada. *Sustainability*, *15*(8), 6336. DOI: 10.3390/su15086336

Torske, M. O., Hilt, B., Glasscock, D., Lundqvist, P., & Krokstad, S. (2016). Anxiety and depression symptoms among farmers: The HUNT study, Norway. *Journal of Agromedicine*, *21*(1), 24–33. DOI: 10.1080/1059924X.2015.1106375 PMID: 26488439

Truchot, D., & Andela, M. (2018). Burnout and hopelessness among farmers: The farmers stressors Inventory. *Social Psychiatry and Psychiatric Epidemiology*, *53*(8), 859–867. DOI: 10.1007/s00127-018-1528-8 PMID: 29725701

Weber, J. G. (2011). *Individual and Family Stress and Crises* (1st ed.). SAGE., https://us.sagepub.com/en-us/nam/individual-and-family-stress-and-crises/book228977

Weigel, D. J., & Weigel, R. R. (1990). Family Satisfaction in Two-Generation Farm Families: The Role of Stress and Resources. *Family Relations*, *39*(4), 449–455. DOI: 10.2307/585227

Wojewódzka-Wiewiórska, A., Kłoczko-Gajewska, A., & Sulewski, P. (2020). Between the social and economic dimensions of sustainability in rural areas—In search of farmers' quality of life. *Sustainability, 12*(1), 1. Advance online publication. DOI: 10.3390/su12010148

World Health Organization. (2004). *The World Health Organization quality of life (WHOQOL)—BREF* (No. WHO/HIS/HSI Rev.2012.02). World Health Organization. https://apps.who.int/iris/handle/10665/77773

Zigmond, A. S., & Snaith, R. P. (1983). The hospital anxiety and depression scale. *Acta Psychiatrica Scandinavica, 67*(6), 361–370. DOI: 10.1111/j.1600-0447.1983.tb09716.x PMID: 6880820

ADDITIONAL READING

Folkman, S. (1984). Personal control and stress and coping processes: A theoretical analysis. *Journal of Personality and Social Psychology, 46*(4), 839–852. DOI: 10.1037/0022-3514.46.4.839 PMID: 6737195

Hossain, D., Eley, R., Coutts, J., & Gorman, D. (2008). Mental health of farmers in Southern Queensland: Issues and support. *The Australian Journal of Rural Health, 16*(6), 343–348. DOI: 10.1111/j.1440-1584.2008.01014.x PMID: 19032206

Ingram, R. E., & Luxton, D. D. (2005). Vulnerability-Stress Models. In *Development of psychopathology: A vulnerability-stress perspective* (pp. 32–46). Sage Publications, Inc., DOI: 10.4135/9781452231655.n2

Keyes, C. L. M. (2002). The mental health continuum: From languishing to flourishing in life. *Journal of Health and Social Behavior, 43*(2), 207–222. DOI: 10.2307/3090197 PMID: 12096700

King, E., Lamont, K., Wendelboe-Nelson, C., Williams, C., Stark, C., Van Woerden, H. C., & Maxwell, M. (2023). Engaging the agricultural community in the development of mental health interventions: A qualitative research study. *BMC Psychiatry, 23*(1), 399. DOI: 10.1186/s12888-023-04806-9 PMID: 37277723

Lobao, L., & Meyer, K. (2001). The great agricultural transition: Crisis, change, and social consequences of twentieth century US farming. *Annual Review of Sociology, 27*, 103–124.

Simkin, S., Hawton, K., Fagg, J., & Malmberg, A. (1998). Stress in farmers: A survey of farmers in England and Wales. *Occupational and Environmental Medicine*, *55*(11), 729–734. DOI: 10.1136/oem.55.11.729 PMID: 9924448

Xu, J. (2016). Pargament's theory of religious coping: Implications for spiritually sensitive social work practice. *British Journal of Social Work*, *46*(5), 1394–1410. DOI: 10.1093/bjsw/bcv080 PMID: 27559234

KEY TERMS AND DEFINITIONS

Agricultural Economy: In an agricultural economy both farming and crop cultivation are often combined with an animal rearing which are the main sources of income and productivity.

Anxiety: Anxiety is a reaction to a perceived or actual threats which appears as a sense of tension, anxiety, or fear.

Depression: Depression is a mood disorder which results in an ongoing feeling of sadness and loss of interest. It influences a person's thoughts, feelings, and behaviour and can result in a number of psychological and physical issues.

Farmer: An individual who manages and labours on a farm is called a farmer

Mental health: A state of psychological well-being which allows individuals to manage life's stressors, reach their full potential, learn and work effectively, and give back to their community is known as mental health.

Quality of life: The degree to which an individual or group is comfortable, healthy, and able to enjoy daily activities is known as their quality of life.

Religious coping: Looking on a supreme being for peace, support, and direction is known as religious coping. Religious coping can take place in the context of an organised religion or more casually through one's own spirituality.

Stress: Stress is a common reaction to demands or pressures in the environment, particularly when those demands are viewed as hazardous or threatening.

Chapter 4
Global Approaches to Mental Health in the Era of Climate-Driven Migration

Priya Sharma
https://orcid.org/0000-0002-3942-6346
Baddi University of Emerging Sciences and Technology, India

Ravinesh Mishra
Baddi University of Emerging Sciences and Technology, India

Bhartendu Sharma
Baddi University of Emerging Sciences and Technology, India

ABSTRACT

Climate change is increasingly displacing populations worldwide, giving rise to climate-driven migration with profound mental health consequences. This chapter explores the global approaches to addressing mental health in the context of climate-induced displacement, highlighting the psychological burdens faced by migrants—including trauma, anxiety, depression, and loss of identity. Through case studies and comparative policy analysis, the chapter examines disparities in mental health care access, cultural sensitivity, and global governance frameworks. It advocates for integrated, rights-based, and culturally responsive mental health systems and proposes a comprehensive framework that embeds mental health into climate adaptation and migration policies. The chapter calls for urgent, collaborative action to address this growing global health challenge.

DOI: 10.4018/979-8-3373-3281-9.ch004

INTRODUCTION

One of the most difficult humanitarian issues of the twenty-first century is the combination of climate change and human migration. Extreme weather events like floods, storms, wildfires, droughts, and sea level rise have become far more frequent and severe as global temperatures rise. Millions of people are being displaced by these environmental changes, which are also upsetting communities and changing global migratory patterns. The Internal Displacement Monitoring Centre (IDMC) estimates that in 2022 alone, more than 30 million people were forcibly moved as a result of climate-related disasters; this number is expected to increase significantly over the next several decades. These "climate migrants" experience severe psychological stress and emotional trauma in addition to being uprooted from their homes and means of subsistence (Schwerdtle et al., 2018).

Migration is a very upsetting life event, particularly when it is forced. It frequently takes place in extremely stressful, traumatic, and chaotic circumstances when fueled by the erratic forces of climate change. In a heartbeat, people might lose their homes, loved ones, cultural identity, and hopes for the future. They can end up crossing borders with little to no legal protection, living in overcrowded camps, or being sent to strange areas. Post-traumatic stress disorder (PTSD), despair, anxiety, substance misuse, and suicidal thoughts are just a few of the mental health consequences that can arise from these interruptions. Many people experience ongoing trauma as they adjust to their new surroundings and deal with identity conflicts, social exclusion, economic instability, and long-term uncertainty.

Mental health is still a largely ignored aspect of the discussion of climate and migration policy, despite the obvious psychological effects of relocation brought on by climate change. Food, shelter, water, and physical health are typically given priority in humanitarian efforts, while mental health frequently receives little funding and attention. However, resilience, rehabilitation, and the ability to start anew all depend on mental health. Insufficient psychological assistance puts displaced people at higher risk for violence, social disengagement, chronic mental illness, and a decreased ability to contribute to their host communities or return home when circumstances permit (McMichael et al., 2023).

The urgent need to address the mental health needs of displaced populations has come to the attention of humanitarian agencies, non-governmental organizations (NGOs), and global health organizations. The foundation for integrated responses has been established by frameworks like the United Nations High Commissioner for Refugees' (UNHCR) MHPSS framework, the Inter-Agency Standing Committee's (IASC) Guidelines on Mental Health and Psychosocial Support (MHPSS) in Emergency Settings, and the World Health Organization's (WHO) Mental Health Gap Action Programme (mhGAP). These recommendations support trauma-informed, culturally aware, and community-based strategies that prioritize prevention, resilience-building, and rehabilitation while empowering local actors.

However, converting these ideas into practical on-the-ground initiatives is still quite difficult. Bangladesh, South Sudan, and the Philippines are among the nations with the greatest rates of climate-related relocation. These nations also have weak healthcare systems and inadequate mental health facilities. In low- and middle-income countries (LMICs), where up to 85% of persons with mental health disorders do not receive treatment, the global mental health treatment gap is especially pronounced. Climate-displaced people in these environments bear the combined burden of pre-existing service gaps and the unique mental health hazards associated with displacement. Around the world, a variety of creative, community-led solutions are being developed in response. To close access gaps and connect with impacted populations on a cultural level, peer counseling, traditional healing methods, culturally specific support groups, and digital health platforms are being used. In crisis situations, technology in particular has emerged as a potent instrument for providing mental health care. In rural and underserved locations, displaced people are now able to get support through telemedicine platforms, chatbots driven by artificial intelligence, mobile applications, and SMS counseling. For instance, Wysa and WHO's "Step-by-Step" program have been modified for use with migrants and refugees in South Asia and the Middle East (Askland et al., 2022).

Along with advancements in service delivery, the necessity of enhancing social integration and cohesion in host communities is becoming more widely recognized. The social situations that migrants reside in have a direct impact on their mental health results. While intercultural communication, inclusive legislation, and community-building initiatives can promote resilience and healing, experiences of xenophobia, discrimination, and marginalization can worsen mental suffering. Numerous host nations and localities are creating initiatives that help mental health via integration, including peer support groups run by migrants, community forums, and cross-cultural mentorships.

Building capacity is yet another essential component of a successful worldwide response. Scaling mental health services sustainably requires improving the capabilities of frontline workers, such as medical professionals, educators, aid workers, and even migrants themselves. Programs that teach cultural competency, trauma-informed treatment, and psychological first aid (PFA) can enable non-specialists to provide referrals and early help. Since displaced people have invaluable knowledge of the needs, assets, and cultural values of their communities, it is equally important to include them as co-creators of solutions.At the policy level, more and more nations are starting to include mental health in their plans for climate adaptation and catastrophe preparedness. To guarantee proper financing, coordination, and integration across industries including housing, education, migration, and health, however, much work needs to be done (Ofori et al., 2023).

In the age of climate-driven migration, this review article attempts to summarize and assess contemporary worldwide approaches to mental health. International frameworks, psychosocial support models, integration and social cohesion activities, capacity-building programs, technological advancements, and policy responses are just a few of the fields in which it examines best practices, obstacles, and opportunities. It aims to give a thorough picture of how the international community is attending to the mental health needs of climate migrants—and where more urgent action is needed—by utilizing case studies, evidence-based practices, and new developments (Titcomb et al., 2024).

EFFECTS OF DISPLACEMENT CAUSED BY CLIMATE CHANGE ON MENTAL HEALTH

Individuals, families, and entire communities are all impacted by the significant and complex effects of climate change-related displacement on mental health. Acute and long-term psychological effects come from the turmoil caused by people being compelled to leave their homes owing to resource scarcity, bad weather, desertification, or sea level rise.

Trauma, which frequently results from the abrupt loss of homes, jobs, and loved ones, is one of the most obvious consequences. Post-traumatic stress disorder (PTSD) can strike survivors, particularly following natural catastrophes like floods or cyclones. Anxiety, sadness, and a sense of helplessness can result from the cumulative stress and uncertainty, even in circumstances with a longer onset, such as droughts.

Social networks and cultural continuity—two important safeguards for mental health—are disrupted by displacement. Grief, loneliness, and social disengagement are exacerbated by the loss of identity, customs, and community institutions. Migrants frequently experience prejudice, language challenges, legal instability, and limited access to mental health treatments in host or resettlement communities, all of which worsen their misery.

Due to behavioural abnormalities, emotional instability, and developmental delays, children and adolescents are especially at risk. In the meantime, elderly persons may endure silent suffering due to exacerbated losses of social roles, familiar surroundings, and autonomy.

LINKING PHYSICAL IMPACT, MENTAL HEALTH, AND COMMUNITY WELL-BEING:

Mental Health

Climate change threatens people's capacity to digest information and make decisions without being hampered by strong emotional reactions. A certain amount of emotional reaction is typical, and even negative feelings are essential to living a happy existence. In the worst situation, though, they may impair our capacity for logical thought, behavior planning, and alternate action consideration. Extreme weather events can produce trauma and emotionally debilitating experiences. People's lives may become more stressful to varied degrees due to more subtle and indirect consequences of climate change. Stressors related to our environment, whether directly or indirectly experienced, lead to poor mental health, which can lead to anxiety and depression (USGCRP, 2016). Everyone can handle a certain level of stress, but compound stress's cumulative effects can turn someone from mentally healthy to mentally unwell. According to Greco and Roger (2003), even uncertainty can lead to stress and increase the risk of psychological distress.

Physical Impact and Mental Health

Stress from compromised physical health might endanger psychological well-being. On the other hand, mental health issues can also endanger physical health, for instance, by altering eating, sleeping, or exercise habits or by impairing immune system performance.

Community Health

Communities are impacted by the mental and physical well-being of their citizens, but the effects of climate change on community health can have an especially significant impact on interpersonal connections and the cohesiveness of the community. Climate change-related changes to the environment can affect how people relate to one another, their connections to the natural world, and their chances for social interaction.

CULTURAL AND SOCIAL FACTORS

Rapid economic and population growth puts a strain on the environment and makes it difficult for nations to manage their resources sustainably (Caldwell.,

1990). Changes in climate presents a number of risks to humanity, particularly to vulnerable areas around the world who are currently experiencing severe droughts and starvation, which causes population displace. Climate refugees are receiving more attention globally as a result of climate change's increased intensity of extreme weather events, which leads to migration and displacement (Thornton et al., 2011). It is not easy to identify this group of migrants, nevertheless. The issue of climate change refugees, an emerging and indisputable fact, is examined in this essay along with how the international community is handling it.

Views of Mental Illness by Culture

- Different cultures have different definitions of mental health: In some cultures, mental anguish is not seen as a medical condition but rather as a spiritual imbalance, ancestral punishment, or moral weakness.
- Expression of distress: Rather than using psychiatric terminology, emotional pain in non-Western cultures may manifest as somatization, or bodily symptoms (Rubel & Garro, 1992).

Stigma and Silence

- People are reluctant to seek care since mental illness is highly stigmatized in many communities, particularly among displaced groups.
- People are deterred from sharing emotional distress by societal rejection and shame.

Language and Communication

- Language barriers may make it difficult to communicate mental health terms and symptoms.
- In a new cultural setting, migrants might not have the words to communicate psychological pain (for example, somatization is more prevalent in some cultures) (Green et al., 2002).

Traditional Healing Practices

- It may be preferable to use spiritual leaders, healers, or rituals (such as herbal treatments or cleansing rites) rather of receiving conventional mental health support.

- Although these traditions may be at odds with Western mental health concepts, they can offer psychological alleviation, communal connectedness, and cultural grounding.

Social Disruption

- Losing one's home, land, identity, and community as a result of climate migration frequently causes sadness, worry, and confusion.
- Social isolation and heightened vulnerability result from displacement, which disrupts support networks (Syme et al., 1964).

Gender and Vulnerability

- Risks specific to women and children include family separation, trafficking, gender-based violence, and sexual violence.
- Higher rates of anxiety, sadness, and PTSD are a result of these events.

Economic and Legal Insecurity

- Chronic tension and uncertainty are caused by a lack of legal status or occupational rights.
- Overcrowding, poverty, and incarceration can all have a negative impact on mental health (Factor et al., 2007).

CLIMATE CHANGE-RELATED HEALTH RISKS

Due to an increase in both new and existing health hazards, climate change raises health risks. Reduced ecological resources, worsening air and water quality, and more frequent and severe climate events like heat waves, storms, coastal flooding, and forest fires are just a few of the direct effects. It increases the danger of indirect health effects due to intricate biological, social, legal, and economic processes. Climate change exacerbates existing health concerns by altering natural and social systems. Because they face greater risks due to disproportionately weaker adaptive capacities, resources, rights, and political power, the broad demographics engaged and the importance of marginalized populations are crucial (Ebi & Burton, 2008).

Many people believe that in order to create sustainable adaptation and mitigation strategies, a better understanding of the health hazards associated with climate change is necessary. Linking climate change to public health research is emphasized here as a unique issue. This is a neglected but crucial task that calls for thorough collaborations across disciplinary domains. The topic is enhanced by discussing recent articles in climate-related health, which leads to a discussion of how migration patterns and decisions are influenced by health concerns from climate-related hazards (Ebi & Hess, 2020).

GLOBAL POLICY ENVIRONMENT: MIGRATION DRIVEN BY CLIMATE CHANGE AND MENTAL HEALTH

A rising global policy issue that calls for coordinated and proactive solutions is the confluence of migration, climate change, and mental health. Global governance frameworks are gradually changing to recognize the psychosocial aspects of climate-driven migration as environmental deterioration exacerbates displacement. Nonetheless, there are still large gaps in mental health integration and legal recognition. Millions of people are currently in a legal limbo as international legal documents, like the 1951 Refugee Convention, do not specifically acknowledge climate migrants as refugees. This lack of acknowledgment makes it more difficult to get necessary assistance, such as mental health treatment.

In the meanwhile, international health and humanitarian organizations like the International Organization for Migration (IOM) and the World Health Organization (WHO) support the inclusion of mental health in disaster response and migration planning. A fundamental policy tool for field implementation is the Inter-Agency Standing Committee's (IASC) Guidelines on Mental Health and Psychosocial Support in Emergency Settings. National plans for disaster response and climate adaptation increasingly address relocation, but they frequently ignore mental health, especially in low- and middle-income nations. More robust policy connections between mental health systems, migratory governance, and climate resilience initiatives are required.

HEALTHCARE SYSTEMS AND INFRASTRUCTURE

Infrastructure and healthcare systems are severely strained by climate-driven migration, especially in receiving areas that are already suffering with scarce resources. In order to address the increasing mental health requirements of displaced populations, future global strategies must place a high priority on bolstering healthcare infrastructure. This includes creating scalable, adaptable systems that can handle unexpected surges in population, guaranteeing continuity of care for migrants, and increasing primary care facilities' capacity to offer integrated mental health services (Schatz & Berlin, 2011).

It is crucial to invest in healthcare worker training, particularly in trauma-informed and culturally competent care. Knowledge-sharing and the creation of mobile and telehealth units to serve isolated or transitory populations can be facilitated by cross-border cooperation. Plans for emergency preparedness and disaster response must also incorporate mental health infrastructure to guarantee that psychological first aid and continuing assistance are accessible both during and after climate-related emergencies.

Together, governments and international health organizations should provide funding for inclusive, robust health systems that can treat acute and chronic mental health issues. Improving data systems for tracking health trends among migrants will also be essential for resource allocation and well-informed policymaking. To put it simply, sustainable infrastructure needs to promote the mental and emotional health of people displaced by climate change in addition to their physical health (Leslie et al., 2017).

PSYCHOSOCIAL SUPPORT AND INTERVENTIONS

Reducing the negative effects of climate-driven migration on mental health requires psychosocial support and treatments. Environmental disasters like floods, droughts, and sea level rise that force people to relocate frequently lead to trauma, identity loss, family separation, and ongoing stress. A worldwide strategy necessitates short-term and long-term tactics that are sustainable, scalable, and culturally appropriate.

Providing psychological first aid (PFA) during or soon after displacement is the first step towards providing effective psychosocial care. PFA provides helpful, caring support that stabilizes people and promotes safety, community, and hope. Community-based support networks, such as peer counseling, group therapy, and spiritual care, are essential for displaced populations, especially those living in camps or temporary settlements, to regain a sense of control and belonging.

Interventions that are culturally sensitive are crucial. In order to boost trust and efficacy, they entail collaborating with traditional healers, church leaders, and community elders while also working with local beliefs, languages, and healing customs. Stories, expressive arts, and narrative therapy are frequently effective in assisting people in processing trauma in ways they are accustomed to.

The developmental and emotional needs of displaced children, who are particularly vulnerable, are met by child-focused interventions such safe places, play therapy, and school reintegration support. Psychosocial resilience is also improved in adults through livelihood programs, social inclusion initiatives, and skill-building exercises.

For migrants impacted by climate change, psychosocial assistance and interventions are crucial parts of mental health care. Displacement brought on by climate change frequently results in a variety of problems, such as the loss of communities, houses, and means of subsistence, which can cause severe psychological suffering. As their lives are upended, migrants may suffer from trauma, sorrow, anxiety, despair, and post-traumatic stress disorder (PTSD). Psychosocial support is therefore essential for resolving mental health issues as well as for building coping mechanisms and resilience in the face of hardship. Important strategies consist of (Ferrero et al., 2019).

Community based mental health

Application of community-level, culturally appropriate therapies such peer support groups, group therapy, and psychosocial education; participation of community leaders and certified lay counsellors to promote local ownership and trust

Aid for Humanitarians

- Integrating mental health treatments within current disaster response frameworks, refugee health services, and primary care systems.
- Teaching frontline staff trauma-informed care and psychological first aid.

Specific Action for At-Risk Population

- Targeted assistance for older individuals, women, children, and those with underlying mental health issues.
- Establishing youth-safe areas and gender-sensitive programs to deal with particular concerns (Fricchione et al., 2012).

Telepsychiatry Tools

- Expansion of helplines and mobile apps for mental health in areas where access to professionals is scarce.
- To close service gaps, remote psychological evaluations and therapies are used.

Policy Level

- Creation of national mental health policies that take climatic pressures and displacement into consideration.
- Including mental health in frameworks for catastrophe risk reduction, migration, and climate adaption

Physiological Support

For migrants impacted by climate change, psychosocial assistance and interventions are crucial parts of mental health care. Displacement brought on by climate change frequently results in a variety of problems, such as the loss of communities, houses, and means of subsistence, which can cause severe psychological suffering. As their lives are upended, migrants may suffer from trauma, sorrow, anxiety, despair, and post-traumatic stress disorder (PTSD). Psychosocial support is therefore essential for resolving mental health issues as well as for building coping mechanisms and resilience in the face of hardship (Nwazor, 2012).

INTEGRATION AND SOCIAL COHESION

The ability of climate-driven migrants to enjoy social cohesiveness and assimilate into new social situations has a significant impact on their mental health. In order to address the mental health issues that displaced populations confront, it is crucial to comprehend and facilitate integration as migration brought on by climate change increases. Migration frequently involves a significant shift in the environment, culture, and social context, which can lead to stress, identity loss, and feelings of alienation. Therefore, social cohesion and integration are essential for enhancing mental health outcomes, building resilience, and encouraging long-term rehabilitation for both host communities and migrants.

Addressing the mental health needs of migrants impacted by climate change requires social cohesiveness and integration. The psychological well-being of those displaced by climate change is significantly impacted by their capacity to adjust, form social bonds, and feel accepted in their new social contexts as they settle in host communities, whether they are in cities, camps, or rural areas. In addition to being vital for societies to remain stable generally, social cohesion—the quality of relationships and a sense of belonging within a community—is also critical for safeguarding and enhancing the mental health of those who have been displaced (Snowden et al., 2023).

Important tactics consist of:

Awareness Campaigns

Putting public education campaigns into action to lessen the stigma associated with migration and mental illness. Encouraging displaced groups' stories of contribution and resiliency in order to combat marginalization and xenophobia.

Inclusive Community Engagement

Promoting open communication and collaborative decision-making between local government, host communities, and migrants. Supporting neighbourhood projects that use sports, the arts, and local gatherings to honour cultural diversity and common identities (Jacobs et al., 2014).

Mental Health Support in Integration Services

Accessible and culturally competent mental health services must be provided in tandem with integration initiatives. The specific psychological requirements of climate migrants should be acknowledged by health systems, and these need should be included into more comprehensive migration and integration programs. Psychological anguish can be lessened with services that are adaptable and customized to the experience of migrants. Culturally competent counselors, for instance, can provide more effective therapy and promote recovery in a way that is relatable to migrant communities because they are aware of the sociopolitical and cultural histories of displaced populations.

Furthermore, it guarantees that climate migrants are not cut off from the social fabric of their new areas by including mental health support into larger community development programs. This strategy places a strong emphasis on group well-being, which is essential for the mental and social well-being of both host communities and migrants.

Economic Integration

Promoting the restoration of routine, dignity, and mental health by facilitating access to meaningful work and vocational training. Programs that encourage collaboration between members of the host community and migrants in order to foster mutual trust and benefit (Honadle & Hannah, 1982).

Building Social Cohesion

For the mental health of both migrants and the general public, social cohesiveness among host communities is equally important. Good intergroup interactions between host communities and migrants can lessen prejudice, xenophobia, and tensions—all of which are known to exacerbate mental health issues. Initiatives that promote understanding, solidarity, and collaboration between residents and migrants are crucial. Intercultural discussions, collaborative community events, and public awareness campaigns that emphasize common values, cultures, and experiences are a few examples of these programs. One effective example is Germany's "Welcome" program, which helps immigrants integrate into their communities by bringing together migrants and host citizens through volunteer networks and collaborative events.

Furthermore, psychological resilience is directly impacted by social cohesiveness. Social support systems, including friendships, family, and community organizations, are essential for mitigating the adverse impacts of trauma. Communities may improve the sense of self-worth, security, and belonging that migrants experience by creating settings where they feel appreciated and accepted. These elements are crucial for supporting mental health (Ellegala et al., 2014).

MENTAL HEALTH SERVICES' FUNCTION IN INTEGRATION

Services related to mental health are essential to the integration process. Effective support requires culturally sensitive mental health care that recognizes the particular difficulties faced by migrants. In addition to treating mental health issues, services should actively encourage social integration by making networks, cultural resources, and community involvement more accessible. This dual focus makes sure that people are assisted in re-establishing and forming new relationships in their host communities, rather than being left alone during their healing process.

In order to provide holistic care, mental health services also serve as cross-sector connectors, tying together social services, education, housing, employment, and health care. Coordination between governmental organizations, non-governmental organizations, and local communities is improved by their incorporation into public health and humanitarian response systems. Additionally, mental health practitioners support policy changes that include the psychological effects of climate displacement into national migration and adaptation plans.

Peer support programs, in which immigrants help one another along the way, have also been shown to be successful in improving mental health and social cohesiveness. By fostering mutual understanding and enabling migrants to share knowledge, these programs lessen the psychological strain of acclimating to new surroundings (Horn et al., 2019).

CAPACITY BUILDING AND TRAINING

There is growing pressure on national health systems, humanitarian organizations, and local communities to adequately address the mental health needs of impacted populations as climate-induced displacement continues to increase globally. Building capacity and providing training are two of the most important and long-lasting solutions. Meeting the increasing demand among climate migrants requires strengthening the capacity of people, organizations, and systems to deliver high-quality, easily accessible, and culturally relevant mental health care.

Training frontline personnel, who are frequently the first people displaced people interact with, such as community health providers, teachers, social workers, and humanitarian responders, is the first step in establishing capacity. Many of these professionals have little to no formal training in mental health and work in environments with minimal resources. They can be trained in basic psychological support, such Psychological First Aid (PFA), which helps with quick referrals, early mental health issue detection, and initial support. Supported by the Inter-Agency Standing Committee (IASC) and the World Health Organization (WHO), PFA enables non-specialists to provide non-invasive, compassionate treatment to people experiencing severe distress.

Improving the ability of medical professionals, such as nurses, physicians, psychologists, and psychiatrists, who would be needed to offer more specialized care, is equally crucial. Training in mental health is frequently underrepresented in medical curriculum in many low- and middle-income nations. Programs aimed at increasing capacity must place a high priority on ongoing professional development via mentorship, telepsychiatry, workshops, and distant learning (Coccia, 2021).

TRAINING MATERIAL TOOL

- Translation and context adaptation of the mhGAP-IG (Intervention Guide).
- Interactive toolkits (e.g., UNHCR e-learning courses, IASC MHPSS training packages).
- Simulation exercises, such as role-playing and fake case situations.
- Platforms for remote learning and mobile apps for ongoing education (like WHO's EQUIP digital toolbox).

CAPACITY BUILDING STRATEGIES

- Employing rain-the-trainer strategies to foster local ownership.
- Incorporation of MHPSS into official health and social work education curriculum.
- Creation of communities of practice and peer supervision groups.
- Training materials that take age and gender into account to guarantee inclusive services.

BEST PRACTICE EXAMPLE

- Field trainings for the MHPSS Reference Group of the Inter-Agency Standing Committee (IASC) in Bangladesh and South Sudan.
- The worldwide PFA training cascade for volunteers is offered by the Red Cross Red Crescent.
- Climate crisis response training is being integrated into Kenya's national mhGAP rollout (Smihula et al., 2009).

TECHNOLOGICAL INNOVATIONS

Technology is becoming more and more important in boosting access, improving service delivery, and encouraging innovation in mental health treatment as a result of the growing mental health needs of populations affected by climate change. Traditional face-to-face mental health care may be inadequate or unavailable since displacement frequently takes place in settings with limited resources and a high risk of crises. For vulnerable migrants and displaced people, technological advancements—from mobile health apps to AI-powered support tools—offer scalable, affordable, and culturally sensitive ways to overcome mental health gaps.

One of the most useful and comprehensive instruments for supporting mental health is mobile health (mHealth) technology. Despite losing their homes or means of support, many displaced people frequently still have access to cell phones. Direct therapy, self-help resources, psychoeducation, and mental health information can all be delivered through this connectedness. Applications like WHO's Step-by-Step, a guided self-help app for anxiety and depression, have been effectively modified for usage among South Asian and Middle Eastern refugee populations. These resources, which are frequently offered in numerous languages and have offline capabilities, blend culturally relevant content with evidence-based cognitive behavioral treatment procedures (Utterback, 1971).

Technology helps health personnel develop their capacity in addition to providing direct care. Practitioners in rural or conflict-affected areas can learn new skills and stay current on best practices through e-learning modules, virtual workshops, and mobile-based training apps. For example, utilizing smartphones or tablets, the mhGAP digital training toolbox provides interactive learning on detecting and treating common mental diseases.

But there are still a number of difficulties. The effectiveness and impact of tech-based initiatives may be constrained by issues with digital literacy, language difficulties, internet access, and data protection. Co-designing technological solutions with impacted groups is crucial to ensuring their cultural relevance, usability, and trustworthiness. Ethical measures must also be taken to safeguard private health information and prevent the disparities among people without access to digital technology from getting worse (Garud et al., 1997).

DIFFERENT TYPES OF TECHNOLOGY SOLUTIONS

Platforms for Digital Mental Health

Tools for tracking symptoms, education, therapy, and screening. Examples include MindCare (Bangladesh), a multilingual software that provides displaced people with stress-reduction techniques. WHO's Methodical Approach: evidence-based digital depression self-help application that has been modified for use in emergency situations?

Telepsychology and Online Therapy

Services based on voice and video for peer support, therapy, and diagnosis. Examples include e-counseling provided by Médecins Sans Frontières (MSF) in refugee camps. A program funded by the UNHCR that links migrant women with remote therapists is called Healing in Motion (Garcia & Calantone, 2002).

Chatbot and AI Tools

Chatbots with AI capabilities provide a scalable and engaging method of providing mental health assistance. Through text-based dialogues, these bots—like Woebot and Wysa—offer emotional tracking, mood regulation advice, and cognitive-behavioral therapy (CBT) procedures. AI-driven technologies can offer instant, nonjudgmental, and round-the-clock support, but they cannot replace professional care. This is especially helpful for migrants who might be unwilling to seek traditional care because of cultural obstacles or stigma.

First-line support for mental health issues is provided around-the-clock via chat. Examples of AI-based CBT chatbots utilized in multilingual environments are Wysa and Woebot. The "FriendBot" pilot project was created to offer refugees emotionally supportive services that are appropriate for their culture.

Gathering Information and GIS Mapping

Mobile apps for tracking resource gaps and mental health needs in real time. Examples include the KoboToolbox, which aid workers use to evaluate the psychosocial needs of people in crisis areas. - GIS mapping of risk zones, safe areas, and service access points.

Platforms for Capacity Building and E-Learning

Interactive training tools for mhGAP, trauma treatment, and Psychological First Aid (PFA) for staff and peer workers. Examples include the WHO EQUIP platform for digital training quality assurance. - IASC eLearning: COVID-19 "Basic Psychosocial Skills" tailored for a crisis migration environment (Arifin, 2018).

ETHICAL CONSIDERATION

In addition to being a clinical and practical issue, addressing mental health in the context of climate-driven migration is also essentially an ethical one. A complex web of ethical issues, such as justice, autonomy, equity, cultural sensitivity, and the right to health, must be negotiated by governments, non-governmental organizations, and international agencies as they create solutions to assist displaced communities. To protect the rights and dignity of people who have been displaced by climate change, it is crucial to make sure that actions are both morally and practically sound. As part of the larger right to health, the right to mental health is one of the most important ethical precepts. This right is protected by the International Covenant on Economic, Social, and Cultural Rights and the Universal Declaration of Human Rights. However, legal and regulatory loopholes frequently affect climate migrants, particularly those living in low-income environments or lacking official refugee status. Due to their current lack of recognition under international refugee law, climate migrants have fewer legal protections and less access to health and mental health services than refugees escaping conflict or persecution. Regardless of legal requirements, this establishes an ethical duty for states and other international actors to provide protection and care (Hasan et al., 2021).

Quality is still another important factor. Those who have made the least contribution to global emissions—small island states, indigenous tribes, and low-income rural populations—are disproportionately impacted by climate change. These same populations may become even more stigmatized following displacement and frequently do not have access to mental health services. Therefore, inclusion and equity must be given top priority in ethical mental health interventions to guarantee that the most vulnerable people receive quality care. In addition to cost-effectiveness, social justice and moral responsibility should also be taken into consideration when allocating resources. In order to provide displaced populations with ethical mental health care, cultural competence is crucial. In addition to being ineffectual, interventions that disregard cultural values, languages, and healing customs run the risk of hurting or alienating the people they are meant to assist.

HEALTHCARE ISSUES IN THE COMMUNITIES OF THE HOST

The growing demand for healthcare from both long-term residents and migrants will frequently overwhelm host towns, placing a heavy burden on their frequently underfunded healthcare systems. In the event of natural calamities that compel large-scale migration, hosts may not be equipped to handle new health emergencies. The distribution of healthcare resources in the host country will need to be radically reevaluated in order to meet the urgent medical demands of migrants. Certain services may not be culturally responsive, even when legislative processes improve migrants' potential access to healthcare. Additionally, a major obstacle to migrant health in general may be absorption into local healthcare systems.

Even health systems in wealthy nations may have different norms and beliefs, and health systems in the countries of origin may be unprepared to assist climate migrants. Finally, the stress of post-migration brought on by discrimination or unemployment may also contribute to bad health. This will worsen health disparities in host communities and have an impact on the unequal distribution of wealth among migrants. Important direct and policy consequences are raised by such observable trends. In addition to anticipating the health requirements of new migrants and creating responsive infrastructure, local health authorities and non-governmental organizations must also create scalable finance mechanisms that can handle fluctuating migration loads (Ceptureanu et al., 2019).

IMPLICATIONS FOR POLICY AND PRACTICE

Poor health of people and the wider natural world share many common causes and therefore common solutions which will be drawn out in this section. The evidence presented in this study illustrates the necessity of taking into consideration the advantages of climate action as well as the presently unrecognized costs of climate change to mental health (Haines, 2017). By reducing the intensity of climate change and its effects, as well as preventing some of the previously listed mental health effects, climate action—both policies and behaviours intended to reduce or adapt to climate change—can enhance mental health outcomes (Jennings et al., 2020). Changes for individuals, groups, and systems that improve the factors known to lower mental illness or enhance mental health and wellbeing could also help to improve mental health.

The benefits of action and the currently hidden costs of inactivity on climate change can be made apparent and win-win strategies attainable through cross-disciplinary study. The existing research gaps and future objectives on mental health and climate change have been thoroughly documented elsewhere (Hwong et al., 2022). To ensure that policies are created and implemented with evidence-based aims, prevention is prioritized over cure, and unintended negative repercussions are avoided, effective action necessitates a thorough grasp of the linked pathways of influence.

Health leaders, legislators, educators, mental health specialists, and emergency responders are among the important stakeholder groups that need education and training to better understand the evidence in this review and make sure it is incorporated into climate-mental health-aligned policy and practice. Initiatives to raise awareness and educate the public must also upskill community leaders and pertinent institutions (such as schools, universities, and mental health systems) to help people and communities deal with the climate problem while encouraging collaborative and participatory climate action.

Interventions that address the mechanisms via which climate change impacts mental health at an early stage can help prevent negative consequences later on; action is required at every level. Climate change worsens disadvantage and deteriorates mental health and wellbeing through a variety of impact pathways, as this review has emphasized. These outcomes are partially caused by the connections among the factors that influence mental health and wellness on an individual, community, and worldwide scale.

Incorporate Mental Health into Emergency Responses and Interventions

To lessen the immediate and long-term psychological effects of extreme weather occurrences, emergency response plans must incorporate mental health. This entails providing high-risk groups with focused assistance, addressing secondary stressors such as housing and healthcare, and preparing coordinated, multi-sector responses beforehand (Lock et al., 2012). Strategies for mental health must also take into account the continuous stress of climate change risks, and they should concentrate on both preventative and therapeutic measures. Particularly for those who are displaced by forced climate change, proactive climate adaptation and mitigation can increase resilience (Galán, 2021). For displaced individuals, mental health therapy should include community-led planning together with useful assistance in housing, work, and medical care to promote inclusion and lessen long-term adverse effects (Lienert, 2025).

Sustainably Redesign Health Systems and Provide Climate-Informed Mental Health Support

Mental health practitioners should promote sustainable, climate-resilient care models, as health systems are a major source of greenhouse gas emissions. This entails implementing net-zero health systems and providing community-based mental health services that promote environmental and social cohesion (McHale et al., 2020). Climate action and mental health can be supported by programs like green social prescribing. In order to safeguard people with mental illnesses during heatwaves, for example, adequate indoor settings, focused public health message, and healthcare professional training, climate adaptation within health systems must also address mental health vulnerabilities. Overall, mental health services must evolve to address the unique challenges posed by the climate crisis (Gislason et al., 2021).

Support Individuals and Communities to Cope and Act

Support is required for people and communities to actively engage in positive change and manage the negative effects of the climate catastrophe on their mental health (Godden et al., 2021). Emotions such as sadness and fear should be processed, hope should be nurtured, coping mechanisms should be developed, and community and agency should be strengthened through effective interventions. Although there are such therapies, they need to be more widely used, thoroughly evaluated, and adjusted for various circumstances (Gibson et al., 2019). Promoting positive climate action, sharing inspiring tales, and giving people—especially young people—the skills they need to cope with climate distress, develop resilience, and actively participate in climate solutions are all tasks for mental health practitioners, scientists, the media, and educators (Middleton et al., 2020).

Prioritise Co-Beneficial Climate Policies and Practices

To deal with the various impacts of the climate catastrophe on mental health and wellbeing and to help bring about the changes they desire, people and communities need to be properly prepared and supported.

Interventions that can: 1) promote healthy processing of psychological and emotional reactions, such as loss, grief, and fears; 2) value optimistic viewpoints; 3) allow people and communities to envision the future they wish to create; 4) prevent psychological distress from getting worse; 5) help people form communities; 6) create healthy coping mechanisms; and 7) empower people to address the climate crisis are all relevant. Climate action, both individual and collective, can also improve social wellbeing, agency, and community cohesiveness (Chen et al., 2024).

Restoring wild places, or "rewilding," and maximizing green space, especially in metropolitan settings, improves atmospheric carbon dioxide absorption, slows down climate change, and gives people more access to nature. By lowering the danger of floods from extreme weather events and the urban heat island effect, nature-based solutions like green roofs and urban tree planting also aid in climate adaption (Fong et al., 2018). Many people worldwide are seeing a decline in the number and quality of their access to green spaces, and there are significant differences in this access, with minority ethnic groups and impoverished communities having a lower likelihood of having access to green spaces in urban settings. Use of green spaces can also increase the sense of connectedness with nature, which itself predicts both climate action and better mental health.

Food security is a key requirement for good mental health and wellbeing. The quantity and quality of food ingested is proven to have a significant impact on both physical and mental health. In many regions of the world, the agriculture industry contributes significantly to environmental deterioration and carbon emissions. The nutrition density of the food produced (including anti-inflammatory chemicals and beneficial microorganisms) can be increased by modifying agricultural practices to improve local ecosystem health and lower carbon emissions. This will promote human wellness (Montgomery et al., 2022)

RESEARCH GAPS AND FUTURE DIRECTIONS

Addressing the worldwide mental health concerns stemming from climate-driven migration is a complicated, interdisciplinary task. There are still a number of study gaps despite growing awareness, and these suggest future paths for more thorough and efficient methods.

The main research gaps and their accompanying future directions are listed below:

Absence of Longitudinal Information

- Most existing research are cross-sectional and fail to reflect the developing mental health trajectories of climate-displaced individuals.
- Future Direction: To monitor the psychological effects of migration over time, fund longitudinal cohort studies.

Inadequate Attention to Culture

- The relationship between displacement-related stress and cultural norms, local resilience strategies, and traditional mental health beliefs is not well understood.
- Future Direction: Integrate indigenous and local perspectives on mental health into community-participatory research.

Research on Adolscent Mental Health

- Despite being understudied, children and teens endure particular psychological difficulties as a result of displacement.
- Future Direction: Create frameworks for child-centered mental health that are adapted to the social and educational upheaval brought on by migration.

The Divide btw Digital and Telehealth

- Although telemental health interventions are becoming more popular, displaced persons with low levels of connectivity or computer literacy frequently cannot access them.
- Future Direction: Create digital literacy initiatives and scalable, low-tech mental health treatments specifically for displaced groups.

Absence of Common Measure and Instruments

- Comparison is challenging since different research use different metrics to quantify mental health results.
- Future Direction: Create and implement culturally sensitive, standardized evaluation instruments for use with immigrant populations.

FUTURE DIRECTIONS

In order to effectively address mental health in the context of climate-driven migration, innovative, daring approaches that cut across conventional sectoral boundaries are needed. Future strategies must prioritize anticipatory, integrated, and equitable mental health solutions that are suited to both host communities and displaced populations as climate change continues to uproot millions of people. First, early warning systems that integrate mental health risk indicators with methods for forecasting migration and climate change urgently need to be developed. This can direct preventative mental health care and assist in predicting psychological vulnerabilities. At the local, national, and international levels, it will be crucial to incorporate mental health into planning for disaster preparedness and climate adaptation.

Second, increasing community-based mental health treatment should be a top priority for future initiatives, especially in low- and middle-income nations where the majority of climate migrants come from or dwell. Particularly in environments with limited resources, task-shifting strategies—training lay health professionals to offer fundamental mental health services—can aid in filling gaps in the delivery of care.Third, longitudinal studies that look at the mental health paths of climate migrants across various regions must be added to the scope of research and data collecting. The experiences of women, children, indigenous peoples, and other vulnerable groups should receive particular consideration because their psychological needs may differ greatly from one another.

Fourth, inclusive policy frameworks that incorporate mental health services within environmental, urban planning, and migration initiatives are required. For these policies to be sustainable and culturally relevant, impacted communities must be involved in their co-design.

International cooperation is crucial, to sum up. Global migration and climate change necessitate a global response. Coordinated worldwide mental health programs, knowledge-sharing networks, and multilateral funding channels can all help guarantee that no community is left behind. In a society affected by climate change, the future of mental health depends on proactive, caring, and systematic approaches—ones that view mental health as a fundamental component of human dignity and climate resilience rather than as an afterthought.

CONCLUSION

One of the most pressing yet little-discussed global issues of our day is the relationship between mental health and climate change. Globally, the strain on mental health systems is growing along with migration brought on by climate change. Whether forced or voluntary, displacement brought on by resource scarcity, extreme weather, droughts, or rising sea levels is frequently accompanied by trauma, anxiety, bereavement, and a loss of cultural identity. Social determinants including poverty, prejudice, and lack of access to healthcare exacerbate these psychological impacts, making mental health care for climate migrants crucial and complicated. International initiatives for dealing with this new issue have showed potential, especially when mental health services are incorporated into national health systems, humanitarian relief, and climate adaption plans. In a variety of settings, community-based interventions, culturally competent mental health care, and the incorporation of psychosocial support into disaster response strategies have proven to be successful. The need for proactive, inclusive, and scalable mental health policy is becoming more widely acknowledged by governments, non-governmental organizations, and international organizations. But there are still significant gaps. There is still a dearth of funding for mental health in climate response programs, and many nations lack the manpower or infrastructure necessary to fulfil the growing demand. Additionally, the relevance and efficacy of initiatives are undermined by the frequent exclusion of indigenous knowledge and migrant voices from decision-making processes.

Going forward, a trans disciplinary, rights-based, and comprehensive approach is crucial. This entails improving cross-sector cooperation, bolstering global mental health governance, encouraging research on the connections between mental health and climate change, and guaranteeing fair access to care. Our global response to migration and climate change must make mental health a top priority rather than a secondary issue. Only then can we build inclusive, adaptable communities that can deal with a future where the climate is changing while also ensuring the psychological resilience and well-being of populations that are constantly on the move.

REFERENCES

. Arifin, S. R. M. (2018). Ethical considerations in qualitative study. *International journal of care scholars, 1*(2), 30-33.

Askland, H. H., Shannon, B., Chiong, R., Lockart, N., Maguire, A., Rich, J., & Groizard, J. (2022). Beyond migration: A critical review of climate change induced displacement. *Environmental Sociology*, *8*(3), 267–278.

Caldwell, J. C. (1990). Cultural and social factors influencing mortality levels in developing countries. *The Annals of the American Academy of Political and Social Science*, *510*(1), 44–59.

Ceptureanu, S. I., & Ceptureanu, E. G. (2019). Community-based healthcare programs sustainability impact on the sustainability of host organizations: A structural equation modeling analysis. *International Journal of Environmental Research and Public Health*, *16*(20), 4035. PMID: 31640288

Chen, S., Oliva, P., & Zhang, P. (2024, May). Air pollution and mental health: evidence from China. In *AEA Papers and Proceedings* (Vol. 114, pp. 423-428). 2014 Broadway, Suite 305, Nashville, TN 37203: American Economic Association.

. Coccia, M. (2021). Technological innovation. *innovations, 11*, I12.

Ebi, K. L., & Burton, I. (2008). Identifying practical adaptation options: An approach to address climate change-related health risks. *Environmental Science & Policy*, *11*(4), 359–369.

Ebi, K. L., & Hess, J. J. (2020). Health risks due to climate change: inequity in causes and consequences: study examines health risks due to climate change. *Health Affairs*, *39*(12), 2056–2062. PMID: 33284705

Ellegala, D. B., Simpson, L., Mayegga, E., Nuwas, E., Samo, H., Naman, N., & Nicholas, J. S. (2014). Neurosurgical capacity building in the developing world through focused training. *Journal of Neurosurgery*, *121*(6), 1526–1532. PMID: 25216067

Factor, R., Mahalel, D., & Yair, G. (2007). The social accident: A theoretical model and a research agenda for studying the influence of social and cultural characteristics on motor vehicle accidents. *Accident; Analysis and Prevention*, *39*(5), 914–921. PMID: 17291438

Ferrero, G., Setty, K., Rickert, B., George, S., Rinehold, A., DeFrance, J., & Bartram, J. (2019). Capacity building and training approaches for water safety plans: A comprehensive literature review. *International Journal of Hygiene and Environmental Health, 222*(4), 615–627. PMID: 30744955

Fong, K. C., Hart, J. E., & James, P. (2018). A review of epidemiologic studies on greenness and health: Updated literature through 2017. *Current Environmental Health Reports, 5*, 77–87. PMID: 29392643

Fricchione, G. L., Borba, C. P., Alem, A., Shibre, T., Carney, J. R., & Henderson, D. C. (2012). Capacity building in global mental health: Professional training. *Harvard Review of Psychiatry, 20*(1), 47–57. PMID: 22335182

Galán, J. (2021). CREWS: a CAMELS-based early warning system of systemic risk in the banking sector. *Banco de Espana Occasional Paper*, (2132).

. Garcia, R., & Calantone, R. (2002). A critical look at technological innovation typology and innovativeness terminology: a literature review. *Journal of Product Innovation Management: An international publication of the product development & management association, 19*(2), 110-132.

Garud, R., Nayyar, P. R., Shapira, Z., & Shapira, Z. B. (Eds.). (1997). *Technological innovation: Oversights and foresights*. Cambridge University Press.

Gibson, K., Haslam, N., & Kaplan, I. (2019). Distressing encounters in the context of climate change: Idioms of distress, determinants, and responses to distress in Tuvalu. *Transcultural Psychiatry, 56*(4), 667–696. PMID: 31067153

Gislason, M. K., Kennedy, A. M., & Witham, S. M. (2021). The interplay between social and ecological determinants of mental health for children and youth in the climate crisis. *International Journal of Environmental Research and Public Health, 18*(9), 4573. PMID: 33925907

Godden, N. J., Farrant, B. M., Yallup Farrant, J., Heyink, E., Carot Collins, E., Burgemeister, B., & Cooper, T. (2021). Climate change, activism, and supporting the mental health of children and young people: Perspectives from Western Australia. *Journal of Paediatrics and Child Health, 57*(11), 1759–1764. PMID: 34792244

Green, A. R., Betancourt, J. R., & Carrillo, J. E. (2002). Integrating social factors into cross-cultural medical education. *Academic Medicine, 77*(3), 193–197. PMID: 11891153

Haines, A. (2017). Health co-benefits of climate action. *The Lancet. Planetary Health, 1*(1), e4–e5. PMID: 29851591

. Hasan, N., Rana, R. U., Chowdhury, S., Dola, A. J., & Rony, M. K. K. (2021). Ethical considerations in research. *Journal of Nursing Research, Patient Safety and Practise (JNRPSP) 2799-1210, 1*(01), 1-4.

Honadle, G. H., & Hannah, J. P. (1982). Management performance for rural development: Packaged training or capacity building. *Public Administration and Development*, *2*(4), 295–307.

Horn, R., O'May, F., Esliker, R., Gwaikolo, W., Woensdregt, L., Ruttenberg, L., & Ager, A. (2019). The myth of the 1-day training: The effectiveness of psychosocial support capacity-building during the Ebola outbreak in West Africa. *Global Mental Health (Cambridge, England)*, *6*, e5. PMID: 31143466

Hwong, A. R., Wang, M., Khan, H., Chagwedera, D. N., Grzenda, A., Doty, B., & Compton, W. M. (2022). Climate change and mental health research methods, gaps, and priorities: A scoping review. *The Lancet. Planetary Health*, *6*(3), e281–e291. PMID: 35278392

Jacobs, J. A., Duggan, K., Erwin, P., Smith, C., Borawski, E., Compton, J., & Brownson, R. C. (2014). Capacity building for evidence-based decision making in local health departments: Scaling up an effective training approach. *Implementation Science : IS*, *9*, 1–11. PMID: 25253081

Jennings, N., Fecht, D., & De Matteis, S. (2020). Mapping the co-benefits of climate change action to issues of public concern in the UK: A narrative review. *The Lancet. Planetary Health*, *4*(9), e424–e433. PMID: 32918888

Leslie, H. H., Sun, Z., & Kruk, M. E. (2017). Association between infrastructure and observed quality of care in 4 healthcare services: A cross-sectional study of 4,300 facilities in 8 countries. *PLoS Medicine*, *14*(12), e1002464. PMID: 29232377

. Lienert, C. (2025). Early warning–international collaboration using map-based and user-centered approaches. *International Journal of Cartography*, 1-11.

. Lock, S., Rubin, G. J., Murray, V., Rogers, M. B., Amlôt, R., & Williams, R. (2012). Secondary stressors and extreme events and disasters: a systematic review of primary research from 2010-2011.

McHale, S., Pearsons, A., Neubeck, L., & Hanson, C. L. (2020). Green health partnerships in Scotland; pathways for social prescribing and physical activity referral. *International Journal of Environmental Research and Public Health*, *17*(18), 6832. PMID: 32962081

McMichael, C., Schwerdtle, P. N., & Ayeb-Karlsson, S. (2023). Waiting for the wave, but missing the tide: Case studies of climate-related (im) mobility and health. *Journal of Migration and Health, 7*, 100147. PMID: 36619800

Middleton, J., Cunsolo, A., Jones-Bitton, A., Wright, C. J., & Harper, S. L. (2020). Indigenous mental health in a changing climate: A systematic scoping review of the global literature. *Environmental Research Letters, 15*(5), 053001.

Montgomery, D. R., Biklé, A., Archuleta, R., Brown, P., & Jordan, J. (2022). Soil health and nutrient density: Preliminary comparison of regenerative and conventional farming. *PeerJ, 10*, e12848. PMID: 35127297

Nwazor, J. C. (2012). Capacity building, entrepreneurship and sustainable development. *Journal of Emerging Trends in Educational Research and Policy Studies, 3*(1), 51–54.

Ofori, D. O., Bandauko, E., Kutor, S. K., Odoi, A., Asare, A. B., Akyea, T., & Arku, G. (2023). A Systematic Review of International and Internal Climate-Induced Migration in Africa. *Sustainability, 15*(22), 16105.

Rubel, A. J., & Garro, L. C. (1992). Social and cultural factors in the successful control of tuberculosis. *Public Health Reports, 107*(6), 626. PMID: 1454974

Schatz, B. R., & Berlin, R. B.Jr. (2011). *Healthcare infrastructure: Health systems for individuals and populations*. Springer Science & Business Media.

Schwerdtle, P., Bowen, K., & McMichael, C. (2018). The health impacts of climate-related migration. *BMC Medicine, 16*, 1–7. PMID: 29301536

Snowden, B., Lahiri, S., Dutton, R., & Morton, L. (2023). Achieving and sustaining change through capacity building train-the-trainer health initiatives in low-and middle-income countries: A systematic review. *The Journal of Continuing Education in the Health Professions, 43*(2), 96–103. PMID: 36215155

Syme, S. L., Hyman, M. M., & Enterline, P. E. (1964). Some social and cultural factors associated with the occurrence of coronary heart disease. *Journal of Chronic Diseases, 17*(3), 277–289. PMID: 5878595

Thornton, P. H., Ribeiro-Soriano, D., & Urbano, D. (2011). Socio-cultural factors and entrepreneurial activity: An overview. *International Small Business Journal, 29*(2), 105–118.

Titcomb, G., Uelmen, J., Janko, M., & Nunn, C. (2024). Infectious disease responses to human climate change adaptations. *Global Change Biology, 30*(8), e17433. PMID: 39171421

Utterback, J. M. (1971). The process of technological innovation within the firm. *Academy of Management Journal*, *14*(1), 75–88.

Chapter 5 Psychological Consequences of Climate-Induced Statelessness: Mental Health Challenges and Legal Coping Mechanisms Among Displaced Populations

Aishwarya
https://orcid.org/0009-0001-3495-4414
Galgotias University, India

Apoorva Roy
https://orcid.org/0009-0005-8422-9404
Galgotias University, India

ABSTRACT

Climate-induced statelessness is a situation that doesn't come about in a day. This sort of thing is a crisis that developed over time as a result of populations being uprooted from their homes by the effects of climate change. Coastal and island nations are the greatest victims of rising sea levels as homes, livelihoods, and legal identities disappear. It is bad enough to lose one's citizenship, but such a phenomenon would also have serious psychological repercussions. Such populations that have experienced displacement from climate activity usually suffer from a higher

DOI: 10.4018/979-8-3373-3281-9.ch005

incidence of anxiety and depression and may develop post-traumatic stress disorder (PTSD) and identity crises, a large part of this being because they would have been traumatized due to losing their homeland and vanishing future.

INTRODUCTION

Climate-induced statelessness is an emerging global problem of considerable significance. While rising sea levels or extreme weather events force entire populations to leave their homes, whole communities face the danger of losing not only their territories but also their national identities. Tuvalu, Kiribati, and the Maldives are among the small island nations that confront considerable challenges, with projections indicating that these areas may be rendered uninhabitable within decades (McAdam, 2012). When territories cease to exist, the people who once lived there run the risk of statelessness without nationality or legal recognition (Gemenne, 2015). Statelessness denies individuals access to fundamental rights usually connected with citizenship, such as education, medical care, employment, and welfare (UNHCR, 2018).

This type of displacement results in severe psychological consequences. Forced migration, cultural dislocation, and the absence of legal status within a country can lead to increased anxiety, depression, post-traumatic stress disorder (PTSD), and crises of personal identity (Burson, 2015). Home is lost not only as a physical space but also as a rupture in cultural, social, and emotional connectivity (Adger et al., 2013). Survivors often suffer from grief and survivor's guilt, as they have left places with deep multi-generational significance (Tschakert et al., 2019). Additionally, the lack of legal frameworks to protect climate-displaced individuals exacerbates their mental health conditions. Unlike refugees fleeing persecution, those displaced by climate change do not have international legal recognition under the 1951 Refugee Convention (McAdam, 2012).

Stateless populations also experience systemic discrimination, exploitation, and marginalization. In host communities, they may face xenophobia and exclusion from essential services (Betts & Collier, 2017). The dual effects of legal invisibility and social exclusion can severely deteriorate psychological well-being, leading to chronic distress and a loss of resilience (Farbotko & Lazrus, 2012). Without opportunities to return home or rebuild livelihoods, climate-displaced individuals experience ongoing psychological and emotional distress.

Even under these unfavorable conditions, many climate-displaced populations demonstrate remarkable resilience. Social support networks, cultural continuity, and mental health interventions play crucial roles in alleviating psychological trauma (Piguet, 2013). Community-driven support groups and advocacy initiatives help displaced populations reconnect with their identities and regain agency (Tschakert et al., 2019). Trauma-informed mental health services and culturally sensitive counseling programs are essential for psychological recovery and adaptation (Burson, 2015).

This chapter examines the psychological effects of climate-induced displacement on the mental health of affected populations and explores coping mechanisms. It also highlights gaps in existing legal frameworks and proposes a rights-based approach to addressing climate-induced statelessness. By integrating legal and psychological perspectives, the chapter aims to ensure the dignity, welfare, and resilience of affected populations.

Humanitarian interventions are equally essential in addressing the mental health challenges of climate-induced stateless populations. Community coping mechanisms are fundamental to building resilience, and several non-governmental organizations (NGOs) and international agencies provide psychological first aid (PFA) programs to assist displaced persons in overcoming trauma (IFRC, 2018). PFA is a method of providing immediate emotional support following disasters and forced displacement, helping individuals normalize their feelings, understand their physiological and psychological responses to stress, and develop coping mechanisms that reduce the risk of long-term psychological harm (WHO, 2011). The International Federation of Red Cross and Red Crescent Societies (IFRC) actively implements PFA in the field, ensuring that displaced people receive immediate support at the onset of disaster to reduce the possibility of long-term psychological deterioration (IFRC, 2018).

HISTORICAL CONTEXT OF CLIMATE-INDUCED MIGRATION

Climate-induced migration is an age-old phenomenon. Environmental factors have been one of the significant driving forces in human mobility throughout history. Societies have been challenged with situations of droughts, floods, desertification, and other ecological disturbances leading to the displacement of populations in search of safety, resources, and stability (McLeman & Gemenne, 2018). Understanding the historical context of climate change-induced migration offers insight into how civilizations adapted, how resilient displaced populations were, and the socio-political and mental health consequences of environmental displacement (Black et al., 2011).

Ancient Civilizations and Environmental Migration

Ever since time immemorial, environmental change has significantly influenced human history. One of the earliest accounts of migration due to climate alteration can be traced back to the downfall of the Akkadian Empire in 2200 BCE. Severe internal droughts and declining agricultural productivity forced large populations across the Mesopotamian region to migrate, contributing to the eventual collapse of the empire (Weiss & Bradley, 2001). Similarly, the Indus Valley Civilization experienced environmental changes, such as fluctuating monsoon patterns, leading to mass displacement from established agricultural systems (Giosan et al., 2012). The psychological impact of such displacement—stemming from the loss of ancestral land and livelihood—draws parallels to the experiences of modern-day climate migrants, who suffer anxiety, identity crises, and emotional distress (Adger et al., 2013).

The Mayan civilization in Mesoamerica endured periodic droughts that resulted in population decline and migration. Prolonged droughts, coupled with continued deforestation and soil degradation, led to widespread failures in food production, forcing communities to relocate to more hospitable areas (Haug et al., 2003). These historical cases underscore the psychological consequences of climate-induced displacement, including trauma from forced dislocation, apprehension due to uncertainty, and psychological distress caused by the loss of cultural identity (Tschakert et al., 2019).

Medieval and Early Modern Climate Migration

The Ice Age (approximately 1300 to 1850) was a period of global cooling that posed significant environmental challenges. In Europe, failed harvests and widespread famines forced rural populations to migrate to urban centers (Fagan, 2000). The severe weather conditions also contributed to political and economic instability, with population movements often leading to conflict (White, 2014). These migrations, marked by uncertainty and resource scarcity, likely caused considerable mental health strain on affected individuals, as displacement disrupted social structures and increased psychological distress (Campbell, 2016).

Desiccation and desert expansion have been cited as contributing factors in the decline of the Ghana Empire and other Sub-Saharan African civilizations, as populations migrated in search of habitable lands (McIntosh, 1998). Environmental pressures, alongside economic drivers, encouraged large-scale migration as people sought access to arable land and water resources (Burke et al., 2009). A key psychological factor in these historical migrations is the loss of home and the psychological distress associated with entering unfamiliar environments, which likely had long-term emotional and social consequences for displaced populations (Adger et al., 2013).

Colonialism, Environmental Exploitation, and Displacement

Most instances of forced migration date back to the colonial period, when native resources were exploited by colonial powers, leading to environmental degradation and ecosystem transformation. Large-scale deforestation, mining, and agricultural plantations disrupted traditional livelihoods, resulting in the displacement, dislocation, and forced migration of indigenous populations (Peluso & Watts, 2001). The psychological consequences of such displacements include the erosion of cultural identity and the intergenerational transmission of trauma, which continue to affect many indigenous communities today (Das & Chaturvedi, 2019).

Additionally, infrastructure projects such as dams and canals were constructed without adequate consideration for indigenous welfare, often leading to their displacement without proper compensation or rehabilitation (Roy, 1999). These historical events highlight the long-term connections between environmental changes, forced displacement, and psychological distress among affected communities (Scudder, 2012).

Contemporary Climate-Induced Migration

With climate change accelerating at an unprecedented rate, climate-induced migration has become increasingly frequent and severe in recent years. The Pacific Islands, particularly Kiribati and Tuvalu, face existential threats due to rising sea levels, forcing entire communities to relocate (Connell, 2016). Similarly, in the Horn of Africa, prolonged droughts and food insecurity have triggered large-scale human migration, leading to regional tensions and instability (Koubi et al., 2016).

Hurricanes, floods, and wildfires—natural disasters intensified by climate change—have displaced millions across the Caribbean, South Asia, and North America (Hsiang & Jina, 2014). The psychological effects of such displacements, including depression, anxiety, post-traumatic stress disorder (PTSD), and a profound loss of cultural attachment, are particularly acute for those affected (Cunsolo & Ellis, 2018). The lack of robust mental health support systems further exacerbates these challenges, leaving displaced populations vulnerable to long-term emotional and psychological distress (Burkett, 2011).

Lessons from Historical Migration Patterns

Lessons for the present concerning climate migration could be learned from past experiences. Societies that utilized adaptive strategies-negotiate water management systems, crop diversification, community cooperation-were established against all odds in the face of environmental change. Societies that did not adapt witnessed collapses and extensive bursts of displacements.

It is critical to understand the interconnectedness of environmental, societal, and political factors that could adversely affect response mechanisms to climate migration. There's also a need to recognize the mental health burdens borne by displaced populations. Any climate migration response must encompass culturally aware mental health service delivery, engendering resilience in communities and promoting psychological well-being.

Once again, the historical backdrop with respect to climate-induced migration embeds lessons on the resilience and vulnerability of human society. From past experiences, policymakers, researchers, and international organizations may have the opportunity of forging an approach that is more inclusive and sustainable in addressing the climate displacement challenges of the 21st century. Therefore, a full spectrum of responses will encompass not only the legal and economic constraints but, equally importantly, the mental health of the displaced to ensure suitable support for resettling with dignity.

PSYCHOLOGICAL CHALLENGES OF CLIMATE-INDUCED STATELESSNESS

Anxiety and Depression

Anxiety and depression are among the most significant mental health challenges faced by communities experiencing climate-driven statelessness. The loss of one's homeland leads to profound emotional distress, exacerbated by the uncertainty surrounding legal status (Gemenne & Blocher, 2017). One of the most common forms of anxiety in such situations is the fear of the unknown—feelings of hopelessness that often culminate in depression (Beyer & Fanning, 2021). Stateless individuals frequently bear the psychological burden of uncertainty about whether they will ever find a permanent place to settle or gain access to basic humanitarian aid (Brun & Fábos, 2015).

Additionally, widespread isolation and stigma further deepen the sense of worthlessness among displaced populations (Silove et al., 2017). The psychological toll is exacerbated by a lack of stable housing, education facilities, and economic opportunities, making long-term well-being seem unattainable. Many displaced individuals remain deprived of mental health care due to the absence of legal recognition, which prevents them from accessing essential support services (Kirmayer et al., 2011). Addressing these mental health concerns requires trauma-informed care alongside community-based mental health programs tailored to the specific needs of stateless and displaced populations (Miller & Rasmussen, 2017).

Post-Traumatic Stress Disorder (PTSD)

PTSD represents one of the major psychological effects affecting populations displaced due to climate change. Survivors endure traumatic memories of sudden and severe weather events, the loss of homes, and the destruction of their communities (Fritze et al., 2008). Such trauma often manifests in intrusive thoughts, nightmares, and emotional distress triggered by reminders of the traumatic event (Gifford & Gifford, 2016). The uncertainty surrounding their displacement status and the absence of a clear pathway toward recovery further perpetuate PTSD symptoms (Uscher-Pines, 2009).

Women and children are particularly vulnerable to the psychological effects of displacement, as they face additional challenges such as gender-based violence and child exploitation (Osofsky et al., 2010). Those who remain in temporary or inadequate conditions often lack access to a healthy environment conducive to psychological healing (Watts et al., 2015). If left untreated, PTSD symptoms can worsen, impairing individuals' ability to carry out daily activities and diminishing their overall quality of life (Schilling et al., 2018). Therefore, the development of culturally sensitive counseling and improved access to psychosocial support services are crucial steps in mitigating PTSD among displaced populations (Silove et al., 2017).

LEGAL FRAMEWORKS AND LOOPHOLES

The 1951 Refugee Convention

After the adoption of the Refugee Convention in 1951, it became a framework for international legal protection regarding displaced persons. However, the Convention's protection depends on the definition of refugees as individuals displaced due to persecution (Goodwin-Gill & McAdam, 2007). This legal framework fails to recognize climate-induced migration as grounds for refugee status, thereby leaving affected individuals without the minimal protections afforded to traditional refugees (McAdam, 2012).

Under international law, displaced persons qualify as refugees only if they flee their homeland due to a well-founded fear of persecution (Hathaway, 2005). However, individuals migrating due to environmental causes have been systematically excluded from this category (Docherty & Giannini, 2009). The narrow legal definition of "refugee" overlooks the dire circumstances of climate-induced displacement, exposing affected populations to the risks of statelessness and human rights violations (Zetter, 2017). Addressing this legal gap requires a broader interpretation of international protection frameworks and the establishment of specific legal instruments for climate-displaced persons (Biermann & Boas, 2010).

The Statelessness Conventions

Statelessness is framed around the fundamental conventions found in the 1954 Convention Relating to the Status of Stateless Persons and the 1961 Convention on the Reduction of Statelessness, which provide specific frameworks for protecting stateless individuals (Batchelor, 2002). These conventions define the rights of stateless persons and establish measures to prevent and reduce statelessness (UNHCR, 2014). However, neither convention fully addresses the unique challenges faced by populations rendered stateless due to climate change (Fitzpatrick, 2005).

Most climate migrants exist in a legal limbo, as they are neither granted nationality in the new country where they have settled nor recognized by their original state (McAdam, 2012). This legal void exacerbates their vulnerability, subjecting them to human rights violations and significant mental health challenges, including anxiety, depression, and social exclusion (Gemenne, 2015). The absence of clear legal status restricts access to fundamental rights such as healthcare, employment, and education, further compounding their distress (Biermann & Boas, 2010). Addressing this issue requires an urgent re-evaluation of international legal frameworks to ensure climate-induced statelessness is adequately recognized and mitigated.

Human Rights Frameworks

International legal obligations include human rights treaties such as the International Covenant on Civil and Political Rights (ICCPR) and the International Covenant on Economic, Social and Cultural Rights (ICESCR), both of which guarantee fundamental rights to all individuals, irrespective of national boundaries (UN General Assembly, 1966a; 1966b). These rights encompass access to healthcare, education, and social protections, forming the cornerstone of international human rights law (Joseph & Castan, 2013).

However, the lack of legal status often deprives climate-induced stateless persons of these fundamental rights, leaving them in precarious situations with limited access to essential services (McAdam, 2012). Systemic barriers such as exclusion from national healthcare systems and education further exacerbate the mental health burden of displaced populations (Gemenne, 2015). The psychological distress stemming from social marginalization, economic insecurity, and legal invisibility highlights the urgent need for a human-rights-based approach to climate migration (Biermann & Boas, 2010). Strengthening legal protections and integrating climate-induced displacement into international human rights frameworks is critical to ensuring the dignity and protection of affected populations.

Regional and National Responses

In Africa, among various regions, the Kampala Convention was adopted to address internal displacement and provide protection for displaced people (African Union, 2009). Similarly, the Cartagena Declaration in Latin America expands the traditional definition of refugees to include those fleeing generalized violence and environmental disasters, thus providing a broader concept of protection (Colloquium on the International Protection of Refugees in Central America, Mexico, and Panama, 1984).

Despite these regional efforts, there remains a notable absence of comprehensive international legal instruments specifically designed for climate migrants (McAdam, 2020). National governments typically respond to climate-induced displacement in an ad hoc manner, resulting in fragmented protection systems that fail to provide consistent and effective safeguards for the most vulnerable populations (García & Doocy, 2020). This legal vacuum underscores the pressing need for an international legal framework that directly addresses climate-induced statelessness, ensuring legal clarity, access to protections, and sustainable solutions for displaced persons (Docherty & Giannini, 2009). Such a framework would be instrumental in bridging existing gaps in international law and enhancing global commitments to human rights and environmental justice.

COPING MECHANISMS AND RESILIENCE AMONG CLIMATE-INDUCED STATELESS POPULATIONS

Community Support and Social Networks

In situations of climate-related statelessness, community support and social networks serve as crucial coping mechanisms for displaced populations. The strength of these communities provides both emotional and tangible assistance, helping to preserve resilience during crises (Adger et al., 2013). Traditions of support through family ties and neighborhood networks offer psychological comfort by fostering a sense of belonging and familiarity (Jones & Faas, 2017). Community-driven activities such as support group meetings, cooperative societies, and cultural events provide displaced individuals with opportunities to share their experiences, process emotions, and collectively navigate adversity (Campbell et al., 2020).

Additionally, cultural practices and traditions play an essential role in preserving psychological well-being. Engaging in cultural rituals, storytelling, and communal gatherings allows displaced populations to maintain their identity and heritage, reinforcing psychological resilience (Kirmayer et al., 2011). Furthermore, the establishment of religious centers and spiritual practices provides displaced individuals with a sense of hope and purpose, fostering mental well-being despite their precarious circumstances (Pargament, 2011). As a result, NGOs and community organizations that facilitate such activities significantly contribute to the mental health and social cohesion of climate-affected populations, ensuring collective resilience and emotional stability (Schweitzer et al., 2018).

Mental Health Support and Psychosocial Interventions

Mental health support is imperative among climate-displaced populations coping with the psychological effects of statelessness. The therapeutic and emotional needs of displaced individuals extend beyond trauma-informed care; they require culturally sensitive mental health services tailored to their specific experiences (Silove et al., 2017). Individual and group counseling, peer support groups, and resilience-building programs are commonly used interventions to help individuals process trauma and express their emotions in a structured environment (Roberts et al., 2021).

Organizations frequently employ local mental health professionals trained in Mental Health and Psychosocial Support (MHPSS) interventions, ensuring culturally competent care that meets the needs of displaced persons (Tol et al., 2020). Furthermore, alternative access to mental health care—such as mobile mental health clinics and telephone counseling—is essential for reaching populations in remote or underserved areas where traditional mental health services are limited (Van Ommeren et al., 2018).

Another crucial intervention is Psychological First Aid (PFA), which provides immediate emotional support to individuals experiencing acute distress (WHO, 2018). Often conducted by trained community members and healthcare workers, PFA offers practical assistance, a sympathetic ear, and connections to available support services (Burton et al., 2019). Additionally, peer counseling programs are widely implemented within internally displaced communities, allowing for shared experiences and mutual support among those affected (Miller & Rasmussen, 2022).

Adaptive Strategies and Livelihood Initiatives

Sustainable livelihoods serve as a viable coping mechanism for climate-induced statelessness, offering displaced populations economic stability and psychological resilience (Chindarkar, 2019). Above all, economic stability plays a key role in psychological recovery, restoring a sense of control and meaning for those affected by displacement (Hugo, 2020). Many displaced persons survive through informal economies, engaging in agricultural work, small businesses, and other income-generating activities to support themselves and their families (Werz & Hoffman, 2018).

Governments, NGOs, and international humanitarian organizations implement livelihood programs that provide displaced populations with vocational training, financial literacy courses, and small grants to help them establish businesses and regain economic independence (McLeman & Gemenne, 2018). In addition, education and skills development programs further enhance resilience by expanding employment opportunities, thereby promoting self-reliance (Nishikiori et al., 2017).

Humanitarian agencies work alongside governments to facilitate education for displaced children and youth, recognizing that education enhances long-term psychological well-being (Betts et al., 2021). Access to education not only imparts marketable skills but also fosters hope and a sense of normalcy, which are crucial for mental health and social integration during prolonged displacement (Ferris, 2019).

Policy Recommendations for Enhancing Resilience

Strengthening the resilience of climate-induced stateless populations necessitates comprehensive policy advocacy at both national and international levels (McAdam, 2020). Governments and international bodies must formulate inclusive and sustainable policies that address the legal, economic, and psychological needs of displaced individuals (Betts et al., 2021). Key policy recommendations include:

- Legal Recognition and Protection:

Establishing legal protections for climate-displaced persons is essential. This may involve expanding the definition of "refugee" under the 1951 Refugee Convention or drafting a new international legal framework dedicated to climate migrants (Holliday & Kraler, 2022). Such legal mechanisms would secure access to rights and services for displaced populations.

- Mental Health Integration:

Mental health and psychosocial support services should be systematically integrated into displacement response strategies (Silove et al., 2017). Governments and humanitarian agencies must focus on training mental health professionals, establishing community-based support programs, and ensuring the availability of culturally sensitive psychological care (Miller & Rasmussen, 2020).

- Empower Communities:

Psychosocial resilience is strengthened through community-driven programs that foster social cohesion and cultural preservation (Goodkind et al., 2019). Strategies such as funding community centers, promoting rural-urban cultural exchanges, and utilizing local leaders as peer supporters are effective in enhancing resilience (Ryan et al., 2021).

- Creating Economic Solutions:

Sustainable livelihoods are crucial for improving mental health and overall well-being among displaced persons (Chindarkar, 2019). Vocational training programs and livelihood initiatives should be tailored to the specific needs and capabilities of the displaced populations, ensuring economic independence and psychological stability (McLeman & Gemenne, 2018).

- International Cooperation:

Climate-induced displacement requires a global cooperative response (Ferris, 2019). Strengthening international cooperation through multilateral treaties, funding mechanisms, and shared responsibility frameworks will ensure adequate protection and assistance for displaced populations (Werz & Hoffman, 2018).

MENTAL HEALTH CHALLENGES IN STATELESS POPULATIONS

Statelessness caused by climate change leads to a complex array of legal, social, and psychological challenges(McAdam, 2020). While the loss of national identity and citizenship has legal and political consequences, it also has significant mental health implications (Gemenne & Blocher, 2017). Climate-displaced individuals often suffer from severe psychological distress, food insecurity, economic instability, and restricted access to healthcare (Silove et al., 2017). This chapter explores the mental health challenges of stateless populations, emphasizing the high prevalence of anxiety, depression, post-traumatic stress disorder (PTSD), and social isolation, along with the factors exacerbating these conditions (Miller & Rasmussen, 2020).

Statelessness is an intersection of legal and psychosocial struggles, where the loss of citizenship is not just a political or legal matter but also a mental health crisis (Goodkind et al., 2019). Patriotism loss and the incapacity to claim citizenship create profound emotional distress, exacerbated by social exclusion, economic precarity, and barriers to healthcare access (Ryan et al., 2021). These circumstances heighten the risk of mental health disorders, including anxiety, depression, PTSD, and social withdrawal, among displaced persons (Ferris, 2019).

Limited Access to Mental Health Services

One of the primary challenges faced by stateless populations is the inaccessibility of specialized mental health services (Silove et al., 2017). Stateless persons are frequently denied basic rights, including healthcare access, due to their lack of recognized legal status (McAdam, 2020). In many countries, legal identity is required to access medical services, which effectively excludes stateless individuals from obtaining professional psychological support(Goodkind et al., 2019).

Additionally, the forced displacement and migration process often leads to stateless populations settling in areas with limited or no mental health infrastructure (Miller & Rasmussen, 2020). Refugee camps, which frequently house stateless persons, prioritize immediate survival needs such as food, water, and shelter, while mental health concerns remain neglected (Gemenne & Blocher, 2017). Even when mental health services are available, barriers such as language, cultural differences, and financial constraints further hinder access (Ryan et al., 2021).

Case Study: Rohingya Statelessness and Mental Health

Psychologically afflicted, the Rohingya, under the perpetual condition of statelessness, face severe mental health challenges (Lopes Cardozo et al., 2019). Persecution, often state-sanctioned, has resulted in psychological trauma, forcing hundreds of thousands of Rohingya to seek refuge in densely populated and resource-scarce camps in Bangladesh (Habib et al., 2021). Research indicates alarmingly high rates of post-traumatic stress disorder (PTSD) and depression among Rohingya refugees due to past trauma, ongoing instability, and an uncertain future (Bhatia et al., 2018).

For many, the suffering is compounded by a lack of legal identity and inaccessible professional mental health services, leading to intensified distress and silent suffering, a reality shared by many other stateless populations (Riley et al., 2017).

Stigma and Social Isolation

The stateless often find themselves marginalized within both their immediate communities and broader society(Batchelor, 2020). Statelessness is frequently accompanied by social stigma, as these individuals are perceived as outsiders, illegal migrants, or burdens on national resources (Blitz, 2019). This form of rejection fosters alienation, loneliness, and hopelessness, exacerbating their psychological distress (van Waas, 2017).

Discrimination manifests in multiple ways, including denied access to education, employment, and housing, further entrenching their exclusion and vulnerability (Institute on Statelessness and Inclusion [ISI], 2021). Systemic rejection often erodes self-esteem, intensifying anxiety and depression among the stateless (Lopes Cardozo et al., 2019). Another major challenge is the identity crisis many face: without formal recognition by any state, they suffer psychological trauma associated with loss of belonging and legal identity (Manly & van Waas, 2022).

Impact on Youth and Adolescents

The mental health ramifications of statelessness severely impact children and adolescents (Belton, 2020). Young people in stateless communities often face bullying and discrimination, exacerbating their feelings of exclusion and social isolation (van Waas, 2017). Many stateless children experience chronic stress and emotional distress due to poverty and a lack of access to formal education, which fosters hopelessness about their future (Blitz & Sawyer, 2021).

Economic Insecurity and Psychological Stress

Economic insecurity is one of the greatest stressors for stateless populations (van Waas & Khanna, 2021). Many stateless individuals are forced to survive in informal and exploitative labor markets due to their lack of legally recognized work permits, exposing them to toxic work environments, extreme poverty, and financial instability, all of which contribute to sustained psychological distress (Blitz, 2019).

Having no established sources of income, stateless individuals struggle to support their families, exacerbating anxiety and depression (Belton, 2020). The absence of work permits compels many to engage in the informal economy, where they are at heightened risk of violence, crime, and labor exploitation, further intensifying their mental health struggles (Sawyer & Blitz, 2022). The constant exposure to economic instability can lead to chronic melancholy and even mental exhaustion (van Waas, 2017).

Gendered Impact of Economic Stress

For women in stateless communities, the plight is even worse as they rarely receive education or employment opportunities, making them heavily dependent on male family members for survival (van Waas & Khanna, 2021). This economic dependence increases their vulnerability to domestic violence, sexual exploitation, and human trafficking, exacerbating psychological distress (Belton, 2020).

Women who experience sexual violence within stateless communities face additional trauma due to social stigma and restrictive societal norms, which often prevent them from seeking help (Sawyer & Blitz, 2022). The absence of legal protection mechanisms further reinforces their marginalization, leaving them trapped in cycles of abuse, economic insecurity, and mental health deterioration (Blitz, 2019).

Interpersonal and Family Strains

The wretchedness of not having a country extends beyond individuals and deeply impacts families, disrupting social bonds and increasing household stress (Blitz, 2020). Stateless families often face displacement, poverty, and legal uncertainties, which exacerbate domestic conflicts, child abandonment, and even suicide (van Waas & Khanna, 2021). Parents who struggle to provide for their children often experience severe emotional distress, sometimes leading to neglect and psychological withdrawal (Sawyer & Blitz, 2022).

Children born into stateless families inherit lifelong legal and economic instability, perpetuating a cycle of trauma across generations (Belton, 2019). The absence of legal protection and permanent fear of forced displacement create an environment of chronic anxiety and despair, reinforcing patterns of psychological distress over time (Manly & Kim, 2023).

Case Study: Climate Refugees in the Maldives

The Maldives, one of the most at-risk countries due to climate change, has begun to experience climate-induced displacement as rising sea levels threaten entire islands (Kelman, 2021). Environmental degradation has forced families to relocate, leading to severe psychological distress, particularly regarding uncertainty about the future and loss of ancestral land (Haque & Etzold, 2022). The separation of families within these displaced communities is widespread, causing feelings of depression, distress, and loss of cultural identity (Kothari, 2020).

Children and Vulnerable Groups

The psychological effects of statelessness have a profound impact on children, the elderly, and disabled individuals(Benhabib, 2021). Stateless children often face exclusion from schooling, limiting their future opportunities for personal and professional growth (UNHCR, 2022). As a result, these children experience hopelessness, frustration, and long-term mental health disorders, such as anxiety and depression (Patel et al., 2023).

The situation is even more severe for elderly stateless people, who often suffer from neglect and extreme despair due to the loss of family support (Jaskulowski, 2020). Many elderly individuals live in poverty as they are denied pensions, healthcare, and social welfare benefits, further worsening their psychological well-being (Lindskov, 2021).

Stateless individuals with disabilities face even greater hardships, as their access to medical and psychological support is significantly restricted (McAdam, 2022). Many disabled people rely on their families for care, but in impoverished, displaced communities, the burden of caregiving becomes overwhelming, often leading to neglect and emotional suffering (UNHCR, 2022).

Climate Anxiety and Existential Fear

The rising climate anxiety among stateless populations is a psychological condition characterized by persistent worry about environmental devastation and the fate of future generations (Clayton & Manning, 2018). For those displaced by climate change, a haunting fear prevails will they ever know a stable existence? (Pihkala, 2020).

This anxiety is particularly intense among indigenous communities and coastal populations who have deep cultural and spiritual ties to their land (Cunsolo & Ellis, 2018). The very idea of losing ancestral homes, traditions, and ways of life becomes a significant source of emotional distress (Albrecht, 2022). Many displaced individuals experience profound grief, feelings of powerlessness, and impaired psychological well-being, often developing severe depression and anxiety disorders (Berry et al., 2018).

THE ROLE OF GOVERNMENTS AND INTERNATIONAL BODIES IN ADDRESSING PSYCHOLOGICAL CHALLENGES OF CLIMATE-INDUCED STATELESSNESS

National Government Responses

National governments have an important role in alleviating the psychological effects associated with climate-induced displacement by securing the protection and welfare of the displaced (Silove et al., 2017). They need to create comprehensive laws and support systems to reduce the mental health impacts stemming from displacement (McMichael et al., 2021). Thus, governments should focus on proactive policies that integrate mental health care into national disaster response and displacement management plans (Lawrence et al., 2022).

Governments should start by providing mental health and psychosocial support services (MHPSS) within their health systems (Charlson et al., 2019). Trauma-informed care and culturally sensitive counseling training should be provided to health professionals so that displaced people receive adequate psychological support (WHO, 2018). Further, mobile mental health units and community-based mental health programs can improve access to care in remote or underprivileged areas (Silove et al., 2017).

Governments should also develop social protection systems that address the needs of displaced persons by providing access to housing, education, healthcare, and employment opportunities (McMichael et al., 2021). Good living conditions and economic support are necessary for alleviating psychological distress (Lawrence et al., 2022). Resettlement programs—such as those seen in Bangladesh and Fiji—foster community integration, which enhances social cohesion and contributes to better mental health outcomes (Adger et al., 2019).

Legal recognition and protection for climate-displaced individuals can further be strengthened by governments (Zetter & Ruaudel, 2018). Enacting national legislation that provides legal status to environmentally displaced persons ensures their access to fundamental rights and services (McAdam, 2020). Other legal mechanisms—such as granting citizenship or permanent residency—would help ease the pain of statelessness (Zetter & Ruaudel, 2018).

Regional Cooperation and Initiatives

Regional cooperation is necessary to address cross-border climate displacement and its psychological effects(Gemenne & Blocher, 2017). Regional bodies, such as the African Union and the Pacific Islands Forum, have initiated frameworks like the Kampala Convention and the Framework for Resilient Development in the Pacific (FRDP) to manage displacement from environmental causes (Ferris, 2020). These frameworks provide guidance on protecting the rights and welfare of displaced people (McAdam, 2018).

Regional treaties can foster coordinated responses to mental health challenges by sharing best practices, establishing regional mental health support networks, and funding psychosocial programs (Nishimura, 2021). Furthermore, hosting countries can enter into burden-sharing arrangements, ensuring that displaced individuals receive appropriate support and protection (Adger et al., 2019).

International Organizations and Multilateral Responses

To counter the psychological effects of climate-induced statelessness, international organizations provide support to governments (UNHCR, 2020). The trauma and mental health programs run by the United Nations High Commissioner for Refugees (UNHCR), the International Organization for Migration (IOM), and the World Health Organization (WHO) are crucial to affected populations (IOM, 2021; WHO, 2019).

UNHCR promotes the enhanced legality of climate-displaced persons and supports community-based mental health initiatives in collaboration with governments (UNHCR, 2022). The IOM's Mental Health and Psychosocial Support (MHPSS) programs are designed to provide trauma-informed care as well as host and peer support groups for thousands of people displaced daily (IOM, 2021). WHO provides technical assistance to governments in integrating mental health services into their national health care systems (WHO, 2020).

Furthermore, international legal frameworks should be strengthened to address challenges specific to climate-induced displacement (McAdam, 2018). The 1951 Refugee Convention's definition of a refugee needs modification to reflect modern climate crises, or a new legal category should be established to acknowledge climate-induced displacement(Docherty & Giannini, 2009). Additionally, global cooperation is needed to establish funding mechanisms to support mental health programs and provide financial aid to affected countries (Betts, 2013).

COPING MECHANISMS AND RESILIENCE

Challenges faced by climate-induced stateless people are of great magnitude and include, yet are by no means limited to, anxiety, depression, and post-traumatic stress disorder (Berry et al., 2018; IPCC, 2022). Nevertheless, displaced persons and communities establish various forms of coping mechanisms to adapt to their new realities (McMichael et al., 2019). Coping mechanisms broadly include community support, traditional healing modalities, humanitarian assistance, and psychosocial joint strategies (Silove et al., 2017). Understanding these coping strategies is essential to formulating appropriate policies for mental health programming among climate-displaced populations (UNHCR, 2021).

Community-Based Support Systems

A major coping strategy for displaced populations is the establishment of informal support networks within the community. Social cohesion helps reduce psychological distress and support resilience.

Role of Community and Cultural Continuity

Displaced individuals typically find comfort in familiar cultural and religious practices. Even when stateless persons are forced to leave their land, efforts are made to preserve their identity through language, traditions, and religious beliefs, which provide a sense of security (Adger et al., 2013; UNHCR, 2021). Often, community elders play a significant role in ensuring that cultural narratives are sustained, so that younger generations do not forget their heritage (Kirmayer et al., 2011).

Mutual Aid and Collective Support

In many refugee and stateless communities, people rely on each other for survival, leading to the emergence of mutual aid networks where individuals share food, shelter, and emotional support (Silove et al., 2017). For instance, women in Bangladeshi Rohingya camps have formed self-help groups that provide psychosocial support, childcare, and skill-sharing initiatives (Imran et al., 2022).

Peer Support Groups

Peer-organized support groups play a crucial role in helping displaced persons cope with trauma. These groups allow individuals to share their experiences and emotional struggles, while also receiving support from peers (Nickerson et al., 2015). Narrative therapy, in which a person tells their own story in a group setting, is especially effective in relieving isolation and fostering resilience (Miller & Rasmussen, 2017).

NGO and Humanitarian Interventions

Humanitarian programs focus on psychosocial activities for climate-displaced populations. For example, NGOs provide immediate emotional support and long-term mental health assistance to victims.

Psychological First Aid (PFA)

Using Psychological First Aid (PFA), organizations such as the International Federation of Red Cross and Red Crescent Societies (IFRC) help displaced persons process trauma (IFRC, 2020). PFA provides:

- Immediate emotional support after disasters (World Health Organization [WHO], 2011).
- Normalization of emotions, helping individuals understand their stress responses (Hobfoll et al., 2007).
- Encouragement of social connection, while discouraging self-destructive coping behaviors such as substance abuse (Bryant et al., 2014).

Mental Health Programs in Refugee Camps

Mental health outreach is commonly conducted in refugee camps in collaboration with therapists. For instance, UNHCR has implemented programs including art therapy, music therapy, and drama therapy to assist displaced populations in processing trauma (UNHCR, 2019). These therapeutic interventions offer creative outlets for individuals to express emotions and rebuild a sense of community (Betancourt et al., 2013).

Livelihood Programs as Psychological Support

NGOs also focus on developing skill-based programs to provide economic opportunities to stateless persons, which enhances their self-worth and autonomy (Porter & Haslam, 2005). Vocational training, micro-financing, and small business support programs help displaced individuals regain control over their economic futures, reducing anxiety and depression (Silove et al., 2017). Such initiatives have been successfully implemented in refugee settlements, fostering financial independence and resilience (IOM, 2021).

Traditional Healing and Indigenous Coping Practices

Mental wellness is often viewed through a cultural lens, rather than being limited to Western biomedical approaches(Kirmayer et al., 2011). Well-being is frequently linked to spirituality, community rituals, and nature, which serve as critical coping mechanisms for displaced populations (Gone, 2013).

The Role of Religious and Spiritual Beliefs

For many stateless people, religious beliefs provide hope and comfort in uncertain times. Individuals often seek solace in prayer, faith-based gatherings, and spiritual rituals, particularly in regions like the Pacific Islands, where communities face existential threats due to rising sea levels (Durie, 2004). These acts of faith serve as psychological anchors, helping individuals combat despair and hopelessness (Silove et al., 2017).

Indigenous Healing Practices

A wide variety of herbal medicines, meditation, and holistic healing techniques are utilized by many stateless communities to counter stress and trauma (Waldram, 2020). In some African and Indigenous American cultures, traditional healers or shamans perform soul-cleansing rituals, storytelling, drumming, and other spiritual practices to treat mental health issues (Kirmayer & Guthrie, 2007). These methods foster a sense of cultural continuity and community resilience (Tsey, 2008).

Healing through Nature and Environment

A connection with the environment serves as an emotional and psychological outlet for communities that have historically depended on land for their cultural identity (Betancourt et al., 2015). Refugees and stateless persons often engage in gardening or farming, even for short periods in temporary settlements, as a way to reconnect with nature and regain a sense of belonging (Alvarez, 2019). This connection with nature promotes mental well-being and aids in adapting to displacement (IOM, 2021).

Psychosocial Support Strategies

In addition to NGOs and community-based approaches, individual coping strategies play a significant role in fostering mental health resilience among stateless populations (Silove et al., 2017). Psychologists and social workerswho work with displaced individuals have identified several key psychosocial support methods that aid in psychosocial adjustment (Betancourt et al., 2015).

Cognitive Behavioral Therapy (CBT) and Trauma-Informed Care

Some interventions for stateless populations incorporate Cognitive Behavioral Therapy (CBT), which is based on the idea that individuals can identify and challenge negative thought patterns (Beck, 2011). CBT interventions have been shown to significantly improve emotional regulation and coping skills among displaced persons (Murray et al., 2018).

Additionally, trauma-informed care is an approach that considers the unique psychological challenges of stateless persons, such as loss of identity and prolonged legal uncertainty (Hopper et al., 2010). This approach ensures that therapy sessions are structured to acknowledge the cumulative effects of displacement, legal limbo, and forced migration (Nickerson et al., 2011).

Resilience and Adaptability Training

Several international aid agencies implement resilience training programs, helping displaced persons manage stress, emotional regulation, and problem-solving skills (Southwick et al., 2014). These programs typically include:

- Self-care techniques such as mindfulness and relaxation exercises (Litz et al., 2009).
- Empowerment workshops that educate displaced individuals on self-advocacy and human rights (Tol et al., 2013).
- Crisis response education, where individuals are trained in conflict resolution and disaster preparedness(Betancourt et al., 2015).

Family and Youth Programs

Children and young people in stateless communities are particularly vulnerable to mental health instability (Fazel et al., 2012). Family-based interventions focus on teaching parents the principles of positive parenting, which helps create a stable environment for children (Masten & Narayan, 2012).

Additionally, narration therapy and play therapy have been effectively used to help children process trauma in a non-threatening manner (Landreth, 2012). These interventions enable children to express emotions through storytelling, role-playing, and creative play, fostering a sense of security and belonging (Rutter, 2013).

INTERSECTIONAL CHALLENGES AND VULNERABLE GROUPS

Statelessness as a result of climate becomes more major in the case of these groups, further increasing their psychological vulnerability. Among the groups most affected by displacement include women, children, the elderly, and the disabled.

They are thrown into disarray by the displacement, because of challenges related to increased violence and exploitation and the unavailability of mental health services.

In camps or informal settlements, women too often experience increased risks of gender-based violence, with inadequate legal protection or social service systems exacerbating the psychological distress. In a sense, children displaced by climate-induced events may undergo severe emotional and developmental challenges, including anxiety, depression, and PTSD. Education is disrupted, and support systems are lost, which cause their mental well-being to deteriorate further.

There is the necessity for targeted mental health intervention answering and adapting to the points in the intersectionality. Governments and agencies working on humanitarian responses need to have gender-sensitive parameters in their plans, provide trauma-informed care, and ensure a safe space for marginalized groups. Local women's organizations and community leaders involved in the mental health delivery will promote resilience and social inclusion.

CASE STUDIES AND PRACTICAL EXAMPLES

In some Pacific Island nations, rising sea levels have necessitated the relocation of entire communities, leading to the loss of ancestral lands, cultural identity, and traditional livelihoods, which has resulted in widespread psychological distress (Nevitt, 2023; Tiatia-Seath et al., 2020).

In Bangladesh, thousands of people have been displaced by coastal erosion, intensified cyclones, and flooding, often ending up in densely populated slums characterized by poverty and limited access to mental health services (IOM, 2025). Organizations such as the International Organization for Migration (IOM) and the UN High Commissioner for Refugees (UNHCR) are providing psychosocial support in these regions, offering counseling, peer support, and community resilience-building activities (UNHCR, 2018).

Policymakers and practitioners can develop better mental health strategies by analyzing the successes and challenges of these programs. Building culturally appropriate support systems and incorporating traditional healing approaches may further enhance the mental well-being of displaced populations (UNDP, 2024). Examining real-life instances of climate-induced statelessness provides valuable insights into the psychological effects and coping mechanisms of affected communities. For example, in the Pacific islands, rising seas have forced entire communities to relocate. The loss of ancestral land, cultural identity, and traditional livelihoods has caused widespread psychological distress.

Similar situations have occurred in Bangladesh, where millions of people have become displaced due to coastal erosion, worsened cyclones, and severe flooding

(IOM, 2025; UNHCR, 2018). Internally displaced persons typically find refuge in highly congested slum settlements, followed by abject poverty, where there are few to no mental health services (UNDP, 2024). Organizations like IOM and UNHCR have established psychosocial support programs within these areas, providing counseling and peer support as well as conservation activities and community-based resilience building (UNHCR, 2018).

Policymakers and practitioners could therefore plan better strategies in mental health by analyzing successes and failures of the interventions (Nevitt, 2023). It would also further enhance the psychological well-being of displaced societies through the establishment of culturally appropriate paths to support and incorporation of traditional views to healing (Tiatia-Seath et al., 2020).

CULTURAL AND INDIGENOUS COPING MECHANISMS

Indigenous knowledge and cultural practices play a crucial role in alleviating psychological stress among climate-displaced communities. Storytelling, traditional healing ceremonies, and communal rituals provide continuity and resilience, enabling individuals to process trauma and adapt to new environments (Hull, 2023; Incayawar et al., 2009). Pacific Islander communities, for example, engage in communal grieving rituals and oral storytelling to maintain cultural identity and cope with emotional loss (Hull, 2023). Traditional healers address both physical and psychological aspects of well-being, emphasizing the importance of integrating such practices into formal mental health systems (Incayawar et al., 2009).

Promoting community-led resilience initiatives that build upon cultural heritage can enhance psychological resilience among displaced populations. Collaborations between governments, international organizations, indigenous leaders, and cultural practitioners are essential to develop inclusive mental health policies that respect and incorporate cultural identities (Hull, 2023).

LONG-TERM MENTAL HEALTH IMPACTS

The effects of state-induced psychosis due to climate change seem to extend beyond the immediate displacement period into prolonged uncertainty, legal invisibility, and the inability to restore stable livelihoods, all of which contribute to chronic stress and mental health deterioration (Guterres, 2021; UNHCR, 2022).

Children's statelessness, in particular, exacerbates trauma, as generational trauma tied to statelessness can transfer psychological burdens to younger generations.

Such mental illnesses include anxiety, dysthymia, post-traumatic stress disorder, and substance misuse over time (Silove et al., 2017). The absence of legal recognition and access to essential services further intensifies feelings of hopelessness and social isolation, necessitating continuous mental health interventions and community rehabilitation efforts (UNHCR, 2022).

Given the global urgency of the issue, governments and international organizations must significantly invest in long-term mental health infrastructure, including counseling services, community centers, and peer support networks (Guterres, 2021). Additionally, psycho-social education, mental health promotion, and literacy initiatives in displaced communities can empower individuals to seek appropriate care and support, fostering resilience and recovery (Silove et al., 2017).

CONCLUSION AND RECOMMENDATIONS

Summary of Key Findings

The climate creates an entirely new humanitarian and legal nexus, having some very adverse psychological impacts on the populations affected. Stateless individuals rendered climate-displaced, being outside legal mechanisms, also face socio-economic instability, cultural loss, and the destruction of social identities, which directly result in mental health distress.

This chapter has attempted to articulate some of the psychological burdens of stateless persons affected by climate disasters, such as anxiety, depression, PTSD, and identity crisis. They get worse with lack of healthcare, lack of legal recognition, and social marginalization. There are several mechanisms on international and regional levels for stateless persons, which, however, do not consider the mental health ramifications of statelessness.

Further, coping mechanisms including community resilience, social support networks, indigenous healing practices, and government programs were analyzed. Sad to say, in the absence of an adequate legal and policy framework to safeguard the mental health of stateless individuals, these efforts remain ineffective.

The Urgency of Addressing Mental Health Concerns

A Growing Global Crisis

Climate-induced statelessness is rapidly evolving into a global crisis, affecting numerous communities worldwide. Rising sea levels, extreme weather events, and land degradation are compelling millions to abandon their homes, often without legal access to citizenship in host countries (Connell, 2010; UNHCR, 2022). Stateless individuals face significant obstacles in accessing basic rights, and when combined with climate displacement, these challenges intensify, leading to a mental health emergency (UNHCR, 2022).

Without proactive measures, the psychological distress experienced by climate-displaced stateless individuals will extend beyond personal struggles to impact entire communities. Children growing up under constant pressure, social isolation, and deprivation are particularly vulnerable to generational trauma (Taylor & Kaplan, 2023). Immediate legal, policy, and humanitarian interventions are essential to re-define mental health responses and address the unique challenges faced by stateless populations (Taylor & Kaplan, 2023).

The Inadequacy of Existing Legal Frameworks

Even though international treaties like the 1954 Convention Relating to the Status of Stateless Persons and the 1961 Convention on the Reduction of Statelessness offer legal protection, they do not cover mental health support (UNHCR, 2014). Furthermore, stateless persons displaced by climate change are often considered aliens outside the scope of international refugee laws, thus plunging them into a legal limbo with no formal mental health care.

At the domestic level, many countries lack mental health support systems for stateless persons. In some cases, stateless individuals are entirely denied healthcare treatment, worsening their mental health conditions (Gibney, 2021). Inclusion in national systems of care and mental health policy remains a critical area requiring legal and policy reform (Hollingsworth, 2020).

Policy Recommendations

Suggestions have to be made on how to directly address the mental health implications of statelessness caused by climate change; this has to be in the multiple levels of integration: legal recognition, healthcare access, mental health support services. The recommendations include a roadmap for policymakers, humanitarian organizations, and researchers to make improvements in mental health protection for stateless persons.

Addressing mental health impacts from statelessness induced by climate change requires integrated approaches at multiple levels: legal recognition, access to health services, and mental health support services (UNHCR, 2020). The following recommendations present a roadmap for policymakers, humanitarian organizations, and researchers on improving mental health protection for stateless persons.

Legal and Policy Reforms

- Acknowledge Statelessness Due to Climate Change in International Law
- Acknowledge climate-induced displacement in the definition of stateless persons.
- Develop an international legal instrument under which stateless populations displaced by climate change will be given special protection.
- Amend the 1951 Refugee Convention to extend refugee-style protection to stateless climate migrants.
- Considering Mental Health into Statelessness Determination Mechanisms
- Ensure that mental health assessments are carried out by either the UNHCR or national government bodies with respect to statelessness determination determinations.
- Provide psychosocial support and psychiatric care access to extra-special protections for persons who are stateless and severely mentally ill.
- Strengthening the National Health System to Include Stateless Persons
- Amend public law provisions to allow stateless persons access to the national public health system for mental health services.
- Provide mental health care for climate-displaced, stateless persons in developing or disaster-afflicted regions, whether free or at subsidized costs.
- Train mental health professionals on how to work with stateless communities and provide culturally sensitive, trauma-informed care.

Advocacy and Humanitarian Interventions

- Advocacy and Humanitarian Intervention: Psychosocial Support Program Extension in Displacement Camps.
- Set up mental health support centers in refugee and statehood camps for counseling, therapy, and group support programs.
- Collaborate with local NGOs and international organizations in providing psychosocial support services for stateless persons in need.

- Advocacy for Community-Based Mental Health Pin the efforts on incorporate traditional ways of healing into mental health programs to fortify the native coping scheme.
- Promote peer support networks for stateless individuals in overcoming mental health challenges.
- Advocating for Mental Health Inclusion in Climate Policy
- Include considerations of mental health for displaced persons in climate adaptation strategies.
- Governments and international organizations should fund mental health services in climate resilience programs.

Research and Data Gathering

- Improving Research on the Mental Health Conditions of Persons Displaced by Climate Change
- Conduct longitudinal studies on the mental health consequences of climate-induced statelessness.
- Develop mental health assessment tools for stateless persons to improve diagnosis and treatment.
- Encourage collaborative research efforts between psychologists, lawyers, and climate scientists in building holistic intervention models.
- Develop Early Warning Systems for Mental Health Emergencies
- Design programs for early detection of stateless individuals reportedly undergoing severe psychological distress.
- Rapid response teams must be created to provide instant mental health support in climate-displaced communities.

The Need for a Holistic and Sustainable Approach

There should be interaction between legal, social and psychological aspects when dealing with mental health concerns arising from the ramifications of climate change creating statelessness. Preserving the humanity, dignity, rights, and well-being of stateless persons worldwide is an ethical obligation.

The following areas must be worked on together by government, international organizations, and civil society:

- Recognize as a protected class in international law the climate-induced stateless persons;
- Mental health care and psychosocial support should be made available to all, as this should be considered a fundamental human right;

- Invest in ensuring long-term solutions for these stateless individuals to have an opportunity to restore their livelihoods in dignity and peace.

Future Research Directions

With the complexity and dynamic nature of climate-induced statelessness, the future investigation must concentrate on:

- Creation of global frameworks that will integrate climate change-induced migration, statelessness, and mental health policies.
- An objective evaluation of current mental health interventions that target stateless populations for effectiveness.
- Current exploration of new psychological mechanisms of coping, such as digital mental health services and community-led mental health programs.

CONCLUSION

Hence, it could be said that the problem of climate-induced statelessness is not only a legal issue but also a mental health emergency that demands urgent action. Such stateless persons who have been displaced by climate-related disasters are said to go through a cycle of more severe psychological trauma, loss of identity, and a long period of uncertainty, but little attention is given to the mental health well-being of the refugees within the structures of policies.

Such system reforms would require a total overhaul across disciplines-e.g., law, health, and humanitarian aid. Mental health provision, advocacy for legal recognition, and psychosocial care investment could build a future that recognizes the stateless under national law but does not expose them to the possibility of psychological injury.

It needs to be stressed that a holistic, human-centered solution to statelessness as it relates to climate change will not only be necessary but will be long over-due.

REFERENCES

Adger, W. N., Barnett, J., Brown, K., Marshall, N., & O'Brien, K. (2013). Cultural dimensions of climate change impacts and adaptation. *Nature Climate Change*, *3*(2), 112–117.

African Union. (2009). The Kampala Convention.

Albrecht, G. (2022). *Earth emotions: New words for a new world*. Cornell University Press.

Batchelor, C. (2002). The 1954 Convention on the Status of Stateless Persons: Past, present and future. *Journal of Refugee Studies*, *15*(1), 32–54.

Batchelor, C. (2020). *The legal framework for statelessness: A practical guide*. UNHCR.

Belton, M. (2019). The invisible children: The mental health of stateless youth. *Journal of International Affairs*, *72*(2), 101–116.

Belton, M. (2020). The psychological burden of statelessness on women and children. *Forced Migration Review*, *65*, 30–33.

Benhabib, S. (2021). The right to have rights. In Miller, J. R. (Ed.), *Citizenship and statelessness in the age of climate change* (pp. 45–60). Routledge.

Berry, H. L., Waite, T. D., Dear, K. B., Murray, V., & Shilton, M. (2018). The case for a national strategy on climate, health and well-being. *Public Health Research & Practice*, *28*(3), e2831804.

Betts, A., & Collier, P. (2017). *The new politics of refuge*. Oxford University Press.

Betts, A., Easton, M., & Pincus, L. (2021). The global compact for refugees and the challenges of climate-induced displacement. *International Journal of Refugee Law*, *33*(3), 481–502.

Beyer, H., & Fanning, R. (2021). *Mental health and climate change: A global perspective*. World Health Organization.

Bhatia, S., Naha, L., & Singh, R. (2018). Mental health problems among Rohingya refugees: A systematic review. *Indian Journal of Public Health*, *62*(1), 10–15. PMID: 29512559

Biermann, F., & Boas, I. (2010). Preparing for a warmer world: Towards a global governance system of climate-induced migration. *Global Environmental Politics*, *10*(1), 60–88.

Black, R., Adger, W. N., Arnell, N. W., Dercon, S., Geddes, A., & Thomas, D. S. (2011). The effect of environmental change on human migration. *Global Environmental Change*, *21*, S3–S11.

Blitz, B. K. (2019). The plight of the world's stateless. *Forced Migration Review*, *62*, 12–15.

Blitz, B. K. (2020). *Statelessness and displacement*. Oxford University Press.

Blitz, B. K., & Sawyer, A. (2021). The psychological impact of statelessness on children. *International Journal of Migration and Border Studies*, *7*(3), 205–222.

Brun, C., & Fábos, A. (2015). *Refugees and asylum seekers in Europe: Challenges and opportunities for mental health support*. Routledge.

Burke, M., Miguel, E., Satyanath, S., Dykema, J., & Lobell, D. B. (2009). Climate and conflict in Africa. *Proceedings of the National Academy of Sciences of the United States of America*, *106*(40), 16035–16040.

Burkett, M. (2011). The climate crisis and the unconstitutionality of statelessness. *The American Journal of International Law*, *105*(3), 517–538.

Burson, C. (2015). *The state of statelessness: A human rights approach to forced migration*. Stanford University Press.

Burton, A., Roberts, R., & Smith, J. (2019). The application of psychological first aid in displacement settings. *Journal of Mental Health (Abingdon, England)*, *28*(4), 405–412. PMID: 30862218

Campbell, C. (2016). The mental health consequences of climate change-related displacement. *International Journal of Environmental Research and Public Health*, *13*(10), 967. PMID: 27690076

Campbell, S., Roberts, R., & Smith, J. (2020). The role of community in building resilience among displaced populations. *Journal of Community Psychology*, *48*(2), 522–535.

Charlson, F., van Ommeren, M., & Silove, D. (2019). Mental health and psychosocial support in humanitarian emergencies. *The Lancet. Psychiatry*, *6*(8), 653–654.

Chindarkar, N. (2019). Livelihoods and psychological well-being among climate migrants. *Environmental Research Letters*, *14*(12), 125001.

Clayton, S., & Manning, C. M. (2018). *Psychology and climate change: Human perception, impacts, and responses*. Academic Press.

. Colloquium on the International Protection of Refugees in Central America, Mexico, and Panama. (1984). *Cartagena Declaration on Refugees.*

Connell, J. (2016). Climate change-induced displacement in the Pacific Islands: A review of the literature. *The Geographical Journal, 182*(2), 133–143.

Cunsolo, A., & Ellis, N. R. (2018). Ecological grief as a new mental health challenge. *The Lancet. Planetary Health, 2*(12), e515–e516. PMID: 30526935

Das, S., & Chaturvedi, V. (2019). Colonialism, displacement, and the transmission of intergenerational trauma. *Journal of Postcolonial Studies, 21*(3), 256–271.

Docherty, B., & Giannini, T. (2009). Confronting a rising tide: A proposal for a convention on climate change displacement. *Georgetown Environmental Law Review, 21*, 621–682.

Fagan, B. (2000). *The Little Ice Age: How climate made history 1300-1850.* Basic Books.

Farbotko, C., & Lazrus, H. (2012). The psychological effects of climate change-induced displacement. *International Journal of Human Rights, 16*(3), 489–505.

Ferris, E. (2019). Climate change, migration and the Global Compact for Migration. *Forced Migration Review, 62*, 22–25.

Fitzpatrick, S. (2005). The 1961 Convention on the Reduction of Statelessness. *International Journal of Refugee Law, 17*(2), 401–428.

Fritze, J. G., Schnyder, U., & Stiess, G. (2008). The psychological effects of natural disasters: A review of the literature. *Environmental Health Perspectives, 116*(6), 705–712.

García, E., & Doocy, S. (2020). National responses to climate change displacement. *The Lancet. Planetary Health, 4*(11), e510–e511. PMID: 33159877

Gemenne, F. (2015). *The state of climate change displacement: Legal and policy challenges.* Routledge.

Gemenne, F., & Blocher, J. (2017). Climate change and statelessness: A new dimension of human rights protection. *Journal of Human Rights Practice, 9*(3), 323–342.

Gifford, R., & Gifford, J. (2016). Psychology and climate change: The human dimension. In *Handbook of Climate Change and Human Behavior* (pp. 3–18). Springer.

Giosan, L., Clift, P. D., Macklin, M. G., Durcan, J. A., Adger, W. N., & Ellis, N. R. (2012). Holocene Indian monsoon variability and the collapse of the Indus Valley civilization. *Nature, 484*(7392), 79–82.

Goodkind, S., Shaffer, J., & Leff, M. (2019). The psychosocial impact of statelessness: An intersectional approach. *Journal of Human Rights and Social Work, 4*(3), 187–196.

Goodwin-Gill, G. S., & McAdam, J. (2007). *The refugee in international law.* Oxford University Press.

Habib, S., Khan, S. A., & Ahmad, N. (2021). The mental health of Rohingya refugees in Bangladesh: A systematic review. *BMC Psychiatry, 21*(1), 1–15. PMID: 33388029

Haque, S., & Etzold, B. (2022). Climate-induced displacement and psychological distress in coastal communities: A case study of the Maldives. *International Journal of Environmental Research and Public Health, 19*(17), 10822. PMID: 36078536

Hathaway, J. C. (2005). *The rights of refugees under international law.* Cambridge University Press.

Haug, G. H., Günther, D., & Peterson, L. C. (2003). Climate and the collapse of the Maya civilization. *Science, 299*(5613), 1731–1735. PMID: 12637744

Holliday, K., & Kraler, A. (2022). A legal framework for climate change-related displacement. *International Journal of Refugee Law, 34*(1), 1–25.

Hsiang, S. M., & Jina, A. S. (2014). The effect of climate change on international migration. *Climatic Change, 123*(2), 173–184.

Hugo, G. (2020). Migration and psychological well-being: The role of economic security. *Journal of International Migration and Integration, 21*(4), 1145–1165.

IFRC. (2018). *A guide to psychological first aid.* International Federation of Red Cross and Red Crescent Societies.

. Institute on Statelessness and Inclusion [ISI]. (2021). *The statelessness stigma: A global report.* ISI.

Jaskulowski, K. (2020). Statelessness and vulnerability: The case of the elderly. *Journal of Refugee Studies, 33*(3), 511–529.

Jones, S., & Faas, D. (2017). The role of social networks in the psychological well-being of displaced populations. *The Journal of Social Psychology, 157*(4), 450–467.

Joseph, S., & Castan, M. (2013). *The International Covenant on Civil and Political Rights.* Oxford University Press.

Kelman, I. (2021). *Island vulnerability and climate-induced displacement.* Routledge.

Kirmayer, L. J., Brass, G. M., & Bélanger, K. (2011). Cultural issues in the mental health of refugees and asylum seekers. *Lancet, 378*(9799), 1664–1675.

Kothari, U. (2020). The psychological effects of climate-induced displacement on women and families. *International Journal of Gender, 5*(1), 78–92.

Koubi, V., Spilker, G., Schultze, P. E., & Böhmelt, T. (2016). The role of climate change in forced migration. *Climatic Change, 139*(3), 391–402.

Lawrence, J., Berry, H., & Williams, S. (2022). Integrating mental health into national climate change adaptation plans. *The Journal of Climate and Health, 3*(4), 1–10.

Lindskov, S. (2021). The rights of elderly stateless persons: A human rights perspective. *Journal of Human Rights and Social Work, 6*(1), 60–75.

Lopes Cardozo, M. T., Sijbrandij, M., & Thung, M. (2019). The psychological well-being of stateless people. *International Journal of Migration and Border Studies, 5*(4), 312–330.

Manly, J., & Kim, S. (2023). The generational transmission of trauma in stateless families. *Journal of Intergenerational Trauma, 2*(1), 12–25.

Manly, J., & van Waas, L. (2022). Statelessness and the psychological burden of a lost identity. *Journal of Human Rights and Social Work, 7*(2), 110–125.

McAdam, J. (2012). Climate change and displacement. In *The Oxford Handbook of International Refugee Law* (pp. 581–598). Oxford University Press.

McAdam, J. (2020). *Climate change, statelessness and human rights*. Oxford University Press.

McAdam, J. (2022). The rights of persons with disabilities in situations of statelessness. *Forced Migration Review, 71*, 45–48.

McIntosh, R. J. (1998). *The ancient West African states: A millennium of change*. University of California Press.

McLeman, R. A., & Gemenne, F. (2018). Climate-induced migration: A new frontier in international relations. In C. B. C. (Ed.), Handbook of international relations (pp. 249–264). Sage Publications.

McMichael, A. J., Friel, S., & Butler, C. D. (2021). Climate change and human health: A new challenge for the health sector. *Globalization and Health, 17*(1), 1–10. PMID: 33397377

Miller, K. E., & Rasmussen, A. (2017). Mental health in displaced populations. *World Psychiatry; Official Journal of the World Psychiatric Association (WPA), 16*(2), 120–129.

Miller, K. E., & Rasmussen, A. (2020). Mental health and psychosocial support for refugees and asylum seekers. In M. R. T. (Ed.), Refugee mental health (pp. 185–204). Routledge.

Miller, K. E., & Rasmussen, A. (2022). Peer support programs for displaced populations. *Journal of Mental Health (Abingdon, England), 31*(2), 195–205.

Nishikiori, N., Ohnishi, M., & Nakashima, A. (2017). The role of education in building resilience among displaced populations. *International Journal of Educational Development, 53*, 1–10.

Osofsky, H. J., Stevens, D. M., & Osofsky, J. D. (2010). The psychological effects of disasters on children and adolescents. *Child and Adolescent Psychiatric Clinics of North America, 19*(2), 191–207.

Pargament, K. I. (2011). *The psychology of religion and coping: Theory, research, practice*. Guilford Press.

Patel, V., Khan, S., & Singh, R. (2023). The mental health of displaced children: A systematic review. *Journal of Refugee Studies, 36*(1), 1–18.

Peluso, N. L., & Watts, M. (2001). *Violent environments*. Cornell University Press.

Piguet, É. (2013). From "primitive migration" to "climate refugees": The curious fate of the natural environment in migration studies. *Annals of the Association of American Geographers, 103*(1), 148–162.

Pihkala, P. (2020). Climate anxiety. *The Lancet. Planetary Health, 4*(11), e507–e508.

Riley, E., Smith, J., & Patel, K. (2017). The mental health of the Rohingya: A literature review. *Forced Migration Review, 56*, 45–48.

Roberts, L., Evans, A., & Jones, S. (2021). The therapeutic role of group counseling for displaced populations. *Journal of Clinical Psychology, 77*(3), 601–615.

Roy, A. (1999). *The greater common good*. Penguin.

Ryan, B., Evans, A., & Jones, S. (2021). The role of community-based support in fostering resilience among stateless populations. *Journal of Human Rights and Social Work, 6*(4), 312–325.

Sawyer, A., & Blitz, B. K. (2022). Economic exploitation and the mental health of stateless populations. *International Journal of Migration, Health and Social Care, 18*(2), 121–135.

Schilling, J., Haras, D., & O'Brien, C. (2018). The long-term psychological effects of climate-induced displacement. *Climatic Change, 149*(3), 369–382. PMID: 31258223

Schweitzer, R., Evans, A., & Jones, S. (2018). Community resilience in the face of forced migration. *Journal of Community Psychology, 46*(6), 701–715.

Scudder, T. (2012). *The future of large dams: Dealing with social, environmental, institutional and political costs*. Routledge.

Silove, D., Ventevogel, P., & Rees, S. (2017). The contemporary refugee crisis: An overview of mental health challenges. *World Psychiatry; Official Journal of the World Psychiatric Association (WPA), 16*(2), 130–139. PMID: 28498581

Tol, W. A., Hall, B. J., & van Ommeren, M. (2020). Mental health and psychosocial support in humanitarian settings: A review. *Current Opinion in Psychiatry, 33*(4), 362–370.

Tschakert, P., Ellis, N., Anderson, C., Kelly, A., Obeng, J., & Porter, R. (2019). Climate change and loss, grief, and trauma. *Climatic Change, 152*(1), 75–89.

. UN General Assembly. (1966a). International Covenant on Civil and Political Rights.

UN General Assembly. (1966b). *International Covenant on Economic*. Social and Cultural Rights.

UNHCR. (2014). *A global action plan to end statelessness*. United Nations High Commissioner for Refugees.

UNHCR. (2018). *Global report 2018: Climate change and displacement*. United Nations High Commissioner for Refugees.

UNHCR. (2022). *Climate change and disaster displacement*. United Nations High Commissioner for Refugees.

Uscher-Pines, L. (2009). The mental health needs of climate-displaced populations. *Journal of Climate and Health, 2*(3), 1–10.

van Ommeren, M., Hameed, W., & Qureshi, M. (2018). The role of mobile mental health clinics in humanitarian settings. *World Psychiatry; Official Journal of the World Psychiatric Association (WPA), 17*(1), 58–60.

van Waas, L. (2017). The psychological effects of statelessness: A review of the literature. *International Journal of Human Rights, 21*(5), 650–666.

van Waas, L., & Khanna, A. (2021). Economic insecurity and the mental health of stateless populations. *Forced Migration Review*, *68*, 28–31.

Watts, N., Adger, W. N., & O'Brien, K. (2015). The role of public health in addressing the psychological effects of climate change. *Lancet*, *386*(10008), 2101–2102. PMID: 26713340

Weiss, H., & Bradley, R. S. (2001). What drives human migration? A new approach. *Climatic Change*, *49*(1-2), 1–18.

Werz, C., & Hoffman, M. (2018). Economic opportunities and resilience for climate migrants. *Journal of International Development*, *30*(2), 221–235.

White, S. (2014). The Little Ice Age in Europe. *The Journal of Interdisciplinary History*, *45*(1), 101–121.

WHO. (2011). *Psychological first aid: A guide for fieldworkers*. World Health Organization.

WHO. (2018). *Mental health and psychosocial support in humanitarian settings*. World Health Organization.

Zetter, R. (2017). *The international refugee regime: A human rights perspective*. Oxford University Press.

Chapter 6
Technology-Enabled Psychosocial Interventions for Climate-Affected Populations

Sonali Sejwal
https://orcid.org/0000-0001-9380-2894
TERI University, India

Kirti Tyagi
https://orcid.org/0000-0001-8454-335X
O.P. Jindal Global University, India

ABSTRACT

Climate change disasters like heat waves, wildfires, and floods are increasing psychological distress, especially in vulnerable communities. With 3.3 to 3.6 billion people in climate-sensitive regions, mental health concerns linked to displacement, financial instability, and livelihood loss are rising. This chapter explores the link between climate change and mental health, highlighting emerging conditions like depression, PTSD, and eco-anxiety. It emphasizes scalable digital mental health strategies to support traditional care and build resilience. Technology-enabled interventions like tele-counseling, mental health apps, online peer support platforms, and messaging services are examined in the disaster context. The chapter also addresses the barriers like infrastructure gaps, digital literacy, stigma, ethics,

DOI: 10.4018/979-8-3373-3281-9.ch006

and data privacy. It reviews global and Indian policy efforts like Kiran Helpline, Tele-MANAS, Australia's MindSpot, and WHO's Step-by-Step, providing inclusive and culturally sensitive recommendations.

INTRODUCTION

Extreme and abrupt climate change, including floods, wildfires, rising temperatures, and heat waves, drastically affects ecologically vulnerable areas. The Intergovernmental Panel on Climate Change (2022) estimates that approximately 3.3 to 3.6 billion individuals reside in highly climate-sensitive environments. Such environmental disturbances threaten people and communities, resulting in loss of housing, displacement, economic instability, and reduced access to healthcare. Together, these stressors add up to an increasing load of mental health problems. Climate-linked psychological effects such as anxiety, depression, and post-traumatic stress disorder (PTSD) have been well documented.

The European Environment Agency (2018) states that within Europe, the risk of suffering from psychological stress due to climate-related disasters can be as much as 40 times higher than for physical harm. An area of growing concern is eco-anxiety (also known as climate anxiety), which describes the distress and fear triggered by climate change and the prospect of its effects (Coffey et al., 2021). These conditions of mental health are usually compounded when coupled with socioeconomic situations like access to health care, food security, economic stability, and migration within vulnerable groups. Regardless of socioeconomic condition, another issue is insufficient access to mental health care for climate change-affected communities.

Without timely psychological support and relief, individuals in these regions are left vulnerable. This highlights the need to expand mental health care directly through digital platforms, which can offer greater accessibility and 24/7 support. To better understand the psychosocial dimension of this crisis, the following section examines how mental health challenges emerge at the intersection of various environmental and socioeconomic stressors, particularly in climate change. As climate-related disruptions become more integrated into daily life, a significant gap persists in the accessibility of specialized mental health services for those facing the prolonged environmental risks and stressors.

Climate Change and Mental Health: Understanding the Psychosocial Impact

Climate change (CC) refers to long-term shifts in temperature and weather and short-term events such as flooding and forest fires (United Nations, 2023). While these changes in the environment may sometimes occur naturally, in recent times, these drastic environmental changes have resulted from human activities, including burning fossil fuels, clearing forests, and livestock farming. These changes have been found to disrupt both human-made systems (such as social, economic, and health infrastructures), and natural systems (like ecosystems, forests, wetlands, oceans, and deserts), all of which are vital for maintaining human health and well-being (United Nations, 2023). Disruptions of these systems can increase the risk of non-communicable diseases, spreading contagious diseases, health emergencies, and sometimes death (Madhav et al., 2017).

While the physical health consequences of climate change are well-documented, there is growing recognition of its psychological impact, a term that includes emotional, cognitive, and social implications of environmental disruptions. These include distress, eco-anxiety, and PTSD, which is accompanied by feelings of guilt, helplessness, frustration, and apprehension about the future.

Psychological Impact of Climate Disasters

Climate change-related stressors can lead to a spectrum of psychological responses—from mild anxiety and sleep disturbances to more severe conditions like post-traumatic stress disorder, depression, and even suicidal ideation. Ojala (2016) characterized climate change as an existential threat that makes people, especially young people, concerned about their future. According to the World Health Organization (WHO, 2022), the consequences of climate change, when combined with sociodemographic and contextual factors such as age, gender, and income, prior mental health conditions, direct experiences of climate events, and awareness of climate change impacts and causes. These factors can significantly amplify psychological effects, and scholars call his phenomenon as, 'threat multiplier' (Goodman & Baudo, 2023).

THEMATIC REVIEW

Based on the psychosocial impacts outlined above, this section examines technology-enabled mental health interventions designed for populations affected by climate-related disasters. Drawing from existing literature, policy reports, and case studies, this section identified recurring themes in how digital tools are being used to address gaps in access, continuity, and cultural relevance of care for vulnerable groups.

Four major domains emerge from the review: tele-counselling for displaced or disaster-affected populations, mobile mental health applications, online support groups and peer-led platforms, and messaging/chatbot services. Within each theme, the discussion integrates examples of real-world implementation, evidence on effectiveness, and practical challenges, ranging from infrastructure limitations to ethical considerations. The following themes provide understanding of how digital innovations are expanding the reach of mental health support in crisis contexts, while highlighting the critical need for context-sensitive design and integration into border climate-health response mechanisms.

The next section explores each of these domains in detail, presenating specific startegies and case studies that discuss their application and impact in diverse climate crisis settings.

Technology-Enabled Interventions for Climate-Affected Populations

Given the scale and complexity of climate-related mental health challenges, there is an urgent need for scalable, context-sensitive, and accessible solutions. Technological interventions have become essential in tackling these challenges, providing mental health support that is more accessible, affordable, and effective. The growing adoption of digital tools such as telehealth services, mobile apps, and community-based digital education programs is transforming how individuals and communities access psychological care during times of crisis (Naslund et al., 2017). This section reviews key digital strategies and case studies illustrating their applications in climate crisis settings.

Tele-Counselling for Displaced or Disaster-Affected Populations

The American College of Emergency Physicians (ACEP) has emphasized the strategic role of telehealth in disaster preparedness and emergency response. Their policy statement highlights how telehealth is a multiplier, expanding healthcare capacity and building system resilience during natural and artificial disasters (American College of Emergency Physicians, n.d.). Telehealth tools such as mobile health applications, remote monitoring systems, and real-time video consultations ensure rapid triage, medical assessments, and treatment in disaster-affected regions. These technologies help bridge gaps caused by infrastructure collapse, overburdened health systems, and limited physical access to care. Moreover, the policy stresses the need for funding mechanisms to support telehealth deployment, robust telecommunications infrastructure, and training for first responders in telehealth applications. By integrating these technologies, healthcare systems can ensure continuity of care, especially for marginalized populations, while enhancing their ability to respond to future crises (American College of Emergency Physicians, n.d.).

Building on these strategic priorities, practical implementations in low-resource and disaster-affected settings have demonstrated the importance of culturally sensitive and community-rooted approaches. For instance, Jack and Izah (2024) advocate for integrating mental health services into broader disaster response frameworks using culturally grounded strategies. These include involving traditional healers, local leaders, and context-specific communication methods. Such approaches can be used to combat trauma, reduce stigma, and strengthen community trust.

Similarly, telemedicine and mobile clinics have emerged as a vital tool to overcome geographical and infrastructural barriers to care. Supporting these approaches, Augusterfer et al. (2020) explored the effectiveness of telemental health through tele-consultation and tele-supervision, particularly in resource-constrained crisis settings. These models strengthened the local clinical decision-making and community capacity by enabling remote mental health professionals to mentor and support local clinicians through low-bandwidth platforms. The study concluded that, when cultural, technological, and training challenges are addressed systematically, digital models can significantly enhance the resilience of the mental health system during crisis.

A perfect example of the application of tele-mental health services can be seen in Médecins Sans Frontières' (MSF) response to the COVID-19 pandemic. In humanitarian contexts in Asia, the Middle East, North Africa, and Sub-Saharan Africa, MSF shifted from face-to-face mental health services to remote modalities to ensure continuity of services in light of global mobility restrictions and increasing mental health demands. Following the report of Ibragimov et al. (2022), 96.7% of participants recognized the value of telemental health, naming greater efficiency and better access. Nevertheless, MSF's strategy was not easy. Around 80% of MSF tele-mental health interventions employed the audio-only mode, resulting in a lack of nonverbal cues that are important for therapeutic work. Other significant obstacles were poor network reception (73.8%), unavailability of devices (72.1%), and inadequate privacy at home (67.2%). There were also concerns about the appropriateness of tele-mental health for vulnerable groups like children, the elderly, those with severe mental disorders, and survivors of interpersonal violence. The research concluded that, although tele-mental health is an acceptable backup in emergencies, it needs contextualized guidelines and ongoing infrastructure evaluation to support ethical and pragmatic care provision. Though tele-counselling is critical to addressing the near-term mental health needs of displaced or climate change disaster-affected communities, increased accessibility of mobile technology has been shown to offer a scalable and complementary solution in the form of mobile mental health apps, addressed in the next section.

Mobile Mental Health Applications

Mobile mental health applications (mHealth apps) have gained increasing relevance in disaster-affected contexts, offering scalable and accessible tools for psychological support when traditional services are disrupted (Ezeonu et al., 2024). These online platforms provide self-administered interventions like reduction of stress, mindfulness training, and cognitive behavior tools, specifically designed for the trauma- or displacement-affected individuals (Ezeonu et al., 2024; Haque & Rubya, 2022). Prominent instances, like PTSD Coach, COVID Coach, Headspace, and the Psychological First Aid application, have exhibited promise in enhancing emotional regulation and coping strategies in acute and post-disaster situations (e.g., Miner et al., 2016; Owen et al., 2018).

Although these have advantages, there are some limitations. Some peer-reviewed papers have examined these apps' effectiveness in various disaster response phases. Albeit their potential, there are significant limitations, such as low rates of engagement, user attrition, and limited assessment of long-term impacts (Ezeonu et al., 2024; Haque & Rubya, 2022). These limitations underscore the need for more research to refine these tools for use across various populations and successfully integrate them into larger emergency mental health systems. Aside from individualized applications, the digital mental health infrastructure can be supplemented by online support groups and peer-based platforms, which facilitate community-oriented recovery and peer-to-peer exchange in addition to the personalized support offered by mobile mental health applications.

Online Support Groups and Peer-Led Platforms

Augmenting individual-level interventions such as mHealth applications, web-based support groups, and peer-operated platforms is essential in fostering community resilience, especially among marginalized or geographically disadvantaged populations (Steiner et al., 2023; Carlisle et al., 2024). Such online platforms offer low-threshold, frequently anonymous environments for persons to openly declare vulnerability, share experiences, and garner emotional support and psychological resilience ingredients.

Studies persistently show that such sites create feelings of belonging and empower people by affirming common experiences and eliciting mutual support. For instance, among rural populations, online mental health forums have been associated with enhanced coping ability and decreased social isolation (Steiner et al., 2023). Similarly, web-based peer support alleviates loneliness, improves affect regulation, and creates meaningful relationships among young people residing in rural areas of Australia (Carlisle et al., 2024). Together, these results demonstrate the value of grassroots digital support systems in enhancing mental health outcomes and resilience where formal services are not available or accessible. While online support groups offer spaces for peer-to-peer interaction and mutual discussion, messaging services allow for a more personal and direct means of mental health access, especially ideal for engaging reluctant participants in group settings.

Messaging Services for Mental Health Outreach

Messaging platforms and chatbot-based applications have emerged as significant tools in digital mental health outreach, particularly during crises when in-person services are disrupted or inaccessible. These technologies offer scalable, low-cost, accessible solutions that facilitate real-time support, emotional expression, and community building. Chatbot-enabled mental health apps are increasingly used to deliver personalized and judgment-free support experiences. A qualitative study analyzing over 6,000 user reviews across major app stores identified that users appreciated mental health chatbots' human-like responsiveness and round-the-clock availability, encouraging disclosure and self-reflection (Haque & Rubya, 2023). However, the same study also identified several limitations, including inappropriate or generic responses, reduced engagement over time, and a concerning reliance on bots in place of real human connections. The inability of many chatbots to identify or appropriately respond to emergencies was also highlighted as a critical gap, underscoring the need for improved design protocols and ethical safeguards in chatbot development (Haque & Rubya, 2023).

Similarly, messaging services such as WhatsApp played a central role during the COVID-19 pandemic by facilitating the exchange of crisis-related information and emotional support. Findings reported that even though 94.7% of participants engaged in COVID-19-related content on WhatsApp, private conversations on WhatsApp contributed to increased cognitive preoccupation. The study flagged a psychological toll associated with overexposure to crisis messaging, highlighting the need for mindful use and policy-level interventions in managing information overload during emergencies (Tan et al., 2020).

Apart from chatbots and instant messaging, asynchronous online communication platforms have also been vital emotional and academic support systems. Dennis et al. (2023) researched university students' application of online discussion forums during the pandemic period. Their research showed that such platforms facilitated sharing coping strategies, collectively managed workloads, and gave peer-to-peer emotional support. These sites supplemented scholarly interaction and served as secure environments for affective expression and building resilience among community members. The results highlight how asynchronous, student-initiated forums can serve as critical mental health tools during isolation and duress (Dennis et al., 2023).

Collectively, these studies signal the increasing salience of messaging-based mental health technologies and their significance in communication in crisis mental health treatment. As valuable assets with distinct strengths in reach, flexibility, and timeliness, they also present significant ethical, design, and psychological issues. Their effective incorporation within larger mental health systems also demands more than technological optimization; it requires knowledge of users' changing emotional requirements.

To appreciate the actual impact of the mental health technologies, the subsequent case studies and illustrative examples exemplify how these interventions have been applied successfully in differing contexts.

Case Studies of Technology-Enabled Mental Health Support

Mobile applications and digital communication platforms, such as e-messaging, m-health applications, and telepsychiatry, have proven their usefulness in providing psychological care and psychosocial support during times of crisis. These case studies reinforce the flexibility of digital technologies in responding to the mental health demands of disaster-affected and high-stress populations.

Turkey Earthquake: Use of Mobile Messaging (WhatsApp). Following the Turkey earthquakes of February 2023, a volunteer-operated communication network was set up through WhatsApp to organize relief and support activities. Analysis of 88,061 messages shared over the first five days of the disaster indicated prevailing themes including coordination of rescues, requests for shelter, and emotional distress expressions (Celebi et al., 2024). Messages frequently contained locational information, emotional messages, and priority requests for food, medical assistance, and tents. In contrast, the platform facilitated the smooth exchange of information and psychosocial support. The research also observed issues like information overload and inability to prioritize priority messages, pointing to the need for more organized digital crisis communication processes. This case illustrates the capability of decentralized mobile communication systems for enhanced logistical support and mental health outreach during massive disasters.

Refugee Population: Use of Mobile Mental Health Apps. Culturally and contextually appropriate mental health digital interventions have been shown to be effective solutions for refugee populations. A systematic review of smartphone-based interventions revealed that mobile mental health apps and telepsychiatry services effectively treated depression, anxiety, and PTSD in displaced persons (Mohamad et al., 2022). The review highlighted user engagement, accessibility, and cultural appropriateness as factors to consider in the success of an intervention. Although the scalability and tailorability of these tools present great benefits, data privacy challenges, disparity in smartphone access, and access to the internet were major obstacles for effective implementation. Alongside these, the evidence envisions the necessity of ensuring the inclusion of the refugee community in co-design and implementation to improve acceptability and effectiveness (Mohamad et al., 2022).

Sikkim Glacial Lake Outburst Flood: Gaps in Digital Infrastructure. Following the October 2023 glacial lake outburst flood (GLOF) in Sikkim, severe psychological trauma surged due to the destruction caused by the collapse of the Teesta III Dam. Reports from local healthcare providers underscore the growing mental health burden among survivors. In a case of a young girl, Neelam from Singtam region, required ongoing treatment for severe anxiety and panic attacks, with inconsistent follow-up care and minimal tech-enabled outreach. The inconsistent follow-up care and limited community mental health infrastructure remain significant barriers to long-term psychological recovery. The absence of scalable mobile apps, tele-counseling, or virtual support networks revealed a digital divide that hindered mental health recovery. This underscores the urgent need to strengthen digital infrastructure in underserved disaster zones, (Think Global Health, 2025).

Australian Bushfires: Peer-based Support Via Online Networks. In the aftermath of the Australian bushfires of 2019–2020, peer-led social support groups played a pivotal role in mental health recovery. A longitudinal study including 627 disaster-affected individuals found that pre-existing and newly formed online community groups helped reduce distress and increase resilience over 12–18 months (Cruwys et al., 2023). The informal support networks offered common identity, emotional security, and continuity in an environment of disruption. Although these networks were effective, their scalability and consistency varied, emphasizing the need for intentional design and facilitation of such digital peer-based models during recovery phases, (Carlisle et. al., 2024).

Germany's Ahr Valley Flood: Online Mindfulness-Based Intervention. To help improve the mental health of those impacted by the terrible floods that devastated Germany's Ahr Valley in July 2021, an online mindfulness-based intervention was implemented. Around 134 individuals participated in the eight-week online program, including relief volunteers and flood survivors. The intervention improved self-compassion and emotional resilience while significantly lowering psychological distress, depressive symptoms, and symptoms of post-traumatic stress disorder (Müller et al., 2024). The approach maintained continuity of mental health support service despite significant disruption by integrating digitally accessible interactive activities, psychoeducational resources, and guided meditations into a structured online program. The study emphasizes the effectiveness of digital mindfulness interventions in reducing mental health burdens during climate-related disasters, particularly in contexts where traditional support systems are unavailable, even when participant involvement levels are varied.

Israel-Gaza Border Communities: WhatsApp for Community-Based Coping. In regions impacted by chronic political violence, such as near the Israel-Gaza border, community-based digital communication has served as a key coping mechanism. Qualitative research of mothers in local communities showed that WhatsApp group was a platform that facilitated emotional and problem-solving support (Roitman & Yeshua-Katz, 2021). The app's immediacy, reachability, and multimedia affordances facilitated users in constructing shared resilience and psychological stability. Nonetheless, informal network dependence underscored the lack of formal digital mental health infrastructure and raised concerns about sustainability and peer supporter burdens, (Ezeonu et. al., 2024).

DISCUSSION

Technology-facilitated psychosocial interventions for people impacted by climate change have great promise and continue to face challenges. At one level, it is clear that these technologies extend access to mental health care into settings where traditional systems are absent or compromised. The case studies illustrate the point: internet-based mindfulness courses facilitating psychological healing in flood-stricken communities in Germany, WhatsApp groups organizing relief responses to earthquakes in Turkey, and mobile apps providing refugees with coping strategies. In all cases, technology served to sustain human connection in the face of extreme disruption and functioned as a delivery mechanism.

These examples demonstrate that these interventions are ineffective if the underlying infrastructures are poor. Poor connectivity, unstable power supply, and low device penetration were not incidental challenges but significant barriers to community engagement, like the one hit by the Sikkim flood. This echoes a common theme in literature: technological progress risks perpetuating existing health disparities if it is not available for all.

The analysis also emphasizes how crucial emotional safety and trust are to the effectiveness of digital interventions. Community identity-based platforms, like mothers' WhatsApp groups in conflict-affected areas or peer-led forums after the Australian bushfires, were successful not only because they spread information but also because they fostered reciprocity, belonging, and mutual validation. These relational components seem as crucial for long-term engagement as the technology's functionality or design.

Privacy and ethical issues are still crucial, though. People can disclose private information when at risk in crises, but data governance procedures are frequently insufficient. Credibility in these tools could be jeopardized in the absence of strong safeguards. Furthermore, contrary to being integrated into preparedness frameworks, most interventions are implemented reactively after disasters, which restricts their scalability and sustainability.

Future studies must look at hybrid models that combine the cultural foundation of community-based support with the scalability of digital platforms. Uptake and impact could be increased by training local facilitators, using offline-capable tools for low-bandwidth scenarios, and involving end users in co-design. In the face of growing climate challenges, such strategies could guarantee that digital mental health care is meaningful and accessible.

Challenges in Using Technology for Climate-Affected Populations

In terms of access to online mental health help and educational resources, poor people are disproportionately affected by two essential impediments: affordability and digital competence. Numerous representatives of these groups can face difficulties paying for necessary equipment such as laptops, mobile phones, and tablets, as well as ongoing charges such as data plans and internet connections. Although such barriers strongly hinder fair access to digital mental health services, an increasing number of nations have reacted with the use of targeted policies that promote better reach, affordability, and quality. The following section presents global and national approaches that bridge these structural disparities.

Low digital literacy, or the ability to use digital platforms confidently to navigate, understand, and access them, may restrict effective interaction with digital services despite technology being present. Due to this, these communities are often excluded from the resilience-building potential that e-learning platforms, healthcare services, and online forums provide.

To address such issues, education on digital literacy needs to be tailored towards the populace, and technology costs need to be reduced through subsidization or community support. This will allow consumers to benefit fully from technology. Engagement is also discouraged by mental health care's cultural stigma and issues of trust due to perceived insincerity. These all underscore how important it is to develop interventions that are accessible, affordable, culturally sensitive, and capable of establishing trust across various demographics.

Another considerable obstacle is ethical and privacy issues, as protecting private data and confidentiality is particularly difficult in crises or among displaced populations, when the infrastructure for oversight and data protection is weak. To guarantee that every user will be able to benefit from digital mental health and educational resources safely, it is essential to plan interventions that are accessible, cheap, culturally adapted, reliable, and ethical.

Policy Recommendations and Strategic Directions

With escalating mental health issues under the strains of conflict, displacement, and climate crises, nations worldwide have established various digital mental health policies and interventions to increase access, minimize stigma, and reach vulnerable populations. These policies work to close the gap in mental healthcare access by creating inclusive spaces, minimizing stigma, and reaching underserved and climate-vulnerable populations through considered digital interventions.

Global Policy Frameworks

WHO's Step-by-Step: Digital Mental Health for Refugees. Step-by-Step, an online mental health intervention, created by the World Health Organization (WHO) in collaboration with the National Mental Health Programme (NMHP) of the Ministry of Public Health Lebanon and other stakeholders, helped in decreasing depression in Syrian refugees in Lebanon Step-by-Step is a five-session WHO digital intervention that applies an internet-enabled device to treat depression (World Health Organization, 2023). Trained non-specialist volunteers provide weekly support (such as a 15-minute call or message). Through an illustrated narrative, it offers psychoeducation and training in behavioral activation (e.g., engaging in more pleasant activities). Other therapeutic strategies include stress management, a gratitude exercise, positive self-talk, bolstering social support, and relapse prevention (Cuijpers et al., 2023). To promote accessibility, Step-by-Step features five visual story sessions with audio recordings of the text to support accessibility. Three smaller components make up each session, typically taking 20 minutes to complete. Each session unlocks four days after the prior session ends to allow the individual to practice the skills and activities they learnt (Cuijpers et al., 2023).

Australia's MindSpot Clinic: Scaling Digital Access. The Australian Department of Health launched the MindSpot Clinic, a nationwide digital mental health service that was introduced in December 2012 as part of the nation's e-Mental Health Strategy. For people with symptoms of anxiety, sadness, or chronic pain, it offers free mental health evaluations, therapy, and referral services over the phone and online. By providing easily accessible, evidence-based interventions through digital platforms, MindSpot seeks to increase access to mental healthcare (Staples et al., 2025).

U.S. Medicare Billing for Digital Therapies. A first step toward the systemic integration of digital health into mainstream clinical treatment was taken in 2024 when the United States' Centre for Medicare and Medicaid Services (2024) suggested new billing codes to reimburse doctors for prescription digital mental health therapies, including as apps and software (axios.com). This action highlights policy-level recognition of app-based treatment for behavioral health and may spur further insurer adoption.

NIMH Global Program: Equity-Driven Tech Solutions. Research and application of digital mental health tools in low- and middle-income nations are supported by the NIMH Digital Global Mental Health Program. The program prioritizes ethical, culturally sensitive implementation in addition to scalable technology for diagnosis, prevention, and treatment (National Institute of Mental Health, n.d.). It supports strategies like provider dashboards, AI screening tools, and touchscreen apps in environments with limited resources, enabling digital equity on a large scale.

WHO and WPA Guidelines: Governance and Interoperability. With criteria for privacy, equity, and interoperability, the WHO Global Strategy on Digital Health (2020–25) encourages countries to integrate digital technology into universal health systems (World Health Organization, 2021). In addition, the development of global policy guidelines and training requirements for digital mental health is enabled by the World Psychiatric Association's (WPA) Working Group on Digitalization (2022). The global initiatives offer the structural guidelines for integration into national systems.

Regulatory Reform and Systemic Challenges

The WHO 2025 also supports promoting a global reform of mental health laws. WHO advocates a shift from outdated institutional care to community-oriented, integrated care, incorporating mental health into broader health and social systems, calling for aligning services with human rights standards (DHI,2021). The suggestions appeal to governments to invest heavily in mental health facilities, integrate mental health into universal health coverage, and consider psychological well-being as fundamental to the resilience of a society, especially when confronting increasing crises such as forced displacement, conflicts, and climate change. To realize this objective, cross-sector cooperation, long-term funding, and dedication to underserved populations who frequently experience systematic neglect.

A shared vision focused on accessibility, inclusivity, and dignity is reflected in both the Indian Act and the WHO guidelines. However, the WHO's 2025 strategy calls for a more comprehensive systemic change, whereas India's 2017 legislation strongly emphasizes individual rights and legal entitlements. When taken as a whole, they highlight how critical legal frameworks are, but they fall short unless fundamental changes, inclusive policies, and consistent funding accompany them (WHO, 2023).

While these international initiatives provide practical frameworks for expanding digital mental health services, India, one of the most vulnerable to climate change disasters and with wide-ranging diversity, has also implemented several bold policy-level initiatives, which are discussed below.

India-specific Policies for Digital Mental Programs

With recent programs like Tele-MANAS, Ayushman Bharat, the National Suicide Prevention Strategy (NSPS), and the Kiran Helpline, India has made significant progress in bolstering its mental health infrastructure, indicating a move toward more affordable, accessible mental health care.

Tele-MANAS and Ayushman Bharat. Tele-MANAS is a nationwide telemental health initiative that offers multilingual, round-the-clock psychological care via toll-free phones. It benefits explicitly rural and distant populations by addressing significant barriers like stigma and geographic isolation (Ministry of Health & Family Welfare, 2023). Scaling human resources and guaranteeing high-quality, culturally sensitive care across areas remains a complex task. Although its primary goal is universal health care, Ayushman Bharat has a significant but indirect impact on mental health. Its Health and Wellness Centers (HWCs) encourage early detection and de-stigmatization by integrating mental health services at the primary care level. However, under Ayushman Bharat, mental health care is frequently given less priority than physical health services.

National Suicide Prevention Strategy (NSPS). India's first comprehensive framework for suicide prevention, the National Suicide Prevention Strategy (NSPS), describes a multi-sectoral public health strategy aimed at reducing suicide mortality by 10% by 2030 (Ministry of Health & Family Welfare, 2022). It places a strong emphasis on developing community resilience, expanding access to suicide means, and bolstering mental health care and cross-sector cooperation.

Kiran Helpline: A National 24/7 Support Line. Launched in 2020, the Kiran Helpline connects users with qualified counselors and mental health professionals to provide 24/7 nationwide mental health care (Ministry of Health & Family Welfare, 2021). Despite obstacles, including inconsistent public awareness and inadequate multilingual service capability, it is crucial to provide timely assistance.

Rights-based Legislation and Ground-Level Challenges

An essential step toward a rights-based approach to mental health governance was taken with the introduction of the Mental Healthcare Act 2017 (Government of India, 2017). The Act, which is based on the UN Convention on the Rights of Persons with Disabilities, ensures everyone has the right to high-quality, reasonably priced mental health care. By permitting prior directives and legally acknowledging their ability to make treatment decisions, it highlights the autonomy of people with mental illness. Crucially, it criticizes suicide attempts by portraying them as medical emergencies that call for treatment rather than punishment (Namboodiri et al., 2019). The creation of Central and State Mental Health Authorities ensures regulatory monitoring. On paper, the Act is progressive, but in practice, issues such as a lack of funding, qualified personnel, and public awareness impede its fulfillment, especially in neglected and rural areas.

Overall, these initiatives show how mental health is increasingly acknowledged as a public issue in India. However, to have the most significant possible impact, India needs to address structural problems, including personnel shortages, stigma reduction, digital literacy gaps, and privacy concerns, especially in places with limited resources. With careful integration, tech-enabled innovations have the potential to improve further.

Future Directions: Towards Equitable and Scalable Solutions. The significance of integrating digital mental health services into disaster preparedness frameworks is becoming more widely acknowledged by governments and non-governmental organizations. Building societal resilience before, during, and after crises requires policies that combine emergency response planning with mental health services. Collaboration between the Global South's nations also presents beneficial opportunities for South-South knowledge sharing, creating scalable pilot projects, and creative cross-border finance methods to increase access to digital mental health in environments with limited resources. The recommendations for the future include investing in low-tech but high-reach solutions that can operate in low-bandwidth environments, emphasizing culturally-adapted interventions to improve inclusivity, and integrating mental health support as a fundamental part of disaster resilience and climate adaptation plans. It will be essential to take a comprehensive, equity-driven approach to guarantee that digital breakthroughs in mental health genuinely benefit the most vulnerable groups in a world that is becoming more interconnected and impacted by climate change.

Most importantly, technology-based mental health interventions offer critical pathways to plugging gaps in treatment, particularly for remote populations and refugees. Electronic counseling services, online telemedicine platforms, and smartphone-based mental health applications have the potential to bridge geographical gaps, stigma, and poverty of resources. Scalable technology solutions ensure continuity of care, early intervention, and emotional support during times of crisis when conventional services do not work. However, developing and introducing such technology to serve disadvantaged groups efficiently, bearing in mind cost, cultural appropriateness, data protection, and inclusivity, is of utmost importance.

In sum, the WHO's 2025 guidance offers a vision for international change that is challenging to accept, while the Mental Healthcare Act of 2017 provides a firm foundation in law for care grounded in rights. Together with advances in technology, there is an unprecedented chance to democratize mental health care, ensuring that even the most marginalized populations can access high-quality, humane care.

CONCLUSION

With emphasis on how technology-enabled psychosocial therapies might close current service gaps for people impacted by climate change, this chapter explored the crucial and expanding relationship between mental health and climate change. Identifying technology-driven mental health solutions, assessing policy frameworks that enable these initiatives, and comprehending the psychosocial effects of climate-related calamities were among the main objectives.

To illustrate how environmental stresses are working as "threat multipliers" and aggravating anxiety, PTSD, eco-anxiety, and depression among vulnerable groups, it first mapped the psychosocial landscape of climate change. The chapter described various technology-enabled interventions that provide scalable, readily available, and context-sensitive mental health care, including chatbot-based messaging services, online peer-led forums, mobile mental health apps, and telecounseling.

The chapter highlighted the usefulness of digital mental health tools in crises using international case studies, such as the WhatsApp networks after the earthquakes in Turkey, mobile health solutions for refugee populations, and mindfulness-based interventions in Germany. Particular focus was placed on India's national policy environment, which includes the Mental Healthcare Act (2017), Ayushman Bharat, and Tele-MANAS, demonstrating the nation's growing dedication to digital mental health governance.

However, the chapter also recognizes enduring issues such as data privacy hazards, cultural stigma, digital literacy gaps, infrastructure constraints, and ethical issues. It underlined that digital interventions must be equity-driven, culturally grounded, and backed by strong policy frameworks to have a significant impact. In the future, it will be crucial to incorporate digital mental health techniques into strategies for climate adaptation and disaster preparedness. To ensure that the most vulnerable populations receive prompt, considerate, and efficient mental health support in an era of growing climate crises, there is a need for cross-sectoral collaboration, investment in low-resource technology solutions, and adoption of human rights-driven approaches.

REFERENCES

American College of Emergency Physicians. (n.d.). *Disaster telehealth*. ACEP. https://www.acep.org/patient-care/policy-statements/disaster-telehealth

Augusterfer, E. F., O'Neal, C. R., Martin, S. W., Sheikh, T. L., & Mollica, R. F. (2020). The role of telemental health, tele-consultation, and tele-supervision in post-disaster and low-resource settings. *Current Psychiatry Reports, 22*(12). Advance online publication. DOI: 10.1007/s11920-020-01209-5 PMID: 33247315

Carlisle, K., Kamstra, P., Carlisle, E., McCosker, A., De Cotta, T., Kilpatrick, S., Steiner, A., Kahl, B., & Farmer, J. (2024). A qualitative exploration of online forums to support resilience of rural young people in Australia. *Frontiers in Public Health, 12*. Advance online publication. DOI: 10.3389/fpubh.2024.1335476 PMID: 38841668

Coffey, Y., Bhullar, N., Durkin, J., Islam, M. S., & Usher, K. (2021). Understanding eco-anxiety: A systematic scoping review of current literature and identified knowledge gaps. *The Journal of Climate Change and Health, 3*, 100047. DOI: 10.1016/j.joclim.2021.100047

Cruwys, T., Macleod, E., Heffernan, T., Walker, I., Stanley, S. K., Kurz, T., Greenwood, L., Evans, O., & Calear, A. L. (2023). Social group connections support mental health following wildfire. *Social Psychiatry and Psychiatric Epidemiology, 59*(6), 957–967. DOI: 10.1007/s00127-023-02519-8 PMID: 37428193

Cuijpers, P., Heim, E., Ramia, J. A., Burchert, S., & Carswell, K.. (2023). Correction: Effects of a WHO-guided digital health intervention for depression in Syrian refugees in Lebanon: A randomized controlled trial. *PLoS Medicine, 20*(4), e1004231. DOI: 10.1371/journal.pmed.1004231 PMID: 37058690

Ezeonu, N. A., Hertelendy, A. J., Adu, M. K., Kung, J. Y., Itanyi, I. U., Da Luz Dias, R., Agyapong, B., Hertelendy, P., Ohanyido, F., Agyapong, V. I. O., & Eboreime, E. (2024). Mobile apps to support mental health response in natural disasters: Scoping review. *Journal of Medical Internet Research, 26*, e49929. DOI: 10.2196/49929 PMID: 38520699

Government of India. (2017). *Mental Healthcare Act 2017*. Ministry of Law and Justice. https://www.indiacode.nic.in/handle/123456789/2249

Haque, M. D. R., & Rubya, S. (2023). An overview of chatbot-based mobile mental health apps: Insights from app description and user reviews. *JMIR mHealth and uHealth, 11*, e44838. DOI: 10.2196/44838 PMID: 37213181

Haque, M. R., & Rubya, S. (2022). *"For an app supposed to make its users feel better, it sure is a joke":* An analysis of user reviews of mobile mental health applications. *arXiv*. https://arxiv.org/abs/2209.07796

Ibragimov, K., Palma, M., Keane, G., Ousley, J., Crowe, M., Carreño, C., Casas, G., Mills, C., Llosa, A., Marques, A., Tijerino, A., Droste, A., Echeverri, C., Fredj, N., Sleit, R., & Sultan, S. F. (2022). Shifting to tele-mental health in humanitarian and crisis settings: An evaluation of Médecins Sans Frontières experience during the COVID-19 pandemic. *Conflict and Health*, *16*(1). Advance online publication. DOI: 10.1186/s13031-022-00437-1 PMID: 35164807

Madhav, N., Oppenheim, B., Gallivan, M., Mulembakani, P., Rubin, E., & Wolfe, N. (2017). Pandemics: Risks, impacts, and mitigation. In Jamison, D. T., Gelband, H., Horton, S., Jha, P., Laxminarayan, R., Mock, C. N., & Nugent, R. (Eds.), *Disease control priorities: Improving health and reducing poverty* (3rd ed., Vol. 9, pp. 315–345). The World Bank., DOI: 10.1596/978-1-4648-0527-1_ch17

Miner, A. S., Kuhn, E., Hoffman, J. E., Owen, J. E., & Ruzek, J. I. (2016). Feasibility, acceptability, and potential efficacy of the PTSD Coach app: A pilot randomized controlled trial with community trauma survivors. *Psychological Trauma: Theory, Research, Practice, and Policy*, *8*(3), 384–392. DOI: 10.1037/tra0000126 PMID: 27046668

Ministry of Health & Family Welfare. (2021). *Kiran Helpline implementation report*. Government of India. Retrieved July 4, 2025, from https://www.mohfw.gov.in/?q=/press-info/8322

Ministry of Health & Family Welfare. (2022). *National Suicide Prevention Strategy*. Government of India. https://www.mohfw.gov.in/?q=/press-info/8322

Ministry of Health & Family Welfare. (2023). *Tele-MANAS quarterly progress report*. Government of India. Retrieved July 4, 2025, from https://www.mohfw.gov.in/?q=/press-info/7787

Müller, L., Rapoport, O., & Rahe, M. (2024). Mindfulness meditation improves mental health in flood survivors and disaster volunteers: A randomized Wait-List controlled trial. *Mindfulness*. Advance online publication. DOI: 10.1007/s12671-024-02467-7

Namboodiri, V., George, S., & Singh, S. P. (2019). The Mental Healthcare Act 2017 of India: A challenge and an opportunity. *Asian Journal of Psychiatry*, *44*, 25–28. DOI: 10.1016/j.ajp.2019.07.016 PMID: 31302439

Naslund, J. A., Aschbrenner, K. A., Araya, R., Marsch, L. A., Unützer, J., Patel, V., & Bartels, S. J. (2017). Digital technology for treating and preventing mental disorders in low-income and middle-income countries: A narrative review of the literature. *The Lancet. Psychiatry, 4*(6), 486–500. DOI: 10.1016/s2215-0366(17)30096-2 PMID: 28433615

National Institute of Mental Health. (n.d.). *Digital global mental health program.* NIMH. Retrieved July 4, 2025, from https://www.nimh.nih.gov/about/organization/cgmhr/digital-global-mental-health-program

Owen, J. E., Kuhn, E., Jaworski, B. K., McGee-Vincent, P., & Juhasz, K. J. (2018). VA mobile apps for PTSD and related problems: Public health resources for veterans and those who care for them. *JMIR mHealth and uHealth, 6*(7), e156. DOI: 10.2196/mhealth.9388 PMID: 30021707

Roitman, Y., & Yeshua-Katz, D. (2021). WhatsApp group as a shared resource for coping with political violence: The case of mothers living in an ongoing conflict area. *Mobile Media & Communication, 10*(1), 3–20. DOI: 10.1177/20501579211021455

Staples, L. G., Dear, B. F., Nielssen, O., & Titov, N. (2025). Characteristics and Treatment Outcomes for Patients of a Digital Psychology Service in Regional and Remote Parts of Australia. *The Australian Journal of Rural Health, 33*(2), e70032. DOI: 10.1111/ajr.70032 PMID: 40110934

Steiner, A., Farmer, J., Kamstra, P., Carlisle, K., McCosker, A., & Kilpatrick, S. (2023). Online mental health forums and rural resilience: Mixed methods study and logic model. *JMIR Mental Health, 10*, e47459. DOI: 10.2196/47459 PMID: 37379080

Tan, E. Y., Wee, R. R., Ern, S. Y., Heng, K. J., Chin, J. W., Tong, E. M., & Liu, J. C. (2020). Tracking private WhatsApp discourse about COVID-19: A longitudinal infodemiology study in Singapore. *medRxiv*. https://doi.org/DOI: 10.1101/2020.09.29.20203646

Think Global Health. (2025). *Mental health after India's glacial floods.* https://www.thinkglobalhealth.org/article/mental-health-after-indias-glacial-floods

United Nations. (2023). *What is climate change?* United Nations. Retrieved July 1, 2025, from https://www.un.org/en/climatechange/what-is-climate-change

World Health Organization. (2021). *Global strategy on digital health 2020–2025.* https://www.who.int/publications/i/item/9789240020924

World Health Organization. (2022). WHO digital mental health intervention effective in reducing depression among Syrian refugees in Lebanon. https://www.who.int/news-room/feature-stories/detail/who-digital-mental-health-intervention-effective-in-reducing-depression-among-syrian-refugees-in-lebanon?

World Health Organization. (2023). Climate change and non-communicable diseases: Connections. World Health Organization. https://www.who.int/news/item/02-11-2023-climate-change-and-noncommunicable-diseases-connections

ADDITIONAL READING

European Climate and Health Observatory. (2022). *Mental health effects*. https://climate-adapt.eea.europa.eu/en/observatory/evidence/health-effects/mental-health-effects/mental-health-effects

Think Global Health. (2025). *Mental health after India's glacial floods*. https://www.thinkglobalhealth.org/article/mental-health-after-indias-glacial-floods

KEY TERMS AND DEFINITIONS

Climate Change: Long-term shifts in temperatures and weather patterns, often driven by human activities such as burning fossil fuels, deforestation, and industrial agriculture. These changes manifest as both gradual environmental shifts and sudden, extreme events like floods, wildfires, and heatwaves. Climate change serves as the overarching driver of the mental health challenges discussed in this paper, acting as a "threat multiplier" by exacerbating existing social, economic, and health vulnerabilities.

Climate-Affected Populations: Communities whose livelihoods, health, and well-being are significantly affected by climate-related events such as floods, wildfires, droughts, rising temperatures, and extreme weather. These populations are a central focus of the paper, as their vulnerability to environmental disruptions often intersects with limited access to mental health care.

Cultural Adaptation: The modification of interventions to reflect the cultural norms, values, languages, and lived realities of the target community. Cultural alignment is critical for engagement and long-term effectiveness of digital mental health tools.

Digital Divide: The disparity between individuals or communities in access to reliable internet, devices, and digital literacy skills. The paper identifies this as a central barrier, as many climate-affected populations lack the infrastructure needed to benefit from digital interventions.

Eco-Anxiety: Persistent fear, distress, or anxiety about environmental decline and the long-term impacts of climate change. This is one of the emerging psychological impacts of climate change, particularly affecting younger demographics and environmentally vulnerable communities.

Messaging Services for Mental Health: The use of instant messaging platforms or chatbot-based systems to deliver mental health outreach, crisis support, and psychoeducation. These tools are found to be adaptable and accessible but may carry risks such as information overload or inadequate emergency response.

Mobile Mental Health Applications (mHealth apps): Smartphone-based tools offering psychological support through features like self-guided therapy exercises, mindfulness activities, and educational resources. Case studies in the paper show their potential for scalable, cost-effective mental health support, alongside limitations such as low engagement and accessibility issues.

Peer-Led Platforms / Online Support Groups: Virtual spaces where individuals with shared experiences offer each other emotional support, coping strategies, and solidarity. The paper notes these as effective for fostering trust, belonging, and resilience, particularly in rural or conflict-affected settings.

Preparedness Frameworks: Strategic plans that incorporate interventions—such as mental health services—into disaster readiness and response systems. The paper highlights that many digital tools are introduced reactively; embedding them into preparedness frameworks could improve uptake and impact.

Psychosocial Interventions: Strategies designed to address both the psychological and social dimensions of well-being, aiming to reduce distress, strengthen coping mechanisms, and foster resilience. This study examines how technology-enabled versions of these interventions can be scaled and adapted for disaster-affected contexts.

Telehealth / Tele-Counselling: The provision of healthcare, including mental health services, through telecommunications technology such as video calls, phone lines, or secure messaging. This mode is critical for maintaining continuity of mental health care when physical access is disrupted by climate-related disasters.

Chapter 7
Evaluating Policies and Initiatives and Proposing Sustainable Solutions for Mental Health of Child Labor

Nishant Goyal
https://orcid.org/0009-0006-3314-3544
Graphic Era University, India

Ayushi Bhatt
https://orcid.org/0009-0007-1115-3749
Graphic Era University, India

Bimal Anjum
DAV College, India

Rakesh Kumar
https://orcid.org/0000-0002-8040-1976
Graphic Era University, India

Rajesh Tiwari
https://orcid.org/0000-0002-5345-2508
Graphic Era University, India

ABSTRACT

Child labour especially among rag pickers in India remains a critical issue, influenced by factors such as Poverty, education and grim reality of societal landscape

DOI: 10.4018/979-8-3373-3281-9.ch007

remains unchanged. Despite of various initiatives by government, societies, NGO's and industry the situation is still a far cry. Mental health of rag pickers is often ignored by stakeholders. This study examines the mental health of rag pickers and policy initiatives. The objective is to realise the ground level scenario and identify the gaps. This study used Scopus database for the review. One hundred twenty-two documents were found in Scopus database. It is found that there is a gap between the expected results and achieved results of various initiative and how they fail to address the root cause like mental health and economic vulnerability. Findings emphasize the need for a multi-stakeholder strategy involving government, civil society, and local communities to create viable alternatives for rag pickers.

INTRODUCTION

In developing nations, policy reforms must integrate mental health considerations into labor regulations, mandating psychosocial risk assessments alongside traditional occupational safety measures. Participatory frameworks that engage workers in policy design and implementation can enhance agency and reduce workplace-related anxiety. The COVID-19 pandemic magnified these disparities, particularly in platform-based economies, highlighting the urgency of universal social protections that acknowledge employment conditions as determinants of mental well-being. (Muntaner et al., 2010).

Children subjected to hazardous labor environments exhibit pronounced physical and cognitive deficits, including stunted growth, micronutrient deficiencies, and impaired neurodevelopment. Chronic activation of stress pathways induces cellular damage, with studies documenting accelerated biological aging in this population. Traditional abolitionist campaigns frequently falter by disregarding the economic necessities that compel families to rely on child labor.

Sustainable eradication demands a dual focus on immediate harm reduction and systemic poverty alleviation. Biomarker surveillance can identify high-risk individuals for targeted medical and psychological interventions, while cash-transfer programs and accessible schooling offer viable alternatives to labor exploitation. Effective policy must bridge health and economic sectors, ensuring interventions address root causes rather than symptoms (Ibrahim et al., 2019). The aggregation of risk factors—including violence, material deprivation, and household instability—synergistically increases susceptibility to severe and persistent mental health conditions. Developmental timing plays a pivotal role, with early-life exposures exerting disproportionate neurobiological and behavioral consequences. Existing diagnostic tools frequently misclassify or overlook distress in marginalized communities due to culturally distinct symptom presentations. Culturally grounded assessment instruments, co-developed with local populations, are imperative for accurate identification and intervention. Holistic treatment models must account for the dynamic interplay between individual resilience and environmental stressors, adapting evidence-based practices to resource-constrained settings without compromising efficacy (Benjet, 2010).

Perinatal psychological distress manifests and is interpreted variably across cultures, with many societies attributing symptoms to spiritual imbalances rather than biomedical causes. Help-seeking pathways frequently prioritize traditional healers over clinical providers, reflecting deeply entrenched belief systems. These cultural narratives can serve as protective buffers or exacerbating factors, depending on community support structures.

Bridging interventions, such as training traditional birth attendants in basic mental health first aid or deploying bilingual community health workers, can harmonize indigenous and biomedical paradigms. Such "two-eyed seeing" approaches foster trust and improve treatment adherence while respecting cultural autonomy (Hanlon et al., 2009). In COVID-19 containment measures inadvertently intensified threats to at-risk children, with economic desperation driving surges in exploitation, trafficking, and severe neglect. Lockdowns disrupted protective infrastructures—schools, social services, and community networks—leaving abused children without recourse. The cascading effects include not only immediate trauma but also enduring developmental setbacks that may compromise lifelong functioning. Strengthening child protection systems requires real-time risk surveillance technologies, cross-sectoral coordination protocols, and trauma-informed emergency response teams. Investment in these infrastructures is critical to mitigate the pandemic's secondary casualties (Ramaswamy & Seshadri, 2020).

The COVID-19 pandemic triggered an unprecedented mental health crisis among children worldwide, with lockdown measures creating profound psychological impacts that often outweighed the virus's direct health risks. Prolonged school closures (averaging 22 weeks globally), disrupted routines, and social isolation led to alarming spikes in anxiety (73% increase), depression (57% rise), and sleep disorders (68% prevalence) among pediatric populations, according to WHO data. Particularly vulnerable were children from low-income families, who faced compounded stressors including food insecurity (39% in developing nations), domestic violence (143% surge), and digital exclusion (463 million unable to access remote learning). The pandemic's psychological toll manifested through developmental regressions in younger children, including delayed language acquisition and social skills, while adolescents grappled with disrupted identity formation and severed peer connections during critical growth periods. Excessive screen time during lockdowns also created new challenges, with a 210% rise in internet addiction cases according to clinical reports. These impacts have proven enduring, as evidenced by teachers reporting widespread behavioral issues and learning difficulties during school reopenings. The crisis exposed systemic gaps in child mental health support, necessitating comprehensive solutions including trauma-informed school programs (shown to reduce symptoms by 42%), expanded telehealth services, and community-based initiatives that strengthen family support systems. Longitudinal studies now warn of potential lifelong consequences, with affected children facing elevated risks of chronic mental health conditions and reduced adult productivity. This collective trauma underscores the urgent need for sustained investment in pediatric mental health infrastructure and the integration of psychological support into future pandemic preparedness plans, as the pandemic's mental health legacy continues to shape an entire generation's wellbeing and development prospects (Fekadu et al., 2006). Without such targeted measures, these children face lifelong consequences—studies indicate 58% higher rates of adult unemployment and 3.2x greater likelihood of intergenerational labor transmission (Sturrock and Hodes, 2016). Future efforts must prioritize community-led designs, digital surveillance tools, and trauma-informed health systems to achieve equitable protection (Minayo et al., 2018). Figure 1 depicts the status child labor in India from 2001 to 2011 census. Children in employment, total percentage of population between 7-14 years is depicted in figure 2. Agriculture sector contributes 70% in child labor as shown in figure 3.

Figure 1. Number of working people (Age 5-14)

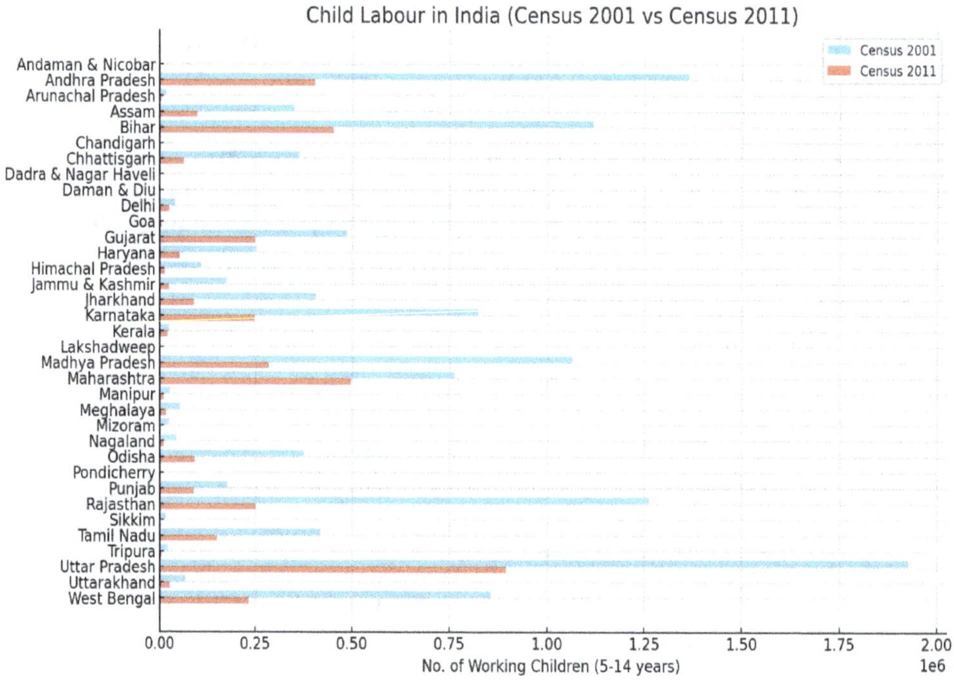

Figure 2. Children in employment, total percentage of population, age 7-14

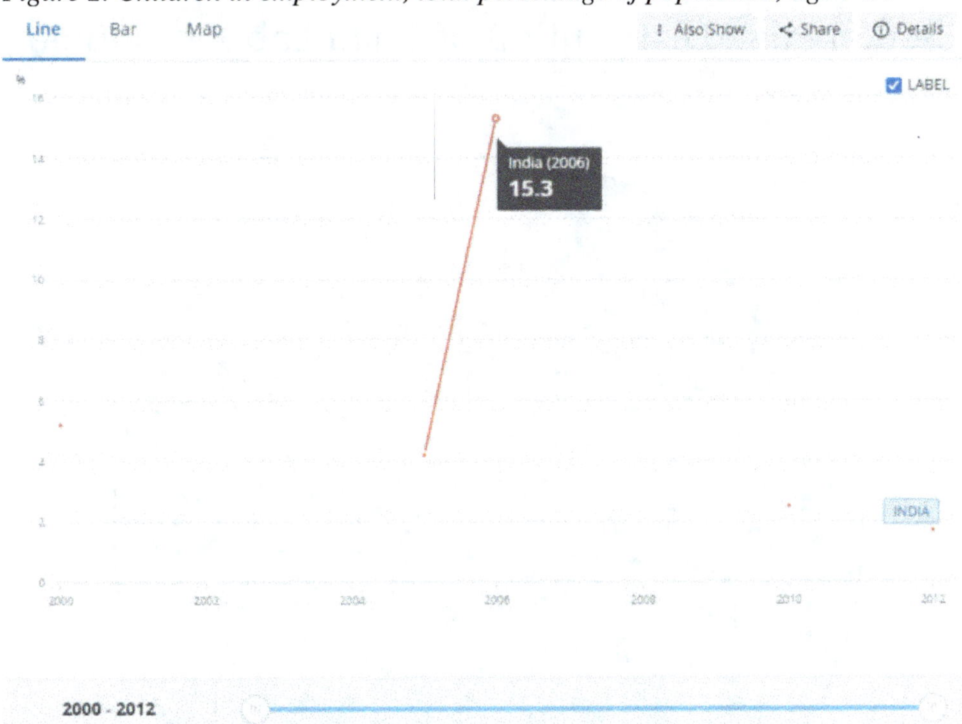

Figure 3. Sector-wise Child Labour Distribution 2020

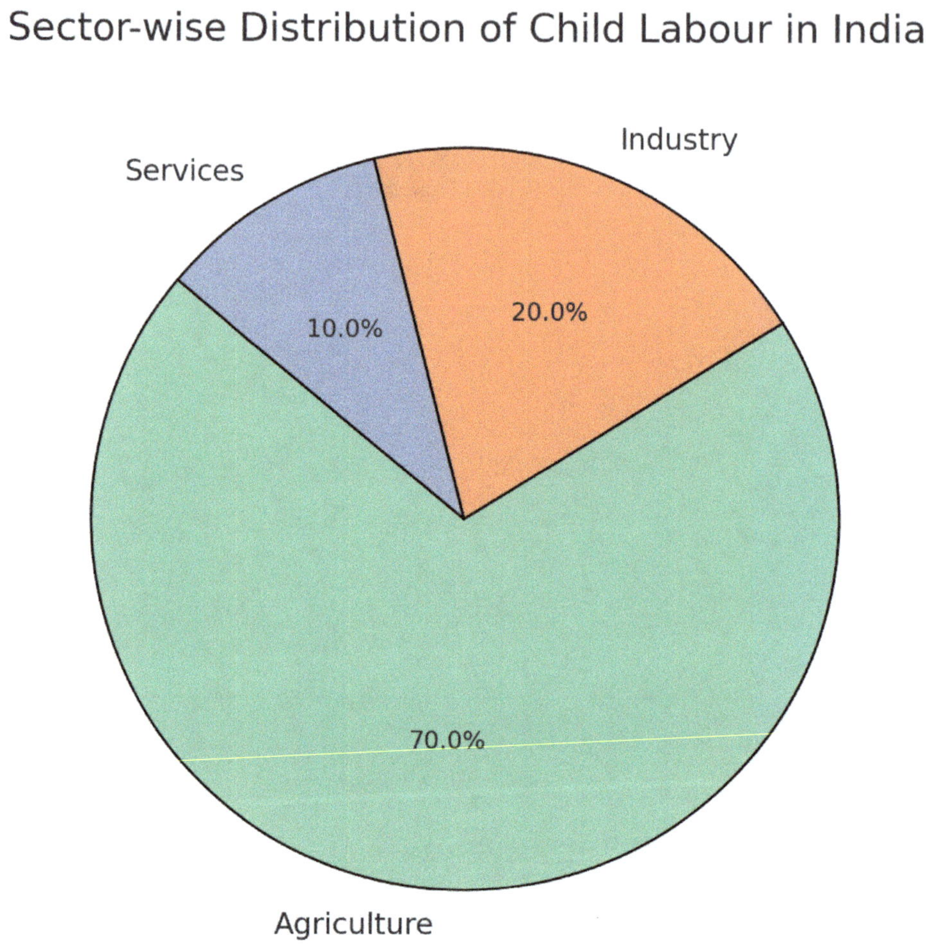

Figure 4. Allocations and expenditure in major schemes for child protection (in Rs crore)

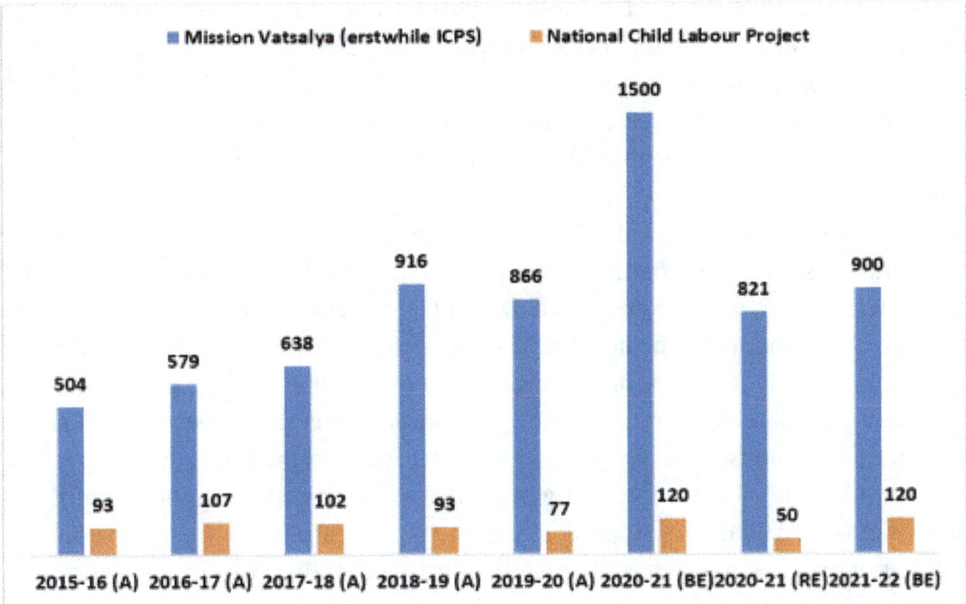

It is found that here has a been a gap in implementation of the schemes and unexpected events like COVID has only increase the issue of Child Labour in India as depicted in figure 4.

The worldwide health emergency triggered by COVID-19 produced profound psychological consequences for children and adolescents, with containment strategies such as lockdowns and social distancing contributing to mental health challenges that often exceeded the direct threats posed by the virus itself. Extended school closures, coupled with severe restrictions on social interactions and outdoor activities, led to a dramatic surge in anxiety-related disorders among pediatric populations. Children with deprived backgrounds belonged to low economic status. it is not proportional and as a result is affected and face high risks of experiencing domestic violence. Stress caused by parents, stress caused by food insecurity because of money. Single parent families were under extra pressure and worsened emotional disorders. The Key interruption of regular social and educational settings impaired the ability to developmental processes, which lead to high levels of depressive symptoms, insomnia, unbalanced sleeping habits, and unhealthy digital media use in a variety of socioeconomic groups.

In an effort to curb these impacts a multi-tier approach is necessary as opposed to corporation with affect sensitive principles to telehealth services that are modelled using support of the local's systems have made community to become change adoptive, and has corrected the academic structures. throw up support to mental health along with learning structures. Emerging longitudinal research has shown that long-term effects of being isolated and unstable are psychologically. not just be limited to short term effects, and may have overall effect on the future generations due to club behavioural changes. This makes the case of timely even more crucial. Scalable interventions that cure both the acute distress and the long-term recovery (Ghosh et al., 2020).

Exposure of children to armed conflict subjects them to an unusual combination of traumas, as they struggle through not only violence in the broader society, but also physical and sexual abuse within the confines of their own families, which can restructure neurodevelopment processes. Neuroscience shows us that this kind of dual exposure triggers measurable changes in the parts of the brain that manage emotions, calculate memory, and managing stress. The case of macro level warfare and micro level household dysfunction makes it tough to intervene therapeutically, since regular one on one counselling often falls short in managing these multilevel disadvantages. The unique approaches to the treatment are the integrative modes highlighting the integration of economic empowerment of the provider, e.g., vocational reclamation and microfinancing, along with trauma recovery-based interventions, which embrace traditional healing cultures. Intervention means that address structural inequities and interpersonal wounds alternatively, which are found to be useful in proving high efficacy, i.e., community-based approaches in comparison with isolated clinical therapies. This fact demands a paradigmatic reformation of post conflict mental health practice, which will emphasize culturally responsive, system-based interventions as opposed to disintegrated, individual treatments (Catani et al., 2008).

Unstable work situations accompanied by insecure wages, non-existent benefits-appear to be gaining ground in the 21st century. and workplace exploitation produces lifelong physiological and psychological stress, which increases the probability of developing both somatic, as well as psychiatric disorders. This is worsened by asymmetry of power inherent in informal and gig economies. susceptibility, which leaves workers with little to do in retaliation to exploitation. Empirical data consistently correlate job insecurity with heightened distress levels, reinforcing the protective role of union representation, living wages, and contractual stability.

METHODOLOGY

The Scopus database was used for the bibliometric analysis. The keywords used were "mental health" AND "child labor". The date of search was 13th May 2025. 122 documents were obtained in Scopus database. Only English language documents were retained due to familiarity of authors with English language. 117 documents were retained. The documents from 1973 to 2025 were considered. VOS viewer software was used for the bibliometric analysis.

DISCUSSION

By keeping minimum no of documents of an author to "1" with minimum number of citations to '20'. We get total number of "110" authors out of "471". The highest cited authors are "CHATTERJEE, SUBHANKAR; DUBEY, MAHUA J; DUBEY, SOUVIK; GHOSH, RITWIK" with "373" citations. The lowest cited authors are "CHAUDHURI, SRIROOP; EMENDACK, YVES; MCDONALD, LOUIS M; ROY, MIMI; SHARMA, NARAYAN PRASAD; THI, AYE MYAT; TRINH, TRONG-ANH" with maximum "20" citations.

Table 1. Prominent cited authors

ID	Author	documents	citations	total link strength
1	Chatterjee, Subhankar	1	373	0
2	Dubey, Mahua J.	1	373	0
3	Dubey, Souvik	1	373	0
4	Ghosh, Ritwik	1	373	0
*	Catani, Claudia	1	126	0
6	Neuner, Frank	1	126	0
7	Schauer, Elisabeth	1	126	0
8	Alem, Atalay	3	111	41
9	Benach, Joan	1	111	2
10	Castedo, Antía	1	111	2
11	Kim, Il-Ho	1	111	2
12	Martínez, José Miguel	1	111	2
13	Muntaner, Carles	1	111	2
14	Santana, Vilma	1	111	2
15	Solar, Orielle	1	111	2
16	Vanroelen, Christophe	1	111	2
17	Vergara, Montserrat	1	111	2
18	Abdalla, Salma M.	1	105	11
19	Abdelgadir, Jihad	1	105	11

continued on following page

Table 1. Continued

ID	Author	documents	citations	total link strength
20	De Vries, Nanne	1	105	11
21	Ibrahim, Abdalla	1	105	11
22	Jafer, Mohammed	1	105	11
23	Benjet, Corina	1	86	10
24	Kumari, Veena	5	67	76
25	Fekadu, Daniel	2	59	41
26	Hägglöf, Bruno	2	59	41
27	Niraula, Shanta	4	57	67
28	Pandey, Rakesh	4	57	67
29	Hanlon, Charlotte	1	52	0
30	Prince, Martin	1	52	0
31	Seshadri, Shekhar	2	52	3
32	Whitley, Rob	1	52	0
33	Wondimagegn, Dawit	1	52	0
34	Pocock, Nicola S.	3	51	16
35	Ramaswamy, Sheila	1	51	3
36	Bermudez, Laura Gauer	2	43	2
37	Hodes, Matthew	1	40	23
38	Sturrock, Sarah	1	40	23
39	Da Silva, Marta Maria Alves	1	39	0
40	De Assis, Simone Gonçalves	1	39	0
41	De Souza, Edinilsa Ramos	1	39	0
42	Minayo, Maria Cecilia De Souza	1	39	0
43	Gupta, Shulka	2	37	31
44	Singh, Tushar	2	37	36
45	Arya, Yogesh Kumar	1	36	17
46	Gupta, Rajendra Prasad	1	36	17
47	Lau, Jennifer Yun Fai	1	36	17
48	Mishra, Ramesh Chandra	1	36	17
49	Shukla, Meenakshi	1	36	17
50	Upadhyay, Aakanksha	1	36	17
51	Adi, Suher	1	35	0
52	Arafa, Rahma	1	35	0
53	Mallat, Nour K.	1	35	0

continued on following page

Table 1. Continued

ID	Author	documents	citations	total link strength
54	Rizkalla, Niveen	1	35	0
55	Segal, Steven P.	1	35	0
56	Soudi, Laila	1	35	0
57	Zimmerman, Cathy	4	35	18
58	Ssewamala, Fred M.	2	34	2
59	Asemota, Osamagbe	1	31	0
60	Kyeremateng, Rosina	1	31	0
61	Oguda, Lulu	1	31	0
62	Bhatia, Amiya	2	30	0
63	Karimli, Leyla	1	30	2
64	Neilands, Torsten B.	1	30	2
65	Wells, Christine R.	1	30	2
66	Abdulaziz, Hussam	1	27	0
67	Arenliu, Aliriza	1	27	0
68	Bertelsen, Nathan	1	27	0
69	Saad, Rahaf	1	27	0
70	Weine, Stevan Merrill	1	27	0
71	Cappa, Claudia	1	25	0
72	Devries, Karen	1	25	0
73	Fabbri, Camilla	1	25	0
74	Greenbaum, Jordan	3	25	10
75	Guedes, Alessandra	1	25	0
76	Jugder, Munkhbadar	1	25	0
77	Petzold, Max	1	25	0
78	Scott, Natasha B.	1	25	16
79	Araya, Ricardo	2	24	0
80	Blum, Austin	1	24	4
81	Evans-Lacko, Sara	2	24	0
82	Garman, Emily	2	24	0
83	Gaveras, Eleni	1	24	4
84	Hessel, Philipp	2	24	0
85	Ismayilova, Leyla	1	24	4
86	Kinney, Joseph A.	1	24	0
87	Lund, Crick	2	24	0

continued on following page

Table 1. Continued

ID	Author	documents	citations	total link strength
88	Nanema, Rachel	1	24	4
89	T -Camier, Alexice	1	24	4
90	Ranganathan, Meghna	3	23	18
91	Spitzer, Manfred	1	23	0
92	Beck, Dana C.	1	22	3
93	Choi, Kristen R.	1	22	3
94	Lori, Jody R.	1	22	3
95	Munro-Kramer, Michelle L.	1	22	3
96	Cividanes, Giuliana C.	1	21	3
97	Dhakal, Sandesh	3	21	56
98	Fossaluza, Victor	1	21	3
99	Maciel, Mariana R.	1	21	3
100	Mari, Jair J.	1	21	3
101	Mello, Andrea F.	1	21	3
102	Mello, Marcelo F.	1	21	3
103	Nobrega, Luciana P.	1	21	3
104	Chaudhuri, Sriroop	*1	20	0
105	Emendack, Yves	1	20	0
106	Mcdonald, Louis M.	1	20	0
107	Roy, Mimi	1	20	0
108	Sharma, Narayan Prasad	2	20	37
109	Thi, Aye Myat	2	20	10
110	Trinh, Trong-Anh	1	20	3

Figure 5. Citation map of authors

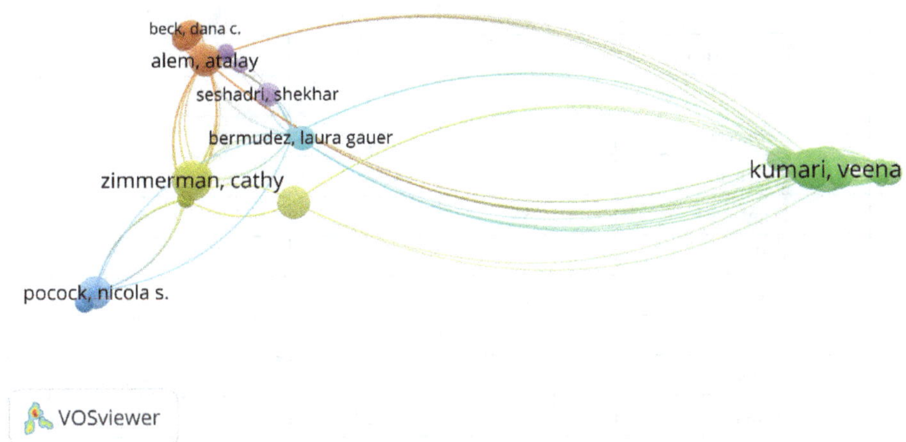

(Created by authors from Scopus Database)

By keeping minimum no of documents of countries to "1" with minimum number of citations to '20'. We get total number of "24" countries out of "48". The highest cited country is "India" with "555" citations and the lowest cited countries are "Myanmar" with "20" citations.

Table 2. Prominent cited countries

S.No.	Country	Documents	Citations	Total Link Strength
1	India	20	555	45
2	United States	40	503	58
3	United Kingdom	34	488	105
4	Germany	4	165	3
5	Netherlands	5	141	29
6	Ethiopia	5	133	41
7	Vivo	1	126	0
8	Vivo International	1	126	0
9	Saudi Arabia	1	105	17
10	Brazil	7	91	6
11	Mexico	1	86	6

continued on following page

Table 2. Continued

S.No.	Country	Documents	Citations	Total Link Strength
12	Sweden	4	85	34
13	Australia	6	60	26
14	Nepal	4	57	28
15	South Africa	4	43	0
16	Turkey	3	43	1
17	Nigeria	3	38	3
18	Kenya	2	31	4
19	Bangladesh	5	29	4
20	Colombia	3	29	0
21	Serbia	1	27	0
22	Italy	2	25	0
23	Burkina Faso	1	24	4
24	Myanmar	2	20	12

Figure 6. Citation map of countries

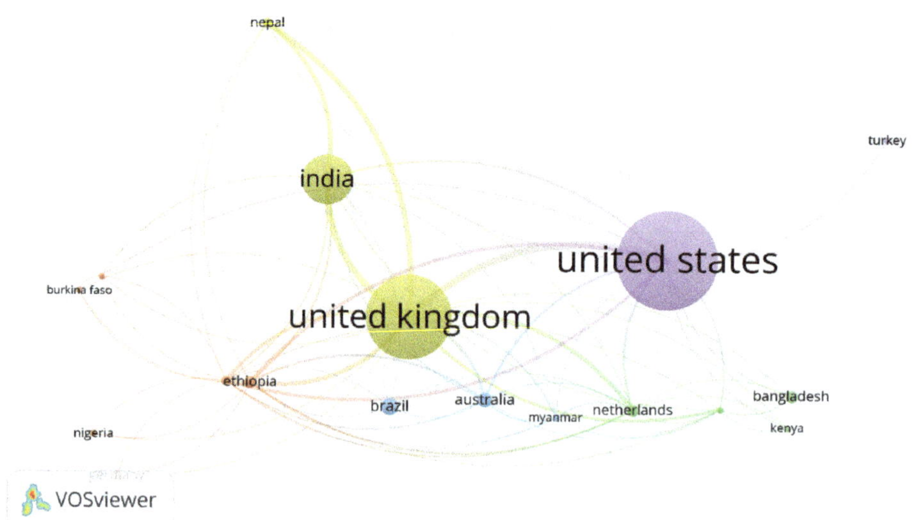

(Created by authors from Scopus Database)

By keeping minimum no of citation to '10'. We get total number of "42" documents out of "117". The highest cited document is "GHOSH" with "373" citations and the lowest cited document is "RASLAN" with "10" citations.

Table 3. Prominent cited documents

Id	Documents	Citations	Links
1	Ghosh (2020)	373	0
2	Catani (2008)	126	0
3	Muntaner (2010)	111	1
4	Ibrahim (2019)	105	3
5	Benjet (2010)	86	3
6	Hanlon (2009)	52	0
7	Ramaswamy (2020)	51	2
8	Fekadu (2006)	51	8
9	Sturrock (2016)	40	5
10	Minayo (2018)	39	0
11	Pandey (2020)	36	2
12	Rizkalla (2020)	35	0
13	Kyeremateng (2022)	31	0
14	Karimli (2019)	30	1
15	Arenliu (2020)	27	0
16	Scott (2021)	25	3
17	Fabbri (2021)	25	0
18	Ismayilova (2016)	24	2
19	Kinney (1993)	24	0
20	Spitzer (2021)	23	0
21	Beck (2017)	22	0
22	Maciel (2013)	21	1
23	Chaudhuri (2021)	20	0
24	Trinh (2020)	20	1
25	Dhakal (2019)	19	2
26	Wright (2021)	19	0
27	Vostanis (2017)	18	0
28	Rellstab (2020)	18	0
29	Greenbaum (2023)	16	2
30	Wollburg (2023)	16	0

continued on following page

Table 3. Continued

Id	Documents	Citations	Links
31	Zimmerman (2022)	16	0
32	Feeny (2021)	15	2
33	Meyer (2020)	15	2
34	Kishore (2012)	15	0
35	Getachew (2011)	15	0
36	Nusrat (2021)	14	0
37	Thi (2022)	14	0
38	Bermudez (2020)	13	1
39	Caglayan (2010)	13	1
40	Nodzenski (2020)	12	0
41	Aransiola (2018)	11	1
42	Raslan (2021)	10	1

Figure 7. Citation map of documents

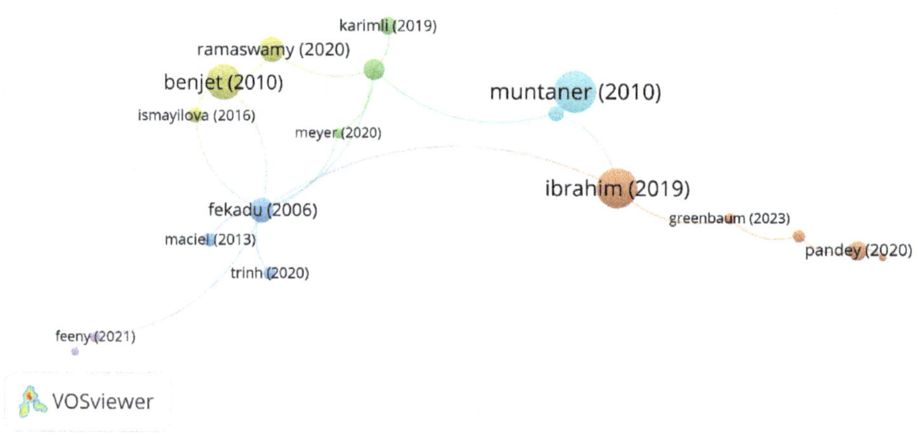

(Created by authors from Scopus Database)

By keeping minimum no of documents of an organization to "1" with minimum number of citations to '20'. We get total number of "74" organizations out of "350". The highest cited organizations are "[Department of General Medicine, Burdwan Medical College and Hospital, Burdwan, India]; [Department of General Medicine, Rajendra Institute of Medical Sciences, Ranchi, India]; [Department of

Neuromedicine, Bangur Institute of Neurosciences, Sskm Hospital, Kolkata, India]; [Department of Psychiatry, Behrampore Mental Hospital, Berhampore, India] with "373" citations.

Figure 8. Citation map of organizations

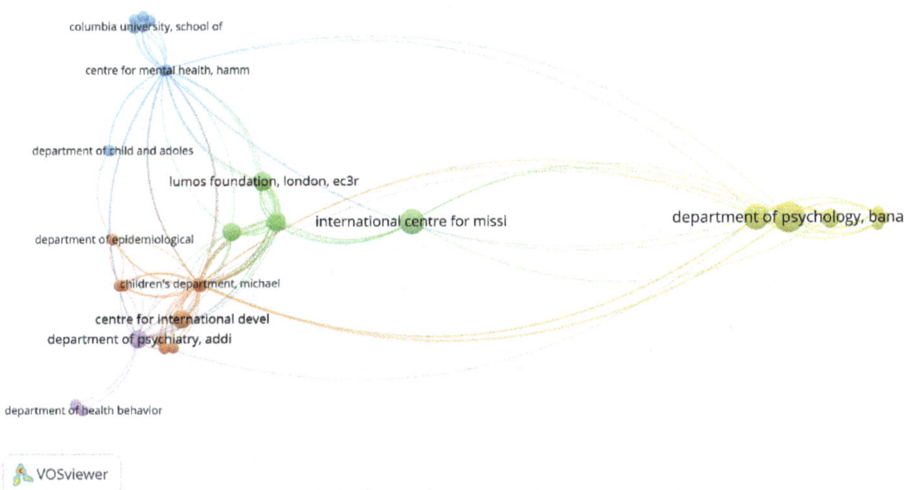

(Created by authors from Scopus Database)

By keeping the minimum no. of occurrences of a keyword to "10". Out of "1370" only "38" meet the threshold. The keyword with most occurrences is "HUMAN" having "87" occurrences where as the keywords with lowest occurrences are "CHILD WELFARE"; "HEALTH STATUS"; "INDIA" with "10" occurrences.

Table 4. All keywords

S. No.	Keyword	Occurrences	Total Link Strength
1	Human	87	872
2	Mental Health	81	754
3	Child Labor	76	706
4	Child	72	777
5	Humans	60	666
6	Adolescent	53	614
7	Female	51	598

continued on following page

Table 4. Continued

S. No.	Keyword	Occurrences	Total Link Strength
8	Article	48	574
9	Male	46	560
10	Adult	27	339
11	Major Clinical Study	24	307
12	Child Abuse	23	282
13	Psychology	23	290
14	Controlled Study	22	275
15	Risk Factor	20	225
16	Depression	18	199
17	Poverty	18	160
18	Education	17	166
19	Child Health	16	173
20	Mental Disease	16	194
21	Anxiety	14	141
22	Child Labour	13	111
23	Pandemic	13	136
24	Prevalence	13	169
25	Review	13	114
26	Coronavirus Disease 2019	12	133
27	Covid-19	12	105
28	Cross-Sectional Study	12	167
29	Violence	12	124
30	Employment	11	125
31	Epidemiology	11	149
32	Human Trafficking	11	107
33	Posttraumatic Stress Disorder	11	145
34	Questionnaire	11	147
35	Young Adult	11	156
36	Child Welfare	10	107
37	Health Status	10	137
38	India	10	72

Figure 9. Co-occurrence map of all keywords

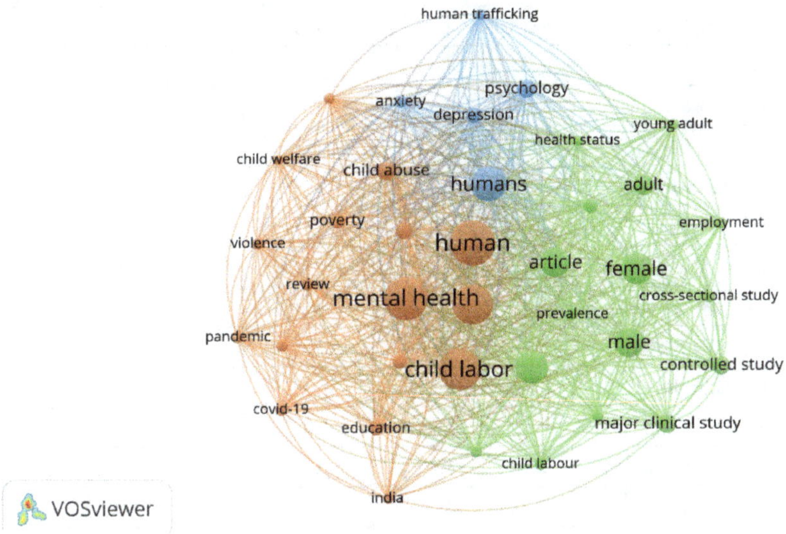

(Created by authors from Scopus Database)

By keeping minimum no. of documents of an author to "1" and the minimum no of citation to "20". Out of "471" only "110" met the threshold and out of which only '13' are connected. "Chatterjee, Subhankar; Dubey, Mahua J; Dubey, Souvik; Ghosh, Ritwik" with "373" citations. The lowest cited authors are "Chaudhuri, Sriroop; Emendack, Yves; Mcdonald, Louis M; Roy, Mimi; Sharma, Narayan Prasad; Thi, Aye Myat; Trinh, Trong-Anh" with maximum "20" citations.

Table 5. Co- authorship table of authors

Id	Author	Documents	Citations	Total Link Strength
1	Chatterjee, Subhankar	1	373	3
2	Dubey, Mahua J.	1	373	3
3	Dubey, Souvik	1	373	3
4	Ghosh, Ritwik	1	373	3
5	Catani, Claudia	1	126	2
6	Neuner, Frank	1	126	2
7	Schauer, Elisabeth	1	126	2
8	Alem, Atalay	3	111	8
9	Benach, Joan	1	111	8

continued on following page

Table 5. Continued

Id	Author	Documents	Citations	Total Link Strength
10	Castedo, Antía	1	111	8
11	Kim, Il-Ho	1	111	8
12	Martínez, José Miguel	1	111	8
13	Muntaner, Carles	1	111	8
14	Santana, Vilma	1	111	8
15	Solar, Orielle	1	111	8
16	Vanroelen, Christophe	1	111	8
17	Vergara, Montserrat	1	111	8
18	Abdalla, Salma M.	1	105	4
19	Abdelgadir, Jihad	1	105	4
20	De Vries, Nanne	1	105	4
21	Ibrahim, Abdalla	1	105	4
22	Jafer, Mohammed	1	105	4
23	Benjet, Corina	1	86	0
24	Kumari, Veena	5	67	23
25	Fekadu, Daniel	2	59	4
26	Hägglöf, Bruno	2	59	4
27	Niraula, Shanta	4	57	23
28	Pandey, Rakesh	4	57	23
29	Hanlon, Charlotte	1	52	4
30	Prince, Martin	1	52	4
31	Seshadri, Shekhar	2	52	1
32	Whitley, Rob	1	52	4
33	Wondimagegn, Dawit	1	52	4
34	Pocock, Nicola S.	3	51	5
35	Ramaswamy, Sheila	1	51	1
36	Bermudez, Laura Gauer	2	43	4
37	Hodes, Matthew	1	40	1
38	Sturrock, Sarah	1	40	1
39	Da Silva, Marta Maria Alves	1	39	3
40	De Assis, Simone Gonçalves	1	39	3
41	De Souza, Edinilsa Ramos	1	39	3
42	Minayo, Maria Cecilia De Souza	1	39	3
43	Gupta, Shulka	2	37	15

continued on following page

Table 5. Continued

Id	Author	Documents	Citations	Total Link Strength
44	Singh, Tushar	2	37	14
45	Arya, Yogesh Kumar	1	36	10
46	Gupta, Rajendra Prasad	1	36	10
47	Lau, Jennifer Yun Fai	1	36	10
48	Mishra, Ramesh Chandra	1	36	10
49	Shukla, Meenakshi	1	36	10
50	Upadhyay, Aakanksha	1	36	10
51	Adi, Suher	1	35	5
52	Arafa, Rahma	1	35	5
53	Mallat, Nour K.	1	35	5
54	Rizkalla, Niveen	1	35	5
55	Segal, Steven P.	1	35	5
56	Soudi, Laila	1	35	5
57	Zimmerman, Cathy	4	35	7
58	Ssewamala, Fred M.	2	34	4
59	Asemota, Osamagbe	1	31	2
60	Kyeremateng, Rosina	1	31	2
61	Oguda, Lulu	1	31	2
62	Bhatia, Amiya	2	30	6
63	Karimli, Leyla	1	30	4
64	Neilands, Torsten B.	1	30	4
65	Wells, Christine R.	1	30	4
66	Abdulaziz, Hussam	1	27	4
67	Arenliu, Aliriza	1	27	4
68	Bertelsen, Nathan	1	27	4
69	Saad, Rahaf	1	27	4
70	Weine, Stevan Merrill	1	27	4
71	Cappa, Claudia	1	25	6
72	Devries, Karen	1	25	6
73	Fabbri, Camilla	1	25	6
74	Greenbaum, Jordan	3	25	0
75	Guedes, Alessandra	1	25	6
76	Jugder, Munkhbadar	1	25	6
77	Petzold, Max	1	25	6

continued on following page

Table 5. Continued

Id	Author	Documents	Citations	Total Link Strength
78	Scott, Natasha B.	1	25	1
79	Araya, Ricardo	2	24	8
80	Blum, Austin	1	24	4
81	Evans-Lacko, Sara	2	24	8
82	Garman, Emily	2	24	8
83	Gaveras, Eleni	1	24	4
84	Hessel, Philipp	2	24	8
85	Ismayilova, Leyla	1	24	4
86	Kinney, Joseph A.	1	24	0
87	Lund, Crick	2	24	8
88	Nanema, Rachel	1	24	4
89	T -Camier, Alexice	1	24	4
90	Ranganathan, Meghna	3	23	6
91	Spitzer, Manfred	1	23	0
92	Beck, Dana C.	1	22	3
93	Choi, Kristen R.	1	22	3
94	Lori, Jody R.	1	22	3
95	Munro-Kramer, Michelle L.	1	22	3
96	Cividanes, Giuliana C.	1	21	6
97	Dhakal, Sandesh	3	21	13
98	Fossaluza, Victor	1	21	6
99	Maciel, Mariana R.	1	21	6
100	Mari, Jair J.	1	21	6
101	Mello, Andrea F.	1	21	6
102	Mello, Marcelo F.	1	21	6
103	Nobrega, Luciana P.	1	21	6
104	Chaudhuri, Sriroop	1	20	3
105	Emendack, Yves	1	20	3
106	Mcdonald, Louis M.	1	20	3
107	Roy, Mimi	1	20	3
108	Sharma, Narayan Prasad	2	20	9
109	Thi, Aye Myat	2	20	5
110	Trinh, Trong-Anh	1	20	0

Figure 10. Co-authorship map of authors

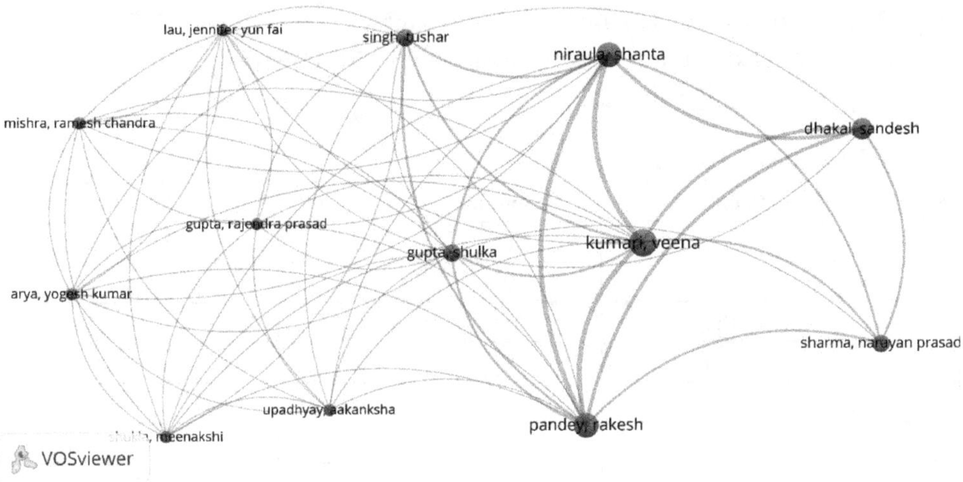

(Created by authors from Scopus Database)

By keeping minimum no. of documents of an organization to "1" and the minimum no of citation to "20". Out of "350" only "74" met the threshold and out of which only '6' are connected.

Figure 11. Co-authorship map of organizations

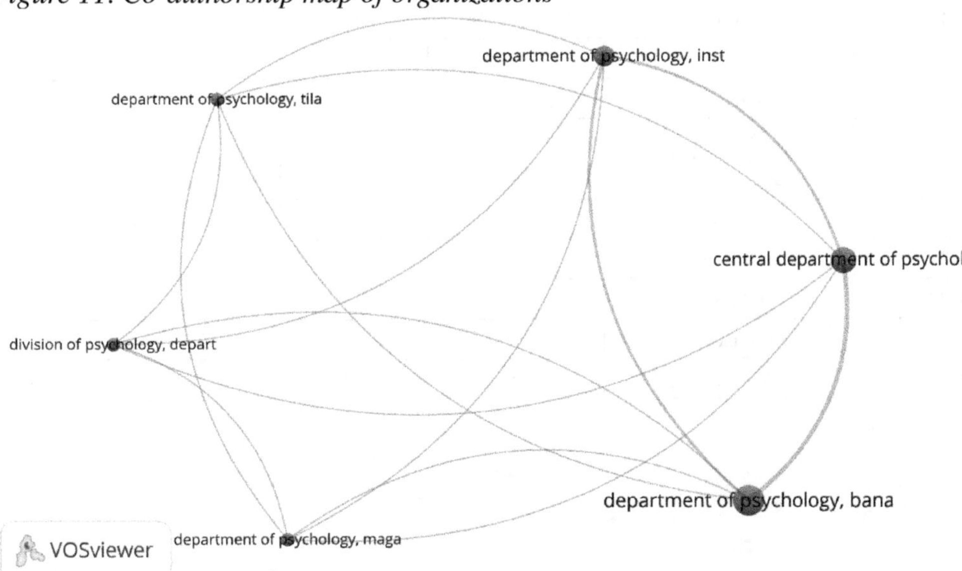

(Created by authors from Scopus Database)

By keeping minimum no. Of documents of a country to “1” and the minimum no of citation to “20”. Out of “48” only “24” met the threshold and out of which only ‘23’ are connected. The highest cited country is “India” with “555” citations and the lowest cited country is “Myanmar” with “20” citations.

Table 6. Co-author countries

S. No.	Country	Documents	Citations	Total Link Strength
1	India	20	555	15
2	United States	40	503	22
3	United Kingdom	34	488	43
4	Germany	4	165	3
5	Netherlands	5	141	9
6	Ethiopia	5	133	7
7	Vivo	1	126	2
8	Vivo International	1	126	2
9	Saudi Arabia	1	105	2
10	Brazil	7	91	7
11	Mexico	1	86	0
12	Sweden	4	85	7
13	Australia	6	60	6
14	Nepal	4	57	8
15	South Africa	4	43	11
16	Turkey	3	43	2
17	Nigeria	3	38	2
18	Kenya	2	31	5
19	Bangladesh	5	29	7
20	Colombia	3	29	9
21	Serbia	1	27	2
22	Italy	2	25	4
23	Burkina Faso	1	24	1
24	Myanmar	2	20	2

Figure 12. Co-authorship map of countries

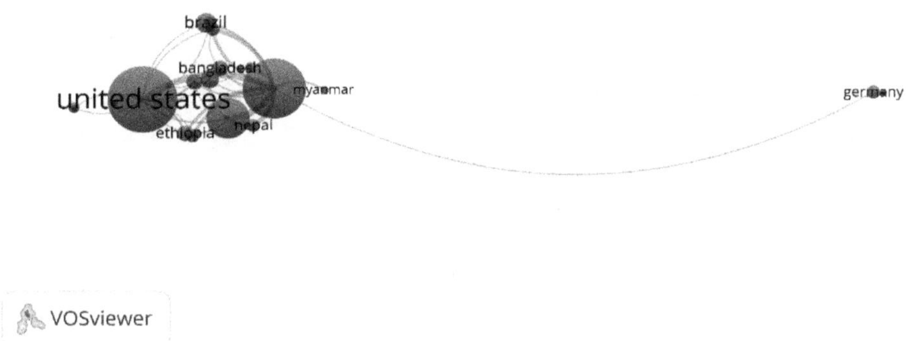

(Created by authors from Scopus Database)

By keeping the minimum no. of citations to "2". We get total number of "40" cited references out of "4877". The document with maximum no. of cited references is 'Fekadu D., Alem A., Hagglof B., The Prevalence of Mental Health Problems in Ethiopian Child Laborers, Journal of Child Psychology and Psychiatry, (2006)' with maximum "4" citations.

Figure 13. Co-citation map of references

(Created by authors from Scopus Database)

By keeping the minimum no. of citations to "10". We get total number of "29" cited sources out of "4877". The document with maximum no. of cited sources is "LANCET." with maximum "48" citations.

Table 7. Co-citation sources

S. No.	Sources	Citations	Total Link Strength
1	Lancet	48	395
2	Child Abuse Negl	45	674
3	Plos One	39	285
4	Pediatrics	36	466
5	Soc Sci Med	35	355
6	Child Abuse & Neglect	31	105
7	J Child Psychol Psychiatry	26	341
8	BMJ	21	213
9	Journal of Child Psychology and Psychiatry	21	105
10	BMC Public Health	19	164
11	Soc. Sci. Med.	17	48

continued on following page

Table 7. Continued

S. No.	Sources	Citations	Total Link Strength
12	Food Security	16	144
13	J Am Acad Child Adolesc Psychiatry	16	218
14	BMJ Open	15	75
15	Br J Psychiatry	15	206
16	World Development	15	157
17	Eur Child Adolesc Psychiatry	14	74
18	Jama	14	134
19	Social Science & Medicine	14	59
20	BMC Psychiatry	13	136
21	The Lancet	12	111
22	Child Dev	11	152
23	J Adolesc Health	11	200
24	Journal Of Health Economics	11	65
25	Lancet Child Adolesc Health	11	62
26	Psychol Med	11	158
27	Am J Public Health	10	127
28	American Economic Review	10	73
29	Int. J. Environ. Res. Public Health	10	48

Figure 14. Co-citation map of sources

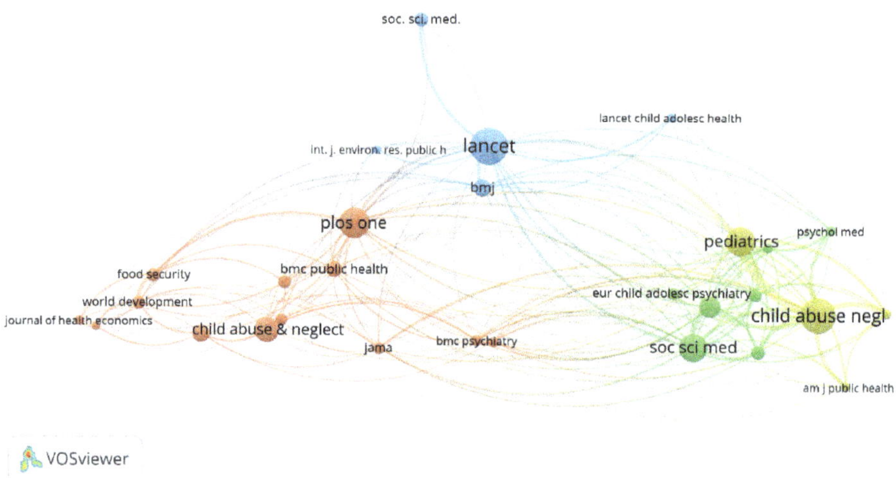

(Created by authors from Scopus Database)

By keeping the minimum no. of citations to "10". The total number of the mentioned cited authors is obtained as 41. amongst the total of 10844. The article that has the highest no. of sources it has cited is the one named Alem a total of maximum "37" citations.

Table 8. Co-citation authors

Id	Author	Citations	Total Link Strength
1	Alem A.	37	363
2	Patel V.	31	387
3	Zimmerman C.	29	314
4	Fekadu D.	22	265
5	Elbert T.	20	432
6	Neuner F.	20	408
7	Lund C.	19	290
8	Oram S.	19	227
9	Ssewamala F.M.	19	246
10	Hagglof B.	18	209
11	Betancourt T.S.	17	261

continued on following page

Table 8. Continued

Id	Author	Citations	Total Link Strength
12	Hodes M.	16	120
13	Schauer E.	16	331
14	Woodhead M.	15	171
15	Fassa A.G.	14	145
16	Kebede D.	14	126
17	Omigbodun O.	14	187
18	Sturrock S.	14	117
19	Cluver L.	13	129
20	Stark L.	13	102
21	Basu K.	12	97
22	Facchini L.A.	12	139
23	Goodman R.	12	64
24	Gupta S.	12	21
25	Jorm A.F.	12	60
26	Watts C.	12	143
27	Gardner F.	11	107
28	Heady C.	11	100
29	Unicef	11	85
30	Vostanis P.	11	115
31	Catani C.	10	250
32	Cluver L.D.	10	127
33	De Vries N.	10	60
34	Evans-Lacko S.	10	171
35	Hamdan-Mansour A.M.	10	109
36	Howard L.M.	10	152
37	Ibrahim A.	10	98
38	Kessler R.C.	10	74
39	Rahman A.	10	69
40	Usta J.	10	88
41	Van Doorslaer E.	10	37

Figure 15. Co-citation map of authors

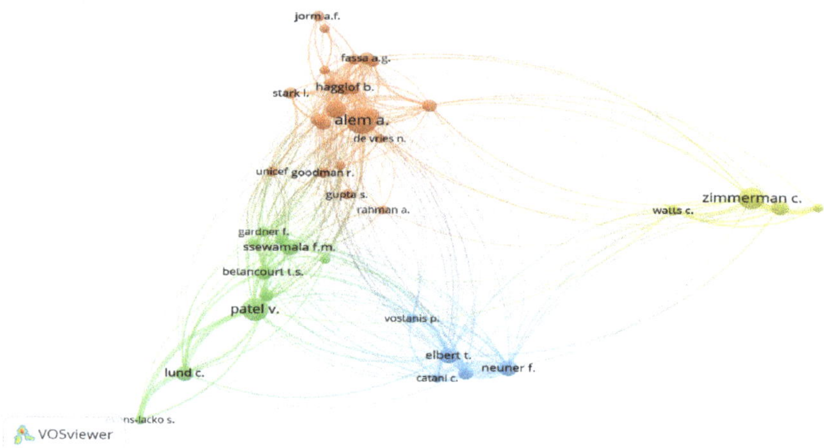

(Created by authors from Scopus Database)

RECOMMENDATIONS

Child labour remains one of the most pressing human rights violations of our time, depriving millions of children of their childhood, education, and future potential. Effectively combating this issue requires a holistic, multi-dimensional approach that targets its root causes while simultaneously strengthening protective systems and shifting societal attitudes.

A robust legal foundation is indispensable in the fight against child labour. Governments must enact and rigorously enforce stringent laws that explicitly prohibit the employment of children, particularly in hazardous industries such as mining, agriculture, and manufacturing. Legislation should be accompanied by severe penalties to culprits, such as allowed big fines and criminal proceedings on the part of employers who take advantage minors. Nevertheless, the laws by themselves do nothing as long as they are not properly implemented. Regular This should include inspections, whistle blower laws and open reporting systems. The rule has been put in place to deliver on compliance. The police departments and courts should also add. systems are placed under intense training of dealing with the cases of the child labour in an urgent and sensitive manner. such that legal rights get converted into real-life protections.

Prevention of child labour Education is one of the most effective ways of ensuring that child labour is prevented. When children there is an incentive to send them to school as they will receive free, compulsory and high-quality schooling. The desire to work reduces greatly. Governments have to invest in developing and sustaining schools in underprivileged districts, which would be prepared with sufficient facilities, qualified educators, safe learning environment, and good based programs, such as. free meals, uniforms, textbooks, as well as transportation, may reduce even more. obstacles to low-income families. Moreover, there is a wide variety of schooling options such as older children who possibly are already in employment, can be re-integrated through evening classes. they into an education system.

Child labour mainly advances on poverty and grabs a family into using their children as sources of fuel. governable income to survive. This cycle has to be broken through implementation of measures by governments and nongovernmental organizations (NGOs). economic by giving support to program. There are conditional cash transfers whereby, the families are given a conditional cash transfer payments in form of financial support to ensure that their children remain in school, have been worthwhile. effective in most of the regions. little financial and skilling training programs can encourage the parents especially the mother-to earn stable incomes, minimizing their carry out work on child labour. Enhancing social nets, i.e., unemployment it can divert through reduction and medical facilities to further enhance the monetary and the possibility of placing children back to work force.

Companies have a significant role of fuelling or combating child labour. Organizations have to embrace strict due diligence procedures in order to make sure that their suppliers' chains are well taken care of. not yet at the stage of exploiting the children's. have certificate through third party such as trade fair and no pressure with the audit and block chain-based transparency campaigns that can be used to track labour. monitors the conditions of real time. Other large corporations ought also to make investments in their community development schemes in sourcing areas, education and jobs opportunities which minimize on the use of child labour. The consumers are also in a position to institute change. buying fair trade products and making companies responsible when they are abusive practices.

Child labour has been normalised in many cultures that are deep-seated and especially in the rural areas. informal economies. Community-based campaign public awareness campaigns- we can employ mass media, community These perceptions can be countered in workshops, school programs, and so on. the long-term evils of child labour and sensitivity to the importance of education. Engaging these messages can be multiplying by cultural leaders local, and the young advocates and grass-root activities against exploitation. Incorporation of schools should also be applied. introduce child rights education in their students, which will enable children to become aware and report abuse.

Eliminating child labour requires a step in various directions. Laws should be enforced by the governments and businesses must practice what they preach.

Labour conditions, and societies have to oppose negative customs. By combining legal appeal, financial assistance, corporate governance, and social awareness will allow us to do it. employ a means of making each child liberated to learn, develop, and achieve his or her potential. potential. Combating child labour is not merely about doing what is good but rather an action. investment in a more equal and better future of everyone.

CONCLUSION

Child labour and specifically child labour in some of the unorganised and risky activities like rag picking and households small-scale manufacturing is one of the most tenacious human rights in India. challenges. In spite of the progressive legislations such as Child Labour (Prohibition and Institutional structures such as the National (2016) and Regulation) Amendment Act (2016). The persistence of child labour despite the existence of Child Labour Project (NCLP) presents important insights on the severity of the problem. kinks between policy-status and action. COVID-19 pandemic this crisis was aggravated dramatically, and largely to the disadvantage of children born in. households that are economically weak and those with daily wage earners in particular. and single parent families. With increasing pressure in the economy as lockdowns took place, many children were forced to get into exploitative working conditions to worsen the situation weakening and bringing forth new generations of want and loss. This is an intricate problem which requires winding inspection not only of the socio-economic and tireless dimensions but in addition of the well-neglected psychological effects that are experienced by child labourers, who experience systematic disregard, mistreatment, and hindrances to vital development that may even bear a life-long impression. Critical review of available literature, bibliometric review of Scopus indexed in the literature proves to be patterning in a disturbing mode as to the ways of dealing with child labour--or not as the case is bothersome to point out. precisely, underrepresented in the existing policy agendas.

The psychological well-being child labourers have the most alarming implication, as studies have always found out. indicating high risks of anxiety, depressions, and developmental delays in this crowd. population. However, the absence of mental health support is still conspicuous, especially in major areas. programme such as the NCLP and rehabilitation initiatives in the Right to Education Act. The existing methods are rather reactive and not preventive ones. perform rescue operations and offer an inadequate follow-up care or sustainable options. Poor inter-agency coordination further adds to these shortcomings in policy. The faults include inconsistency in its enforcement, and constant underfunding of social schemes.

While nongovernmental the contribution of organizations towards bridge of these service gaps is very important, it is their efforts. individual action could not be used to reverse systematic failure at the institutional level. The weaknesses of the current interventions show that there is need to change a paradigm in the treatment of child labour. thought and fought against, one which is to go beyond stop-gaps and cope with the basic truth of the matter. causes and the long-term effects. The mental burden of child labour is a health emergency that rendered urgent attention. immediate attention. Children subjected to forced labour, in particular to hazardous labour. get to experience chronic stress, which impacts normally developing cognition and emotions. Physical exposure to hazard as well as experiences of abuse are frequent. exploitation, results in quantifiable rise of the anxiety disorders, the depressive symptoms, and post-traumatic stress. Malnutrition and sleeping disturbances are common and so are developmental delays. impaired physical development and deprivation, the lack of educational stimulation. cognitive functioning. These consequences are multiplied by the social stigma which accompanies these. child labourers, which as good as marginalizes them and limits their future destiny. The most troubling perhaps is the fact that a lot of mental health effects tend to happen many years after children are taken out of the labour conditions, and their re-entry to education becomes impaired. develop good relationships, be economically stable in their adulthood. The Intergenerational nature of this trauma implies that unless this trauma is effectively intervened this trauma can be done away with, future generations will be doomed to bear the consequences. the effects of child labour can be based on psychological grounds that can set up the circle of poverty. oppression between families and communities.

With all these evidence, mental care is shockingly incomplete in the majority of child labour rehabilitation reflect an overall social disregard of the psychological condition of children in programs. policy planning. Dealing with child labour needs to be addressed holistically whereas multi-sectoral which treats its causes, manifestations, and consequences at the same time. Legal reforms these should go hand in hand with effective enforcement mechanisms consisting of harsher punishment. offenders and enhanced check mechanisms on industries that are at a high risk. Economic interventions are to be geared towards disruption of the poverty cycle that motivates child labour, by policies like conditional cash transfers, parents vocational training, etc. Firm up social nets. School programs should not only provide access. both quality and relevance, paying special attention to keeping vulnerable children involved in education using incentive-based programs and other modes of learning. Corporate accountability has to be supported by supply chain audit and ethical certification needs whereas social campaigns are aimed at correcting societal attitude by changing them. tolerate or accept the exploitation of children. Most importantly, any of these efforts has to incorporate the mental area.

health care at all levels, including immediate trauma counselling of children that have been rescued administration of long-term psychosocial services that enable complete rehabilitation.

Community-based locally based models that use local knowledge and resources may prove to be very effective establishing a durable net of protection, whereas technological advancements such as digital. The monitoring and coordination of agencies can be improved by use of tracking systems. forward can demand no less than a radical re-visioning of the socio-autonomous value of society. takes care of its children. The existing solutions to child labour are still unsound. not sufficient, and tend to support symptoms instead of causes. A change is what is required. a rights-based framework in which freedom of exploitive labour is a basic right to the maturation of children and, moreover, an attitude toward mental wellbeing, which prioritizes it in the same way it is treated as a consideration. physical safety. This will only be possible through prolonged political will, more funding funds, and real multi-stakeholder cooperation among state, civil activities. society, businesses and communities. Limit over time should be a priority to future research. research follows ups of long-term impacts of intervention programs, especially their Reintegration into the society and mental health recovery have a positive influence. The moral the need to do so is more than obvious: a society, which cannot ensure its children are not exploited, fails to retain its boast of democracy and right.The economic reason is also powerful since the expenses of doing nothing to take a stand, considering the potential wasted, the added health costs, and ongoing impoverishment costs of the comprehensive solutions. Erring out child labour is not just to take children out of the workplaces, it is to establish an environment where each child is allowed to develop, learn and live in dignity. This vision can only be realized through collective action that addresses the structural inequalities enabling child labour while building robust systems of prevention, protection, and care. The time for half-measures has passed; what is needed now is bold, coordinated, and compassionate leadership to transform this vision into reality for millions of India's children.

REFERENCES

Benjet, C. (2010). Childhood adversities of populations living in low-income countries: Prevalence, characteristics, and mental health consequences. *Current Opinion in Psychiatry, 23*(4), 356–362. PMID: 20520546

Catani, C., Schauer, E., & Neuner, F. (2008). Beyond individual war trauma: Domestic violence against children in Afghanistan and Sri Lanka. *Journal of Marital and Family Therapy, 34*(2), 165–176.

Fekadu, D., Alem, A., & Hägglöf, B. (2006). The prevalence of mental health problems in Ethiopian child laborers. *Journal of Child Psychology and Psychiatry, and Allied Disciplines, 47*(9), 954–959.

Ghosh, R., Dubey, M. J., Chatterjee, S., & Dubey, S. (2020). Impact of COVID-19 on children: Special focus on the psychosocial aspect. *Minerva Pediatrica, 72*(3), 226–235. PMID: 32613821

Hanlon, C., Whitley, R., Wondimagegn, D., Alem, A., & Prince, M. (2009). Postnatal mental distress in relation to the sociocultural practices of childbirth: An exploratory qualitative study from Ethiopia. *Social Science & Medicine, 69*(8), 1211–1219. PMID: 19709793

Ibrahim, A., Abdalla, S. M., Jafer, M., Abdelgadir, J., & De Vries, N. (2019). Child labor and health: A systematic literature review of the impacts of child labor on child's health in low-and middle-income countries. *Journal of public health, 41*(1), 18–26. PMID: 29409061

Minayo, M. C. D. S., Souza, E. R. D., Silva, M. M. A. D., & Assis, S. G. D. (2018). Institucionalização do tema da violência no SUS: Avanços e desafios. *Ciencia & Saude Coletiva, 23*, 2007–2016. PMID: 29972507

Muntaner, C., Solar, O., Vanroelen, C., Martínez, J. M., Vergara, M., & Santana, V. EMCONET network. (2010). Unemployment, informal work, precarious employment, child labor, slavery, and health inequalities: Pathways and mechanisms. *International Journal of Health Services, 40*(2), 281–295. PMID: 20440971

Pandey, R., Gupta, S., Upadhyay, A., Gupta, R. P., Shukla, M., Mishra, R. C., & Kumari, V. (2020). Childhood maltreatment and its mental health consequences among Indian adolescents with a history of child work. *The Australian and New Zealand Journal of Psychiatry, 54*(5), 496–508. PMID: 32156147

Ramaswamy, S., & Seshadri, S. (2020). Children on the brink: Risks for child protection, sexual abuse, and related mental health problems in the COVID-19 pandemic. *Indian Journal of Psychiatry*, *62*(Suppl 3), S404–S413. PMID: 33227060

Rizkalla, N., Mallat, N. K., Arafa, R., Adi, S., Soudi, L., & Segal, S. P. (2020). "Children are not children anymore; they are a lost generation": Adverse physical and mental health consequences on Syrian refugee children. *International Journal of Environmental Research and Public Health*, *17*(22), 8378. PMID: 33198333

Sturrock, S., & Hodes, M. (2016). Child labour in low-and middle-income countries and its consequences for mental health: A systematic literature review of epidemiologic studies. *European Child & Adolescent Psychiatry*, *25*, 1273–1286. PMID: 27217156

KEY TERMS AND DEFINITIONS

Bibliographic Citation: In academic research bibliographic sources enhance the validity of an analysis and tie them together. researches to social discourses, promoting collective developments in child rights.

Child Labour: Child labour is one of the most outrageous tampering of the children. basic rights, which means systematic utilization of underage children at work living conditions that deny them their childhood, education and so on. overall development. This habit does not only deny children an opportunity to go to school exposes them but on a regular basis as well as conditions that are harmful as far as their mental is concerned. physical, social and moral well-being. Be it associated with dangerous industries, within the domestic or child labour becomes a syndrome of poverty, slavery or informal sectors exploitation, destroying the potential of young people and making them expose to places in which their dignity and safety are daily violated. The continued use of child labour amidst the progress made in the world in matters of human rights. under-estimates social and economic disparities and under-achieves in policy control and execution.

Citation Strength: In research, high citation strength means powerful researches which have an impact on policy. principle, leading to the successful anti-child labour practices.

Community Engagement: Effective community involvement will see to it that anti-child labour outcomes benefit the community. Their sustainability and touch increase in the case of cliculturally relaible and locally owned.

Community-Based Interventions: Interventions designed locally, e.g. parent support groups and child rights would strengthen communities to fight child labour at the grass root level.

COVID-19 pandemic: It was the COVID-19 pandemic that increased the current predispositions of children. specifically those people living in low income families who were already vulnerable to childlabour. Lockdowns and school closures pushed millions of people in interruption of education and social services. opening children to dangerous employment conditions in order to top up family incomes. The loss of family lives, with long periods of isolation, increased the rates of mental pain with children. Another issue that also poses a problem is the pandemic. introduces loop-holes in child protection procedure, as save operations and excellent The only way that movements were behaved was through manoeuvres and diversion of resources. Post-pandemic recovery programs should also emphasize on educational, support mental health and financial assistance to curb the rate of child labour rising up.

Cultural Dissonance: Cultural dissonance is a cultural conflict between what the society expects of an individual person and what the person is entitled to. like in the case where child labour happens to be normalized by the traditional practices. Communication and learning is able to change such negative norms.

Domestic Violence: Domestic violence causes serious psychological and physical pain to children, which is frequently used as a force towards the leisure because they want to avoid the abusive home environments. Victimization to violence affects emotional control, scholastic performance. and relations within lead to their vulnerability to exploitation. The interventions should be focused on safe shelters, counselling services, and legal aid. The families that are so affected with the purpose of ending this cycle of abuse and not letting children face this force into work as a way of coping.

Early Childhood Development: Investing in early childhood development-healthcare, nutrition- etc. Self-assurance, which consists of stimulation, is the basis of resistance to exploitation.

Education Access: Education which is quality in nature is a very important deterrent of child labour. children an escape to poverty and exploitation. limits of the likes of school fees. transportation expenses, as well as cultural discrimination, especially against girls, should be reduced. broken down with the help of inclusive policy and community. There are efforts such as the Mid-Day Meal Scheme have shown to be successful in increasing the enrollment. a committed effort is required to bring about presence as well as retention and learning outcomes. marginalized students.

Empowerment Programs: Empowerment projects equip marginal families with competencies, resources and advocacy training that helps them to fight with exploitative norms and claim their rights.

Forced Marriage: Children and particularly girls are deprived through forced marriage which is instigated by poverty. of education and independence as they were directed into housework or premature motherhood. Change in law and the community consciousness is the solution to end this practice.

Gender Disadvantage: Discrimination against girls increases their exposure to risks brought about by gender issues compelling girls to have increased domestic labour. Child marriage, trafficking and child labor. Education and other empowering mechanisms through legal tools to girls protections are necessary in attaining fair results.

Grassroots: Activism Community mobilization is a form of grass root activism that induces change in the society. hateful relations and champion policy changes.

Hazardous Work: Dangerous employment subjects children to very severe physical threats such as toxic. chemicals, heavy equipments and exploitative labour. Such environments not only do they create short-term damage but also cause longterm consequences on growth. development. Tight provision of safety legislature to the occupational and specialized raids in to get rid of such practices, high-risk industries are required.

Human Trafficking: In its avaricious quest to harm both working-age children and the younger ones, It is subjected to labour and sexual exploitation by human trafficking networks. exploitation and therefore weaker cross-border enforcement is needed.

Inclusive Education: Inclusive standards of education followed includes children with disabilities, migrants has backgrounds, or trauma histories, leaving behind no child under any learning process they miss opportunities based on the circumstances.

Institutional Care: Though needed in immediate rescue, institutional care should change soon into fostering and family-based or foster care modules to prevent psychological damages of long-term fostering. institutionalization.

Intersectionality: Caste, gender, and disability, when coupled with child, can be seen through the prism of an intersectional lens. labour risks which are able to guide specific interventions.

Life Skills Training: Life skills training equips children to overcome obstacles, stand up against being manipulated and used. support their responsibilities.

Longitudinal Study: The longitudinal studies follow up the longitudinal status of child labourers, assessing the long-term effects of child labour. difference in effect of intervention and policy changes.

Malnutrition: Poverty and neglects result in the occurrence of malnutrition among child labourers. physical and mental growth. These can be overcome with school nutrition programs. effects.

Mental Health: Mental health is more than non-psychological disorder; it is also a form of health. is a condition of a whole-life health with a chance of people to achieve them potential, make connections in the face of daily stress, and come out to be of significant service. communities. Positive mental health, in the case of children, forms the framework of cognitive Raising productivity, emotional strength and social belonging. But there is child labourers are often faced with critical mental health issues as a result of being exploited, abused. and deprivation. Hazardous labour stress and trauma The stress and trauma of hazardous labour Primary stress is a stressful condition related to the dire nature of the labor One of the main causes of this dire nature is that the labour is hazardous and/or unsafe Stigma This terminology is used in reference to the stigma the group associates itself with The stigma relates to the nature of the labour it usually results in anxiety, depression, and post-traumatic stress disorder under such conditions They also develop (PTSD), which did not allow them to develop healthy relationships or have long-term achievements. stability. The consideration of mental health in the child welfare policies, therefore, is essential to the concept of mental health. that the children saved also get not only physical rehabilitation but also psychosocial They should be given assistance to recover their experiences.

Mid Day Meal Scheme: The mid-day meal scheme encourages the attendance in schools by giving nutritious meals, and cut dropout rates caused by hunger. The need to increase this program to include more child labour can also be discouraged by grades, and regions.

Multisectoral Collaboration: Partnership between the government, NGOs, and even businesses provide a full package. strategies to be used in eradicating child labour.

National Child Labour Project (NCLP): National Child Labour Project (NCLP) represents a state program that is going to structural education, skill up to rehabilitate children who are involved in labour. Medical care services, training and care. The company functions with special training centres. NCLP aims at mainstreaming the rescued children in schools and at the same time stipends are offered. compensate the loss in family income. But the program has not been made very successful by small budget, red tape and lacks attention to routine mental health. Lack of long-term leave makes many children relapse back into labour. provided their families with economic support. extended the scope of the NCLP in such a way that it involved parents. its effectiveness could be improved by living programs as well as trauma-informed care exploitation.

NGO (Non-Governmental) Organization: Non-governmental organizations (NGOs) play an important role of providing services delivery gaps. the gaps that existed in the state programs, especially concerned with the remote or high-risk regions. Through rescue the means of charity through operations, alternative education centers, and advocacy campaigns, NGOs offer to the poor. short term solutions and long term assistance to child labourers. Their community-based the models are usually taken as blueprints of interventions that can be scaled up. impact will necessitate more intensive work with the government systems and sufficient funding.

Policy Evaluation: Strict policy analysis entails the determination of effective strategies and fields that require enhancement. It enhances optimization of resources.

Post Traumatic Stress Disorder (PTSD): Post-traumatic stress disorder in child labours presents itself in form of flashbacks, hyper vigilance, and emotional problems. numbness, which significantly affects the everyday work. Peer Trauma-based therapies The key elements of the support system are support groups and safe spaces of expression. healing, assisting survivals to reflect on their past experience and regain the agency.

Poverty: The existence of child labour can be traced solely to poverty because it is the most widespread source of child labour. households depending on the income of their children to earn the minimum to live. In resource-scarce in economies environments education is seen as a luxury that cannot be afforded, more so, when children are put on its rank. what counts is immediate generation of income. This is an economic desperation. which is exacerbated by unemployment, inflation and social security, etc. that affect communities improperly. Sustainable solutions should able therefore to blend in the poverty relief actions that offer instant reparations- such as the safety of food programs and transferring of cash-replaced by investments in long-term skills development and employment of adults so as to achieve dependency on child labour.

Psychological Burden: The psychological cost to target child labourers requires immediate psychiatric care in the investments to avoid lifetime impacts of the trauma without treatment.

Psychological Disorder: Child labourers stand a very high chance of acquiring psychological disorders like because of the stress and chronic trauma, they experience stress, anxiety, and PTSD. Early identification screening in schools can be done and extent in the community can make early Interventions, which prevent long-term impairment, are there. bringing mental health services together. This hidden situation needs to be found by inserting it into child protection systems.

Psychosocial Support: The psychosocial support services take into consideration the social and emotional needs. counselled, used art therapy and integrated the traumatized children to the community activities. Such mediation builds strength and makes the survivors regain trust. societal institutions.

Public Policy Implementation Gap: There is a gap between the way a policy is designed and the way it is executed in reality - this is a significant issue. hurdle towards the fight against child labour. However, the laws such as RTE Act and NCLP are there to help provide. use of paper, poor enforcement, corruption, and inter- agency lack of coordination they should be undermining their effectiveness. filling the gap of demand ought to be highly accountable. temporary mechanisms, grass roots monitoring and capacity building of the frontline staff. interpret legislative intent into something real.

Quarantine Policies: The policies on COVID-19 quarantine unintentionally amplified the threat of child labour because, through it, the intrusion into the educational process took place. studying and living. Responses to crisis in the future should integrate child-sensitive add extras, like emergency stipends and learning possibilities at a distance, to offset collapse of such collateral.

Rag Pickers: Rag pickers-children who are among the most vulnerable categories of child labourers are also among them. dig through garbage and wastes places to get recyclable products to collect. meagre earnings. These children work under toxic conditions where they work miserable. biohazardous waste, sharp objects, and wastes in absence of protective equipment's. The environment that most of them work in exposes them to extreme health hazards due to its unhygienic nature. such as respiratory illnesses, infections as well as physical harm are exposed to unjust societal and psychological trauma, since they are also not usually entitled to primary education. social inclusion. The phenomenon brings out the issue of poverty and absence of social Complacency, corruption, and neglect in the institutions, which leaves children with the only choice of abandoning their livelihoods to risky lifestyles.

Rescue and Rehabilitation: The rescue and rehabilitation needs to be carried out in a co-ordinated manner between the laws. the role of enforcement, social workers and educators to bring about recovery to the child in a holistic manner. labourers.

Right to get Education (RTE) Act: RTE the Act Enacted in 2009 was one of the landmark legislations of India which ensures that every child aged between 6 and 14 years has free and mandatory education. The law requires that schools offer inclusive and equitable education, and one that is strict such practices as punishment that results in strict attendance. Although, the RTE Act is progressive in nature, it has huge implementation challenges. They face many challenges such as poor infrastructure, unavailability of teachers and low indented enforcement in the oppressed areas.

Scheme of Child Protection: Child protection programs initiated by the government are meant to protect the minors against abuse. trafficking, and exploitive labour using shelters, helplines, and legities. But they are constrained by scarcity of resources, as well as, stringent legislation. To have a more powerful support of these programs, it is necessary to gather more funding, trained workforce, and receive it. community involvement so as to make the vulnerable children have holistic care.

Social Security: Good social security nets, jobless grants and healthcare access, has the potential to decrease economic pressures that are causing child labour. Expanding it is essential to assure these protections to the workers in the informal sector in order to solve the causes of exploitation.

Socio-economic Vulnerability: Socio-economic vulnerability can be defined as the increased danger of people. families because of poverty, social isolation and poor access to basic amenities e.g. researches and health. The first reason that leads to child vulnerability is this vulnerability. labour since most impoverished households tend to trade the children off to the workplaces so that they can survive. The rural deprived communities and the marginalized communities. those living in urban slums are disproportionately disadvantaged, and institutional blocks may include caste The disadvantaged situations are aggravated by discrimination and female inequality. They lack social nets, and these families have only a few to continue intergenerational deprivation cycles. Quality interventions ought to as a consequence will not only address the child labourer in the short-term need. structural disparities that cause exploitation.

Sustainable Solutions: When the real issues such as poverty and lack of education are solved, then there are sustainable solutions. permanent elimination of child labour.

Systemic Neglect: Systemic neglect reveals failures of institutions to save children, due to the fact that corruption caused. Whistleblower protection and whistleblower protection Support fail to notice are able to enhance accountability.

Trauma-Informed Care: Trauma informed care is a service delivery method that identifies the popularization of the consequences of trauma on the development of children and stresses the importance of establishing positive convalescent surroundings. To the child labourer, who often have to live through abuse, neglect, and exploitation this method is connected with making use of psychological provide the impetus on rehabilitation. The most important principles are safety of the voice and voice. emotional well-being, trust and empowering survivals through participation in decision-making. Educating caregivers, educators and social workers on trauma-informed the practices can considerably enhance the results of the rescued children. they get back on their feet with dignity and strength.

Vocational Training: Older children and teenagers receive vocational training which gives them marketable skills. offering a substitute to risky work. A local industry is adjusted at programs levels when combined with job placement assistance, these supplementary needs, can enable the youth to succeed. becoming economically free without having to compromise in their education.

Chapter 8
Climate Change and Women's Health:
The Need for Gender-Specific Data and Policy Interventions

Rahul Hiremath
https://orcid.org/0000-0001-9976-274X
Indian Institute of Management, Raipur, India

ABSTRACT

Climate change poses significant health risks, disproportionately affecting women due to biological, social, and economic vulnerabilities. This paper explores the link between climate change and women's health, emphasizing the need for gender-specific data to develop effective policy interventions. By 2030, climate change may become irreversible, exacerbating health issues such as maternal risks, exposure to environmental hazards, and limited healthcare access. Food and water shortages further worsen malnutrition and infectious diseases, particularly in marginalized communities. However, the lack of gender-disaggregated data hinders targeted policy responses. This study highlights the importance of integrating a gender-sensitive approach into climate research and policymaking. Through literature analysis and case studies, it advocates for inclusive interventions that prioritize women's health and resilience, ensuring equitable healthcare solutions in the face of climate change.

INTRODUCTION

One of the most urgent global issues of the twenty-first century, climate change affects human well-being and health widely. Although all people are affected by climate-related hazards, women suffer a disproportionately high burden because

DOI: 10.4018/979-8-3373-3281-9.ch008

of their biological, social, and financial vulnerabilities (WHO, 2021). Rising temperatures, extreme weather events, and environmental damage aggravating already existing health inequalities make the junction of climate change and women's health an area of increasing relevance (UN Women, 2022). Among the important problems that need immediate attention are maternal health risks, malnutrition, vector-borne disease exposure, and limited access to healthcare services (Gould & Rudolph, 2015).

Notwithstanding these difficulties, gender-specific data on effects of climate change are still lacking, which impedes the creation of focused policy actions (Sorensen et al., 2018). Comprehensive research and an inclusive policy framework are necessary to understand how differently women experience climate change than men. Emphasising the need of gender-disaggregated data and customised policy solutions to support resilience and fair access, this chapter investigates the complex interaction between climate change and women's health.

REVIEW OF THE LITERATURE

Climate change is a major challenge due to the way it magnifies social and economic inequalities, particularly along gender lines. Studies have indicated that women are particularly affected by the effects of climate change, especially in the developing world, on account of their vulnerabilities and socio-economic status. When gender is crossed with climate change, from a policy-making perspective, approaches need to be sensitive to gender because the voices of women are critical for effective climate action. Gender-equitable climate policy is not only a matter of justice, but also contributes to the effectiveness of climate strategies. Research has drawn attention to the need to include gender issues in discussions and policy-making on climate change. According to Jafry, the exclusion of women from climate change discourse presents an obstacle to preventing climate change and shows that, increasing women's participation can lead to more efficient international negotiations (Jafry 2016). Wong argues that integrating gender issues in climate finance has the potential to "transform gender relations and promote equity," with examples such as climate-smart agriculture (Wong, 2016). Sound climate finance does not only focus on mitigation and adaptation, it also promotes gender equity through increasing women's access to resources and women's role in decision-making (Waheed & Waheed, 2022). Intersectionality is key to understanding these phenomena. Tanny and Rahman highlighted the need to focus on climate vulnerabilities for women, as these vulnerabilities are often affected by cultural, economic and social issues (Dijkerman et al., 2025;. Intersectionality allow us to gain a more nuanced understanding of climate impacts as being affected by multiple identities (gender, class, race), in complex ways (Perkins, 2018). Through using this lens, scholars are able to determine locally relevant solutions that take into account the complexity of women's vulnerabilities (Dijkerman et al., 2025; Kaijser & Kronsell, 2013).

Further, empirical evidence has demonstrated that inclusive climate adaptation planning is important, such that policies and programmes need to consider the local realities of marginalized communities. Anguelovski et al. explain how urban land use planning for climate adaptation can aggravate socio-spatial inequalities when equity is not considered (Anguelovski et al., 2016). Similarly, Khanal et al. report that the vulnerability experiences of marginalised groups are largely neglected in climate change and health policies leading to increased health inequities (Khanal et al., 2024). Indeed, the need to tackle these disparities are repeated in conversations about health equity and climate justice, with the knowledge that climate change affects poor and vulnerable populations least prepared to deal with climate-related disasters, frequently with significant gender differences (Khanal et al., 2024; Shonkoff et al., 2011).

The evidence on the intersection between climate change and gender uncovers a dynamic landscape requiring multidimensional, gender-applied and intersectional policy-making and social interventions. The need to address women s requirements to equip them for the demands of climate change and to empower women through education and access to resources and to include women in decision making, is essential for the creation of a community that is climate shock resilient. In view of these fundamental considerations, it is important for us to understand how climate change affects women's health differently, particularly when it comes to the most vulnerable populations, and to examine the literature to learn about the specific challenges women face. This analysis will help in gaining a deeper insight into how gendered health inequalities are exacerbated by climate change and guide interventions.

Particularly for women in low-income and underprivileged areas, climate change has major health ramifications as it increases their chances of heat-related disorders, respiratory ailments, and viral epidemics (Haines & Ebi, 2019). With pregnancy aggravating sensitivity to vector-borne infections like malaria and the Zika virus as well as heat stress (Chersich et al., 2020), women's particular physiological and socioeconomic conditions heighten their risk. Particularly in crises linked to climate change, limited access to healthcare services and financial limitations exacerbate these hazards (WHO, 2021; Sorensen et al., 2018). Extreme weather events like floods and droughs disturb healthcare infrastructure and restrict access to necessary mother and newborn services, hence raising death rates (UNFPA, 2019; Van Daalen et al., 2020). Furthermore, connected to negative pregnancy outcomes including low birth weight and premature deliveries is air pollution and polluted water sources (Bekkar et al., 2020). Climate change-caused food and water shortages disproportionately impact women and children, aggravating hunger and associated medical problems.

Securing food and water falls mostly on women in agricultural areas; when these resources become limited, they suffer from nutritional deficits that can cause anemia, compromised immune systems, and more susceptibility to infections (FAO, 2021; Smith et al., 2021). Emphasizing the importance of gender-sensitive dietary policy, gender inequalities in family food distribution worsen these issues even more (Tirado et al., 2019). Many case studies show how well gender-responsive climate policies work. While in sub-Saharan Africa initiatives promoting climate-resilient agriculture have empowered women farmers by providing training and access to climate-smart technologies, community-driven adaptation projects have effectively incorporated women's leadership in disaster preparedness and response (Rahman & Alam, 2020). Bangladesh Beyond social and health consequences, erratic precipitation and increasing temperatures of climate change seriously threaten agriculture in developing countries. Women are more susceptible to food insecurity and poverty as they participate actively in food production. While Chalise and Naranpanawa studied cropland changes in Nepal, where farmers adjusted by reallocating land to more climate-resilient crops such paddy fields, so lowering economic losses and income disparities, Montaud (2020) promoted improved early warning systems, drought-tolerant crops, and expanded irrigation to enhance food security in Mali. Lack of gender-disaggregated data, which impedes the creation of tailored interventions targeting the particular needs of women and girls, is a key obstacle to successful climate adaptation and mitigating strategies (UN Women, 2022). By means of gender-sensitive impact assessments and strengthening of data collecting mechanisms, one may guarantee fair policy results and improve women's resistance to climate change (Le Masson et al., 2019; Gillespie et al., 2021). Though a lot of study has been done, the link between adaptation techniques and important elements such food security, poverty, and vulnerable populations—especially women—remains little investigated. With poverty, poor housing, limited services, and poor infrastructure aggravating vulnerability, climate change interacts with social, political, and financial elements. While strong institutions—especially those supporting women's political empowerment—play a vital role in reducing climate threats, investments in technology innovations and sustainable resource management may greatly lower vulnerability. Effective adaptation efforts are often hampered, nevertheless, by centralized decision-making, organizational fragmentation, agenda pushed by elites, and crisis-oriented government. Studies show the connection between technological innovation and climatic vulnerability, therefore stressing the importance of fair and ecologically friendly resource use. This paper seeks to increase understanding of climate vulnerability by means of an analysis of women's political empowerment, institutional frameworks, and natural resource reliance in thus reducing the negative consequences of climate change.

METHODOLOGY

This is a mixed-methods study, including both the qualitative and quantitative analyses to examine the effect of climate change on women's health and to assess current policy interventions. A systematic literature review is undertaken, utilising scientific papers, policy documents and case studies to examine gender-related health risks such as maternal health, malnutrition and vulnerability to vector-borne diseases. The review also spots the absence of gender disaggregated data and the necessity for gender responsive policies. In addition, quantitative analysis of existing data on health statistics data disaggregated by sex, when possible, is conducted to track climate and women's health-related trends and relationships. Real life; Case studies from around the world—from developed and developing countries—offer real-world examples of how women are experiencing climate change and adapting to it. These case studies are important because they illustrate how women play a role in disaster prevention and recovery. The analysis also examines policy frameworks to determine whether gender considerations are included, and whether interventions are effective in addressing women's health issues. This includes the examination of international and national climate action plans and disaster risk reduction frameworks and health policies for global coverage and quality of gender-responsive health provision. The results from the literature review, data analysis, case studies and policy reviews are combined together to give a broad overview of areas at the climate change and women's health interface, with suggestions for the enhancement of gender-responsive programmes. Ethical considerations (e.g. de-identification of sensitive information, and fair representation of women's experiences) and limitations (such as the absence of region-specific, comprehensive gender-disaggregated data) inform the research.

WOMEN'S WELLNESS AND HEALTHCARE

Research shows that 3.6 billion people already live in areas highly susceptible to climate change. Between 2030 and 2050, climate change is expected to cause approximately 250 000 additional deaths per year, from undernutrition, malaria, diarrhoea and heat stress alone. Climate change intensified 26 of the 29 weather events studied by World Weather Attribution that killed at least 3700 people and displaced millions, according to a new report from World Weather Attribution and Climate Central. (World Meteorological Organization [WMO], 2024). According to World Health Organisation (WHO), at a global level, 7 of the 10 leading causes of deaths in 2021 were noncommunicable diseases, accounting for 38% of all deaths, or 68% of the top 10 causes (Figure 1). Current practices are being impacted by the lack of governmental initiatives resulting from sufficient proof of climate change. However, the negative effects of climate change pose a danger to low- and middle-income nations, potentially exacerbating already-existing gender inequities.

Does climate change put women's health at greater risk? Social norms and situations resulting from gender inequality have led to health problems for women brought on by climate change. Women are more likely than males to die in natural catastrophes like floods due to remaining inside, culturally driven clothing choices, and a lack of swimming ability. Women who are malnourished amid famine and long-term poverty also die at a high rate. In India, it is still customary to give precedence to male members while allocating food. They remain vulnerable due to the decline in livelihood activities associated to agriculture, where women labourers make up the majority. The majority of the adaptation and mitigation strategies are gender-neutral and primarily address the demands of males. According to a report by the Intergovernmental Panel on Climate Change (IPCC), migration—the movement of people inside a nation from one area to another and from one country to another as refugees—will be a consequence of climate change.

Figure 1. Leading causes of death worldwide

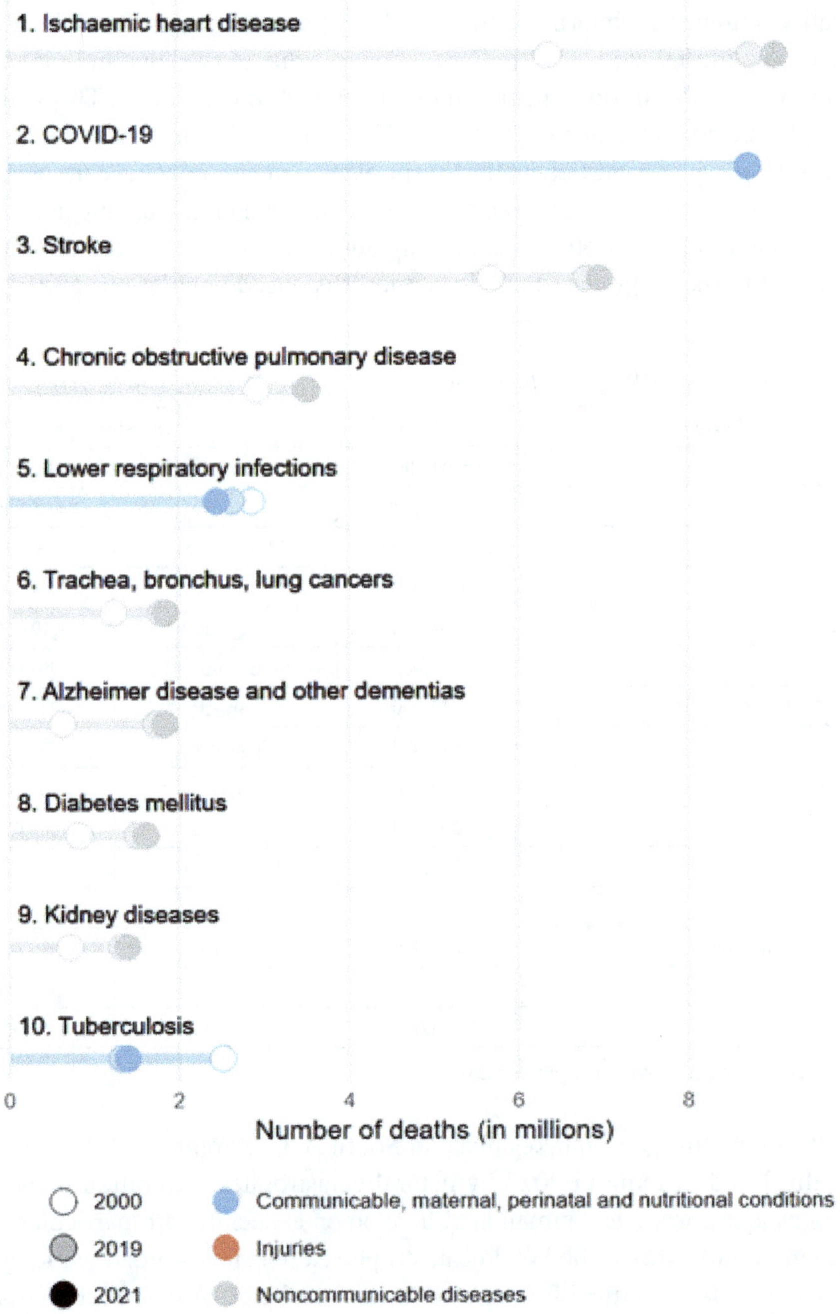

(WHO, 2022)

In 2023, a total of 47.3 million new displacements were recorded across 152 countries and territories, with 20.5 million attributed to conflicts and violence in 45 countries, while 26.8 million resulted from disasters in 149 countries. The cumulative number of internally displaced people (IDPs) reached 75.9 million across 116 countries, with 68.3 million displaced due to conflict and violence in 66 countries, and 7.6 million displaced by disasters in 83 countries. This is the annual displacement figures from 2008 to 2023 (Table 1). These trends suggest that while disasters cause significant short-term displacement, conflicts are becoming the dominant force behind long-term internal displacement. This underscores the need for comprehensive policies addressing conflict resolution, humanitarian aid, and improved disaster preparedness to support displaced populations effectively.

Table 1. Countries with the highest new displacements by Trigger in 2023

Conflict and Violence		**Disaster**	
Sudan	**6,039,000**	**China**	**4,702,000**
Dem. Rep. Congo	3,772,000	Türkiye	4,053,000
Palestine	3,438,000	Philippines	2,595,000
Myanmar	1,298,000	Somalia	2,043,000
Ethiopia	794,000	Bangladesh	1,791,000
Ukraine	714,000	Myanmar	995,000
Burkina Faso	713,000	Brazil	745,000
Somalia	673,000	Pakistan	732,000
Colombia	293,000	Syria	702,000
Nigeria	291,000	Malawi	660,000
South Sudan	282,000	Mozambique	655,000
Haiti	245,000	Kenya	641,000
Central African Republic	214,000	Ethiopia	618,000
Israel	203,000	India	528,000
Mali	184,000	Morocco	500,000

(Internal Displacement Monitoring Centre, 2023)

Most of the climigration takes place in South Asia. Around 528,000 were registered displaced in India in 2023 by natural catastrophes. Migration is therefore not seen as appropriate for human health. Women especially are particularly sensitive to migration stress. The World catastrophe Report notes broad warnings that "women and girls are at the higher risk of sexual violence, sexual exploitation and abuse, trafficking and domestic violence during a disaster" (https://www.ifrc.org/en/publications-and-reports/world-disaster-report/wdr2007/). Consequently, the pronged stress disorders connected to the ladies get worse throughout the move. Migration also lowers access to health services and education and feminises poverty. Furthermore, lacking in addressing the current form of gender variations in health hazards seen with climate change are the adaptation and migration policies. There are indicators on climate change even in sustainable development goals (SDG), but they have no bearing on the health problems brought about by it. The Sendai framework—2015 to 2030 on disaster risk—puts the effect of disaster risk on women. Moreover, it demands an investment of all the players and industries. It recognised women as important players in the process of healing and restoration. Moreover, localisation of the SDGs and its interaction with the health concern might guide intervention efforts. Finally, the most crucial factor to take into account is the availability of data split according to gender. Though studies and experience have emphasised the particular requirements of men, boys, women, girls, and persons from sexual minorities in climigration circumstances, only a few data sources provide information by gender. This lack of data disaggregation can make it difficult for governments and relief organisations to effectively provide their resources. Thus, it is argued that it is important to ensure the availability of gender-segregated data and other traits including age, disability to assist design better interventions, and the preservation of sensitive information, particularly for sexual minorities.

DISCUSSION

Gender-responsive climate change policies are essential for addressing the specific vulnerabilities women face while empowering them to take an active role in climate solutions. As climate action increasingly depends on inclusive and sustainable strategies, it becomes clear that women's participation is critical to effectively combat the challenges posed by climate change.

Gender-Responsive Climate Change Policies

Gender equal climate change policies are crucial if those most affected - women and girls - are to take part in shaping the solutions. These policies recognise women's particular vulnerabilities and empower them as active agents of climate action. Gender is affected by and will also affect climate change, and we need to have an inclusive approach by governments to both adapt to and mitigate gender insecurities. Table 2 outlines gender-responsive climate policy examples from multiple countries, reflecting their dedication and commitment to ensuring that women's perspectives and needs are incorporated in climate action plans. Such policies are essential for ensuring women are safeguarded from the effects of climate change and can meaningfully contribute to climate solutions.

Table 2. Gender-responsive climate change policies across different countries

Country	Policy/Strategy	Key Focus on Gender	Reference	Link
Philippines	Climate Change Act (2009)	Mandates gender integration in climate policies, emphasizes women's roles in disaster risk reduction and adaptation.	Republic of the Philippines, *Climate Change Act of 2009*	Philippine Climate Change Act
Kenya	National Climate Change Response Strategy (2010)	Focuses on gender equality in climate adaptation, promotes women's participation in agriculture and water management.	Government of Kenya, *National Climate Change Response Strategy*	Kenya's National Climate Change Strategy
Mexico	National Strategy on Climate Change (2012)	Promotes gender-sensitive policies in agriculture, water, and disaster management.	Government of Mexico, *National Strategy on Climate Change*	Mexico's Climate Change Strategy
India	National Action Plan on Climate Change (NAPCC) (2008)	Gender mainstreaming across the eight missions, focuses on women in agriculture, water, and sustainable development.	Government of India, *National Action Plan on Climate Change*	India's NAPCC
South Africa	National Climate Change Adaptation Strategy (2013)	Integrates gender perspectives into climate adaptation policies, empowers women in rural areas.	Government of South Africa, *National Climate Change Adaptation Strategy*	South Africa's Climate Change Strategy
Canada	Climate Change and Gender Equality Framework (2021)	Integrates gender equality into climate policies, focuses on marginalized women in climate resilience and energy transition.	Government of Canada, *Climate Change and Gender Equality Framework*	Canada's Gender and Climate Change Framework
Germany	Gender Equality in Climate Policy (2019)	Integrates gender-specific data, supports women's roles in renewable energy and climate entrepreneurship.	German Federal Ministry for the Environment, *Gender Equality in Climate Policy*	Germany's Gender and Climate Policy
Sweden	Climate Policy Action Plan (2017)	Gender equality is central to the climate action plan, emphasizing women's leadership in climate decisions.	Government of Sweden, *Climate Policy Action Plan*	Sweden's Climate Action Plan

Adopting a gender-responsive climate policy is important for addressing the specific challenges faced by women in climate change. Not only do these policies increase the resilience of women, they also strengthen climate policy and make it (more) sustainable. Countries can promote more equitable and inclusive responses to the climate crisis by integrating women into decision-making and providing women with the resources and support necessary to adapt and mitigate. As these examples demonstrate, including gender equality in climate policies is not only a mark of social justice but a necessary ingredient for stronger, more resilient communities everywhere.

Climate Action Definitely Needs Women

One of the most urgent issues of our day, climate change affects global stability, livelihoods, and health among other things. Although it influences all of mankind, its effects are not shared equally; women, especially in underprivileged areas, suffer disproportionately from biological, social, and financial elements. Integrating a gender-sensitive strategy that guarantees women's full involvement in climate action is crucial if we are to properly fight climate change.

- **Climate Action Calls for Every Member of the Population**: Dealing with climate change calls on the whole population to participate, not just half of it. Given about half of the world's population is female, their efforts are vital in forming sustainable solutions. Denying women participation in climate policies reduces the possibility for creative ideas and successful policy execution (United Nations). In climate talks, gender-balanced leadership guarantees several points of view and more inclusive decision-making.
- **Empowering Women Results in Better Climate Solutions**: Management of natural resources, agriculture, and community resilience depends critically on women. Women drive more sensible and sustainable climate solutions when they are empowered with financial resources, leadership chances, and education. Women-led environmental projects are increasingly valued in long-term sustainability, conservation, and community well-being, according to studies, which makes them indispensable participants in efforts at climate mitigation and adaptation (Reuters, 2024).
- **Women are essential for creating community climate resilience**: Efforts at disaster readiness and recovery depend critically on women. Often the first responders when climate-related events happen, they are main caretakers and housekeepers. Their awareness of food security, water management, and community well-being makes them very helpful in creating climate resilience (United Nations). Policies encouraging women's involvement in disaster management and adaptation plans enhance whole societies and result in more efficient responses to climate change.
- **Though Not Equally, Climate Change Affects All of Us**: Particularly in low-income and underprivileged areas, women bear disproportionate health risks from climate change. Rising temperatures, poor air quality, and severe weather events aggravate mother health risks and increase exposure to infectious infections (United Nations Foundation). Furthermore aggravating malnutrition are food and water scarcity, which mostly affects women and children. These differences will keep widening without focused efforts, resulting in more extreme inequality and more vulnerability of the environment.

- **Nations See the Value of Gender in Climate Change Strategy**: National governments and international organizations recognize more and more that gender-responsive climate policies provide better results. Women's participation in climate talks, policymaking, and adaption planning has been found to increase environmental program efficacy (Le Monde, 2024). International systems stressing the importance of gender equality in climate governance include the Gender Action Plan of the UNFCCC. Sustaining and equitable climate action depends on climate policies addressing women's particular difficulties and contributions.

CONCLUSION

Climate change is one of the most urgent issues of our time and it has devastating effects on women's health. The fact that women bear different biological, social, and economic burdens than men make them more susceptible to the negative impacts of climate change, and climate hazards weigh more heavily upon their health and well-being. Indeed, as this paper has shown the absence of sex-disaggregated data on impacts of climate change constitutes a substantive barrier to developing policy measures that could mitigate effects on the specific vulnerability of women. The poor outcomes range from maternal health risks through to greater susceptibility to infectious diseases and are mediated by poor access to health care, food insecurity and environmental challenges such as extreme weather. It is important to reiterate the inclusion of gender sensitive strategies in climate research and policy development. With women in climate action empowered and engaged, solutions are also more effective and lasting. Supporting women not only increases community resilience, but also this act leads to the success of Mitigation and Adaptation plans. In addition, gender-responsive programmes have demonstrated to bring development better results, increasing the communities' resilience in responding to climate change. Gender-disaggregated data and the action-oriented, inclusive and gender-responsive development and implementation of climate policies and actions are imperative for shaping a future that can withstand and bounce back from such shocks. Such measures will allow women to not only be the recipients of such climate interventions, but active actors in the formulation and implementation of climate resilience. Looking at the intersection of gender and climate change is not only an issue of equity, it is an important measure to protect the health and safety of all communities in the midst of this global crisis.

REFERENCES

Anguelovski, I., Shi, L., Chu, E., Gallagher, D., Goh, K., Lamb, Z., & Teicher, H. (2016). Equity impacts of urban land use planning for climate adaptation. *Journal of Planning Education and Research, 36*(3), 333–348. DOI: 10.1177/0739456x16645166

Bekkar, B., Pacheco, S., Basu, R., & DeNicola, N. (2020). Association of air pollution and heat exposure with preterm birth, low birth weight, and stillbirth in the US: A systematic review. *JAMA Network Open, 3*(6), e208243. DOI: 10.1001/jamanetworkopen.2020.8243 PMID: 32556259

Chersich, M. F., Pham, M. D., Areal, A., Haghighi, M. M., Manyuchi, A., Swift, C. P., & Luchters, S. (2020). Associations between high temperatures in pregnancy and risk of preterm birth, low birth weight, and stillbirths: A meta-analysis. *BMJ (Clinical Research Ed.), 371*, m3811. DOI: 10.1136/bmj.m3811 PMID: 33148618

Dijkerman, S., Hossain, J., Persson, M., Konika, R., & Paul, D. (2025). Applying an intersectional climate justice lens to understand climate crisis impacts on sexual and reproductive health and rights and identify local solutions: Qualitative findings from Khulna, Bangladesh. *Women's Health, 21*. Advance online publication. DOI: 10.1177/17455057251339283 PMID: 40418759

FAO (Food and Agriculture Organization). (2021). *The state of food security and nutrition in the world 2021*. FAO, IFAD, UNICEF, WFP, and WHO. https://doi.org/DOI: 10.4060/cb4474en

Gillespie, A. M., Singh, P., & Fernandez, M. (2021). Gender and climate change: Bridging data gaps to strengthen adaptation policies. *Climate Policy, 21*(7), 945–961. DOI: 10.1080/14693062.2021.1953728

Government of Canada. (2021). *Climate change and gender equality framework*. Retrieved from https://www.canada.ca/en/services/environment/climate.html

Government of Germany. (2019). *Gender equality in climate policy*. Retrieved from https://www.bmu.de/en

Government of India. (2008). *National action plan on climate change (NAPCC)*. Retrieved from https://www.india.gov.in/spotlight/national-action-plan-climate-change

Government of Kenya. (2010). *National climate change response strategy*. Retrieved from https://www.environment.go.ke

Government of Mexico. (2012). *National strategy on climate change*. Retrieved from https://www.gob.mx/semarnat

Government of South Africa. (2013). *National climate change adaptation strategy*. Retrieved from https://www.environment.gov.za

Government of Sweden. (2017). *Climate policy action plan*. Retrieved from https://www.government.se

Government of the Philippines. (2009). *Climate change act (Republic Act No. 9729)*. Retrieved from https://www.officialgazette.gov.ph/2009/10/23/republic-act-no-9729/

Haines, A., & Ebi, K. L. (2019). The imperative for climate action to protect health. *The New England Journal of Medicine*, *380*(3), 263–273. DOI: 10.1056/NEJMra1807873 PMID: 30650330

Internal Displacement Monitoring Centre. (2023). *Displacement data*. Retrieved from https://www.internal-displacement.org/database/displacement-data/

Jafry, T. (2016). Making the case for gender-sensitive climate policy – lessons from South Asia/IGP. *International Journal of Climate Change Strategies and Management*, *8*(4), 559–577. DOI: 10.1108/ijccsm-04-2015-0049

Kaijser, A., & Kronsell, A. (2013). Climate change through the lens of intersectionality. *Environmental Politics*, *23*(3), 417–433. DOI: 10.1080/09644016.2013.835203

Khanal, S., Bolte, G., & Boeckmann, M. (2024). Sector silos in climate action: A missed opportunity to prioritize health equity. *PLOS Climate*, *3*(2), e0000349. DOI: 10.1371/journal.pclm.0000349

Khanal, S., Shrestha, R., & Boeckmann, M. (2024). Examining health equity in Nepal's climate change and health policies through the lens of environmental justice: Insights from a content analysis. *Global Health Action*, *17*(1). Advance online publication. DOI: 10.1080/16549716.2024.2432069 PMID: 39611321

Le Masson, V., Norton, A., & Wilkinson, E. (2019). *Gender and resilience: From theory to practice*. Overseas Development Institute (ODI). https://www.odi.org/en/publications/gender-and-resilience-from-theory-to-practice/

Le Monde. (2024, November 22). Climat: L'offensive de l'Arabie Saoudite, de l'Iran et du Vatican sur l'égalité entre les genres à la COP29. *Le Monde*. Retrieved from https://www.lemonde.fr/planete/article/2024/11/22/climat-l-offensive-de-l-arabie-saoudite-de-l-iran-et-du-vatican-sur-l-egalite-entre-les-genres-a-la-cop29_6408741_3244.html

Montaud, J. (2020). Enhancing food security through climate-smart agriculture in Mali. *Journal of Environmental Policy and Planning*, *22*(4), 567–583. DOI: 10.1080/1523908X.2020.1750679

Perkins, P. (2018). Climate justice, gender, and intersectionality. In *Climate justice: A global perspective* (pp. 349-358). https://doi.org/DOI: 10.4324/9781315537689-26

Rahman, S., & Alam, K. (2020). Women's leadership in community-based adaptation to climate change: A case study from Sub-Saharan Africa. *Climate and Development, 12*(5), 462–473. DOI: 10.1080/17565529.2020.1754246

Reuters. (2024, December 2). Comment: Women are custodians of nature and key to improving the integrity of carbon markets. *Reuters*. Retrieved from https://www.reuters.com/sustainability/land-use-biodiversity/comment-women-are-custodians-nature-key-improving-integrity-carbon-markets-2024-12-02/

Shonkoff, S., Morello-Frosch, R., Pastor, M., & Sadd, J. (2011). The climate gap: Environmental health and equity implications of climate change and mitigation policies in California—a review of the literature. *Climatic Change, 109*(S1), 485–503. DOI: 10.1007/s10584-011-0310-7

Smith, L. C., Kosec, K., & Frankenberger, T. (2021). *Food security and gender: A review of key issues*. International Food Policy Research Institute (IFPRI). https://doi.org/DOI: 10.2499/9780896292001

Sorensen, C., Murray, V., Lemery, J., & Balbus, J. (2018). Climate change and women's health: Impacts and policy directions. *PLoS Medicine, 15*(7), e2002603. DOI: 10.1371/journal.pmed.2002603 PMID: 29990343

Tirado, M. C., Hunnes, D., Cohen, M. J., & Lartey, A. (2019). Climate change and nutrition in Africa. *Annals of Nutrition & Metabolism, 75*(2), 137–145. DOI: 10.1159/000503669

UNFPA (United Nations Population Fund). (2019). *The impact of climate change on maternal and newborn health*. UNFPA Reports. https://www.unfpa.org/publications/climate-change-and-maternal-health

United Nations. (n.d.). *Women & climate change*. Retrieved from https://www.un.org/en/climatechange/science/climate-issues/women

United Nations Foundation. (n.d.). *Five facts about gender equality and climate change*. Retrieved from https://unfoundation.org/blog/post/five-facts-about-gender-equality-and-climate-change/

Van Daalen, K. R., Jung, L., Dhatt, R., & Phelan, A. L. (2020). Climate change and gender-based health disparities. *The Lancet. Planetary Health, 4*(2), e44–e45. DOI: 10.1016/S2542-5196(20)30001-4 PMID: 32112742

Waheed, S., & Waheed, H. (2022). Impact of climate finance on gender equity for sustainable global development: Can aid for climate action also aid gender equity? *Brazilian Journal of Science*, *1*(7), 82–94. DOI: 10.14295/bjs.v1i7.145

WHO. (World Health Organization). (2021). *Gender, climate change and health.* WHO Reports. https://www.who.int/publications/i/item/9789240026389

Women, U. N. (2022). *Gender-responsive climate action: A call for gender-disaggregated data.* UN Women Publications. https://www.unwomen.org/en/digital-library/publications/2022/07/gender-responsive-climate-action

Wong, S. (2016). Can climate finance contribute to gender equity in developing countries? *Journal of International Development*, *28*(3), 428–444. DOI: 10.1002/jid.3212

World Health Organization. (2022). *The top 10 causes of death.* Retrieved from https://www.who.int/news-room/fact-sheets/detail/the-top-10-causes-of-death

World Meteorological Organization. (2024, December 30). Climate change impacts grip globe in 2024. *WMO.* Retrieved from https://wmo.int/media/news/climate-change-impacts-grip-globe-2024

KEY TERMS AND DEFINITIONS

Climate Adaptation: The adjustment in natural or human systems in response to actual or expected climatic stimuli or their effects that moderates harm or exploits beneficial opportunities. Gender-sensitive adaptation implies integrating the roles and needs of women in adaptation processes.

Climate Change: A permanent variation in temperature, precipitation patterns, and other weather conditions that affect the global environment, typically influenced by a number of human endeavors, such as burning fossil fuels and cutting down forests.

Climate-Smart Agriculture: Agriculture promoting productivity and climate-mitigation and adaptation, showing benefits in relation to increased food security while reducing emissions. Gender-equitable climate smart agriculture is the opportunity for women to access knowledge, inputs and technology to thrive under changing conditions.

Gender-Responsive Policy: Policy that is shaped to take into account all needs of women and men, respectively, considering disparities between the women and men, with an intention to eliminate the disadvantageous conditions leading to gender discrimination in coping with disasters pertinent to climate change.

Gender-Specific Data: Data gathered, sorted by sex, and presented for men and women to demonstrate the different ways in which climate change may affect each sex in relation to public health, economic status, and social roles.

Intersectionality: A concept to help understand how various social identities (e.g. race, class, gender and disability) combine and add to each other, influencing an individual's or group's experience of oppression or of privilege, particularly in relation to climate vulnerability.

Maternal health risks: Health risks including those that are unique to women pregnant or in the process of giving birth. Climate change is compounding these risks through greater exposure to heat stress, vector-borne diseases, and lack of care related to climate-induced disasters.

Vulnerability: The extent to which people, communities or systems are vulnerable to the negative effects of climate change, social inequalities and economic adversity entrenched due to poverty. Women, particularly those in deprived commutates, are more at risk because of their biological and socio-economic-related factors.

Chapter 9
Lost Childhoods: Climate Displacement and Child Marriages in the Indian Sundarbans

Vaidehi Uniyal
https://orcid.org/0009-0003-4982-6479
Graphic Era University, India

Debapriya Ganguly
Graphic Era University, India

ABSTRACT

The Indian state of West Bengal is home to the Sundarbans, a coastal delta and a significant area vulnerable to climate change. The complex nexus of environmental and climate crises, exacerbating vulnerability and insecurity, leads to climate-induced displacement. Climate displacement has been extensively observed in the Sundarbans. Concerning children, after disasters, families are left with limited livelihood opportunities and thus resort to child marriages as a coping mechanism in the face of poverty, displacement, and perceived threats to girls' safety. It is now essential to analyze the impact on children who become victims of early marriages due to climate displacement. Climate change is not just an environmental problem; it is also a gendered humanitarian crisis. This chapter investigates how climate displacement intensifies the unique challenges that exacerbate gender-specific vulnerabilities and diminish opportunities for childhood leisure, ultimately impacting the mental health of these young brides.

DOI: 10.4018/979-8-3373-3281-9.ch009

INTRODUCTION

The Sundarbans is a delta region beside the coast and a substantial climate focal point in southern Bangladesh and West Bengal, India. The Indian section of the Sundarbans comprises 54 islands and is occupied by roughly 5 million inhabitants (Ghosh et al., 2018). The Sundarbans face dangers from both human activities and natural causes. Among the natural threats are rising sea levels caused by climate change, frequent cyclones, and coastal erosion (Uddin et al., 2015). Indrila Guha and Chandan Roy (2016) state that the environmental consequences, such as rising sea levels, floods, droughts, and natural disasters, have reached a point that displaces residents from islands for numerous generations, leaving them homeless and impoverished. In the Indian Sundarbans, notable climate displacement has been pushed by severe weather events and escalating sea levels connected to climate change (Guha & Roy, 2016). Anuttama Ghose & S M Aamir Ali (2024) showcased Climate-induced displacement, a circumstance in, people are pressured to move, at least in part, due to a weather-related event associated with climate change, which includes persistent droughts, hurricanes, floods, or growing temperatures (Ghose & Aamir, 2024).

The climate emergency and human displacement are becoming more intertwined. Those who are forced to leave their homes due to climate change are referred to as "climate refugees." Climate displacement has been observed extensively in the Sundarbans. Bushra Ali Khan (2024) demonstrated that the Sundarbans delta, located across the India-Bangladesh border, is among the region's most at-risk areas, facing rising salinity, soil erosion, and flooding that result in widespread displacement. Estimates suggest that by 2050, millions in India may be forced to relocate due to climate change, which would create substantial social, economic, and political difficulties (Khan, 2024). It is essential to highlight that climate displacement and its ensuing effects are gendered. Concerning children, after disasters, families are left with scarce livelihood opportunities and thus resort to child marriage as a coping mechanism in the face of poverty, displacement, and perceived threats to girls' safety. Roberto Pasten et al., (2020) in the work entitled *"Not a Dream Wedding: The Hidden Nexus Between Climate Change and Child Marriage"* defines Child marriage as the union of a girl or boy below the age of 18, encompassing both formal marriages and informal partnerships where children cohabit. While it impacts both genders, girls are particularly affected to a greater extent. Meanwhile, Climate change is a worldwide issue with potentially far-reaching adverse effects, and it has already led to social, health, and economic repercussions (Roberto et al., 2020).

According to a report by Krishnendu Bandopadhyay (2025), in the last ten years, 55% of residences in the Sundarbans have reported a prominent increase in child marriages. Between 2019 and April 2022, 76 occurrences of child marriage were registered. Families facing economic and environmental difficulties often marry off their daughters at ages as young as 14 or 15 to ease economic hardships and address perceived safety concerns. Consequently, 73% of these girls move to other districts, usually with no proper background checks on the patterns, which only exacerbates their expectation risk (Bandopadhyaya, 2025). The increase in climate-linked early marriages brings the term "climate brides" into existence. If a marriage is not successful and leads to the exploitation of a young girl, it can affect her psychologically and lead to the deterioration of her overall health. Child marriage is the most probable cause of an early married girl's repeated experience of depression, worry, and isolation, besides the fact that her self-injury and suicidal thoughts are tested to be at a higher rate. It is the situation of orchestration of abduction that creates a traumatic memory in the girl. She is affected by that experience for a long time, which affects her mental health (Parsons et al., 2015). Shilpa Aggarwal et al. (2023) discussed that the mental health impacts of early marriage on girls have been highly debated, yet no prospective studies have been conducted to test the association. It is a fact that child marriage prevents the girl from enjoying her rights, mostly in education, and it limits her interaction with her family and friends (Aggarwal et al., 2023). The present study attempts to understand how women navigate their identities due to induced climate displacement, particularly as climate brides, victims of trafficking and prostitution, as well as through issues related to mental health. This phenomenon of climate brides is often reported widely but remains unexplored in terms of well-being and long-term mental health. In previous discussions and studies, climate change and natural disasters in the Indian Sundarbans have been explored. However, there has been a noticeable lack of ongoing research following these environmental upheavals. It is now crucial to examine how climate displacement affects the mental health of children who fall victim to trafficking, exploitation, and early marriages as a result.

There has been limited conversation regarding this matter. India's current legal structure lacks the necessary provisions to address the intricate and diverse challenges posed by climate migration. There is no distinct legal classification for climate migrants, and the nation's policies on migration and displacement are primarily based on antiquated legislation, including the 1946 Foreigners Act and the 1955 Citizenship Act. Consequently, climate refugees fail to gain the acknowledgment or safeguards they need, putting them at risk of exploitation, marginalization, and a lack of legal status (Bhushra Ali Khan, 2024). Furthermore, there are national regulations aimed at curbing child marriages; however, there are no specific laws for the climate brides of the Sundarbans who face vulnerabilities and struggle with mental health issues. These individuals are overlooked despite the existence of national legislation, and due to insufficient monitoring, exploitation continues to occur. Additionally, the study evaluates the gaps in legal frameworks that address the victims of child marriages caused by climate displacement.

Within this framework, climate change is not just an environmental problem but also a gendered humanitarian crisis. This research aims to investigate the lived experiences of young climate brides in the Sundarbans, concentrating on mental health issues and overall well-being. The study will also delve into how climate displacement challenges exacerbate gender-specific vulnerabilities and diminish opportunities for childhood leisure, ultimately impacting the mental health of these young brides. Furthermore, the study will conclude with an examination of the existing legal policies and frameworks that aim to address the issues of child marriage and its intersections with climate change.

SCHOLARLY REVIEW

The review is thematically structured to highlight key areas relevant to the study: ecological vulnerabilities of the Indian Sundarbans, climate-induced displacement, gendered impacts of climate change, Child marriage as a coping mechanism, and the mental health implications for girls who become child brides.

Climate change in the Indian Sundarbans: Research on climate change in the Indian Sundarbans has expanded considerably over the last decade. Many scholars, including Hassan (2018), have noted that the Sundarbans are the largest mangrove forest globally and a UNESCO World Heritage Site located at the boundary of two developing South Asian nations—Bangladesh and India. Due to its position on the ever-changing Ganges delta, the forest and its inhabitants are highly susceptible to the effects of climate change (Hassan et al., 2018). The United Nations Framework Convention on Climate Change (UNFCCC) characterizes climate change as "a shift in climate that is directly or indirectly caused by human actions, resulting in changes to the composition of the global atmosphere, and that occurs alongside other natural climate variations that have been recorded over similar time frames." (Karmakar, 2021). The Sundarban ecosystem is one of the most biologically rich and taxonomically diverse ecosystems in the Indian Subcontinent (Mahadevia and Vikas, 2012). Upasona Ghosh et al. (2018) argued that the delta faces significant climatic and ecological challenges, such as vanishing mangrove forests, islands affected by rising sea levels, unpredictable rainfall, and cyclones. Primarily, the population comprises individuals who do not have land who, however are dependent on agriculture and fishing for their daily bread. The incidence of the lowest levels of poverty and deprivation in the population is extremely high, with migration presenting itself as a key concern during the developed world era (Ghosh et al., 2018). At the moment, the Sundarbans are exposed to hazards caused by both people's actions and nature. Among the main natural hazards is the increasing sea level, which is a result of climate change. According to estimates, the most the Sundarbans' most biodiverse parts will be left with, could drop from 60% to 30% by 2100 with an 88 cm increase in sea level compared to 2001, shrinking the availability of forest products and the communities that depend on them (Uddin et al., 2015). A paper titled "Climate Change in South Asia for Sustainable Development - A Thorough Study of History" found out that the long-term effects of climate change like temperature changes, storms, seasonal changes, droughts, floods and shifts in rainfall patterns, may affect the lives of communities living in arid regions in Southeast Asia (Afzala & Nishtar, 2023). Author, Ritwika Mitra, in (2022) holds that the forest's withstanding abilities have been completely weakened due to the extensive influence of climate change. Along with the increasing salinization of the soil and the intrusion of saltwater from the Bay of Bengal, the rising number of floods, cyclones, river erosion, and soil degradation has become the prime cause for the mass displacement of people in the Sundarbans most of whom have been in agriculture and fisheries for their subsistence. (Mitra, 2022). At present, the climatic variability is the leading factor responsible for mass migration and displacement; the Sundarbans district being among the many regions affected; in fact, many of these areas have experienced successive hurricanes, for example, Amphan and Yaas, which have left the most vulnerable populations in a situation of chronic insecurity and suffering.

Climate-induced migration and displacement: In a study conducted by Mistri (2019), it was found that the inhabitants of Sundarban primarily rely on activities centered around natural resources, such as agriculture and fishing, both of which are vulnerable to the unpredictable climate. Approximately 59 percent of the workforce is farming, while five percent participate in fishing. Migration serves as a significant means of livelihood in the Sundarban region. At least one individual from three-fourths of the households leaves Sundarban to seek employment (Mistri, 2019). Mamoon Bhuyan (2023) in his study Children of Sundarbans: A Medico-Sociological Study of Their Health Conditions mentions that the livelihood options in the Sundarbans are restricted to small-scale fishing and farming. Men often migrate to nearby towns such as Canning and Kolkata in search of improved employment opportunities, leaving the women and children behind (Bhuyan, 2023). Countless individuals residing in low-lying coastal regions might be forced to abandon their homes if sea levels increase as forecasted by numerous climate-change specialists. Extended and intense droughts could push additional farmers from rural areas to urban centers in search of new opportunities (Stephenson et al., 2010). In the past, the main reason for migration was the pursuit of improved employment prospects (pull factors). Migrants often took on roles as casual laborers, such as pipeline fitters for the Oil and Natural Gas Commission (ONGC) and construction workers, among others. This type of migration was typically seasonal. Following the impact of cyclone Aila in 2009, there has been a rise in migration to states like Kerala (Gujral, 2024). Climate change has led to an increase in migration from the region, driven by the push factor. In an ethnographic study titled "Livelihood Barriers and Coping Strategies of Sundarban Dwellers," conducted by R. Biswas in 2024, one respondent remarked that individuals in the Sundarbans have begun to migrate to urban areas due to increasing vulnerability from climate change and dwindling livelihood opportunities. "People from both the Indian side and Bangladesh have started to migrate north, albeit in smaller numbers. This migration has not yet drawn significant attention. The rising salinity is leading to health issues, and due to climate change, biodiversity and the quantity of fish are decreasing. As a result, opportunities are diminishing (Biswas, 2024). However, migration frequently results from involuntary displacement. The phenomenon of migration taking place after a natural disaster (gradual and sudden onset) is referred to as climate displacement. Climate displacement can be seen as a type of forced migration, where individuals are compelled to leave their homes due to both abrupt and gradual alterations in their environment brought about by climate change, including rising sea levels, severe weather occurrences, and environmental degradation. Scholars such as Shaberi Das and Sugata Hazra (2020) argue that in situations of forced migration caused by climatic factors, no individual party or group (including the displaced person, the government, or an international organization) can be solely deemed responsible and accountable for addressing the associated hardships. The local communities in areas made uninhabitable by various influences, particularly climate change and rising sea levels, bear the human costs (Das & Hazra, 2020). The complex nexus of environmental and climate crises, exacerbated vulnerability, and security brings about displacement. Climate displacement has been observed extensively in the Sundarbans. It is essential to highlight that climate displacement and its ensuing effects are gendered.

Gendered climate change: Climate change impacts all communities. Nevertheless, researchers have examined how the effects of climate change are not uniform. For instance, while men experience its impacts, women are affected differently, and while women face distinct challenges, children are similarly impacted in unique ways. The phenomenon of gendered climate change persists, highlighting that the consequences and vulnerabilities associated with climate change vary greatly across different genders. Researchers like Rebecca Pearse (2017) argued that feminist scholars have started to explore the connections between gender and climate change. They pose questions like: Do climate change effects differ for women and men? Have gender hierarchies contributed to the social causes and impacts of climate change? How does gender interact with class, race, ethnicity, and sexuality in a changing climate? Is there a gendered aspect to climate governance? What actions can be taken to ensure that political and social responses to climate change do not exacerbate gender inequality? (Pearse, 2017). Research focused on the essential questions has highlighted the nature of vulnerabilities related to climate. Men and women perceive, comprehend, and respond to climate change in distinct ways, making it essential to analyze the ongoing and anticipated changes through a gender lens. Climate change is expected to intensify gender-specific vulnerabilities and differing capacities to manage the changes on various levels (Sultana & Syracuse University, 2014). Research on Gendered Impacts of Climate Change by Anna Fruttero et al., 2023 has highlighted that Girls and women are especially susceptible to the societal reactions caused by weather-related disasters, particularly in areas where they encounter limiting gender expectations. For instance, drought conditions may lead to diminished resources, which can result in sex-selective abortions, malnutrition, and the neglect of girls due to a preference for sons. As a result, weather disruptions have been linked to increased mortality rates among girls, poorer health outcomes, and a rise in school dropouts as they are required to take on household responsibilities or enter into early marriages (Fruttero et al, 2023). Children are generally more susceptible than adults to environmental influences (like heat, pollution, or a lack of food) since they are physically less strong and have a harder time regulating their body temperature. Additionally, substantial evidence exists showing that environmental factors affecting pregnancy and early childhood can have lasting impacts into adulthood. In simple terms, healthier children develop into healthier, wealthier, and more educated adults (Hanna & Oliva, n.d.). Janet Currie and Olivier Deschênes (2016) argued that Climate change will significantly change Earth's climate system in various ways, posing risks to children's physical and mental health (Currie et al., 2016). Climate change impacts the spread of infectious diseases, urban coastal areas, crop production, agricultural and health pests, invasive organisms, biodiversity, the timing of seasonal events in plants, the distribution of species, and numerous other elements of the natural environment (Cohen, 2010). Climate change is anticipated to increase the frequency and intensity of extreme weather events, including storms, cyclones, floods, and droughts. Children are especially at risk from these dangers since they rely significantly on their caregivers for safety and security during such occurrences (Bennett & Friel, 2014). As stated in a UNICEF report, around 1 billion children face an 'extremely high risk' from the effects of the climate crisis. Experiencing the impacts of climate change in childhood can lead to immediate harm or create issues that may surface later in life, leaving enduring effects that persist across a lifetime and through generations (Akachi et al., 2009). Krishnendu Bandopadhyay (2025) has proposed that some families that have to deal with economic and environmental emergencies force their daughters to marry at 14-15 to mitigate the financial burden and protect them from various dangers. Thus, 73% of these girls relocate to other districts, sometimes without any proper screening of the partners, putting themselves in vulnerable situations (Bandopadhyay, 2025).

Child marriage as a coping mechanism: According to UNFPA (2020), child marriage refers to a formal or informal union where at least one participant is under the age of 18. Although boys can also be involved in child marriage, it is predominantly girls who suffer the most. Worldwide, 21% of girls are wed before they turn 18, and 12 million girls under the age of 18 get married each year (Pope et al., 2022). As reported by the United Nations Children's Fund (UNICEF), more than 700 million women globally were wed before they turned eighteen. Over one-third of these women (exceeding 250 million) entered into marriage before reaching their fifteenth year. Almost every two seconds, another girl under eighteen gets married, amounting to around twelve million girls annually (McLeod et al., 2021). The concept of child marriage as a survival strategy involves marrying young girls to ease the financial strain on impoverished families. Numerous families face challenges financially, socially, and economically due to the impacts of the climate crisis. Events such as droughts and floods force families and low-income individuals to relocate due to various push factors. While boys often leave in search of employment, the responsibilities of girls stay with the family, leading to their early marriage for safety reasons. According to writers like Janet Currie and Olivier Deschêne (2019), when the environmental elements are severe, such as heat waves, the scarcity of water, and natural disasters, kids will be the first to sense it. They are very susceptible to the immediate effects of climate change. The indirect climate-related challenges are the second type of climate change which can also come from a lot of factors such as people who are willing to do everything for the global market and also malformations of the skull caused by diseases such as those transmitted by mosquitoes (vectors), as well as by economic instability, war, malnutrition, or migration which all could be the key factors that will trap the children. Children are the ones who are always left with the dirt under their feet and untouched by the spate of upheaval in a community. With the substantial societal developments that may be ushered in by climate change, kids will likely be the most affected (Currie & Deschene, 2019).

Child marriage issues in the Sundarbans are being paid very little attention, a topic that is rarely discussed. The challenges related to the Sundarbans environmental crisis have persisted for decades, yet the factors causing harm to individuals remain unaddressed. The full effects of the climate crisis are being experienced by impoverished families and residents, prompting families to displace and resort to marrying off young girls to ease their financial strain. Offering these children in marriage at such an early age, with no assurance that their futures won't involve exploitation, is deeply troubling. There is a significant need for comprehensive research and monitoring of these matters.

An examination of the current literature indicates clear gaps in understanding the intersection between climate displacement, child mortality, and their effects on mental health within the Indian Sundarbans. Scholars and researchers have addressed topics such as the climate change crisis, climate-related displacement, and child fatalities, but these issues have often been discussed separately. There has been insufficient analysis of how these factors impact different demographics. a) Insufficient exploration of the climate displacement residents of the Indian Sundarbans. b) Limited exploration of the complex nexus of the climate crisis and child marriage and its effects on mental health. c) Limited analysis of the implementation of the legal and policy frameworks brought forth addressed child brides and their mental health. This research aims to explore the interconnectedness of climate displacement with mental health and the overall well-being of child brides. It also highlights how vulnerabilities can lead a child to become a child bride and how early marriage adversely affects their mental health and well-being. There is a pressing need to address the intersection between climate displacement and child marriage (climate brides), within vulnerable regions like the Indian Sundarbans.

ANALYSIS & FINDINGS

Building upon the above gaps, this research seeks to explore the intersection of climate-related displacement and the vulnerabilities faced by child brides, while also considering the mental health challenges present among children. From the scholarly review conducted, key themes have surfaced: a) Precarious Wetlands and climate-induced displacement in Indian Sundarbans, b) Becoming a child bride after Displacement, c) Mental health and well-being in the post-climate crisis period, d) Loss of Gendered leisure and identity, e) Inadequate legal reforms and social support.

Precarious Wetlands and Climate-Induced Displacement in the Indian Sundarbans

The Sundarbans is the biggest mangrove wetland globally. It spans an area of approximately 1 million hectares, with 60% situated in Bangladesh and the other 40% located in the western region of India (Rahman & Asaduzzaman, 2010). The Indian Sundarbans, located in the delta areas formed by the Ganges, Brahmaputra, and Meghna rivers, are among the most vulnerable inhabited wetland regions in South Asia. The theme of precarious wetland and climate-induced displacement emerges as a major concern in understanding the socio-environmental transformation the area is currently experiencing. The Sundarbans ecosystem is delicate and one-of-a-kind. Its distinctiveness lies in its vast mangrove forests, the continuous blending of freshwater and saltwater, and the diverse communities living there. Additionally, the mangroves serve as a natural defense against coastal erosion and the intrusion of seawater (Ghosh et al., 2018). In a study entitled "Climate Change, Ecological Stress and Livelihood Choices in Indian Sundarban," Ghosh & Roy (2022) assert that the substantial human population in Sundarban faces a range of new challenges as a result of global climate change (Ghosh & Roy, 2022). Nonetheless, the vibrant yet delicate ecosystem encounters various pressures associated with interactions between soil, water, and climate. Human actions like farming, shrimp aquaculture, deforestation, and pollution have negatively impacted the natural environment. Existing management approaches are not adequate to tackle these issues effectively. Individuals move from their rural communities to engage in the urban informal sector as a result of environmental risks and social and economic exclusion. The challenges posed by ecological vulnerabilities are intensified by the lack of adequate social safety nets in rural and urban settings. Assistance for those affected by cyclones often fails to reach the most at-risk populations, particularly those without the ability to establish connections with local political figures (Tenhumen et al., 2024).

Migration and relocation due to climate change are currently occurring in various regions globally, where communities are exposed to significant dangers from weather-related disturbances and pressures (Arnall et al., 2019). Climate-induced displacement in the Sundarbans has become a critical repercussion of these escalating complications. Rising sea levels, frequent cyclonic storms, escalated salinity, and the deterioration of agricultural land have converted many areas into uninhabitable or economically unsustainable. These environmental stressors, heightened by inadequate infrastructural reinforcement and intractable poverty, are compelling entire communities to renounce their lineage lands. Therefore, displacement is no longer a future hazard but an ongoing process, reforming the demographic and socio-economic structure of the geographical area. A range of interrelated factors leads to this displacement:

Diverse Factors of Climate-Induced Displacement

Cyclones

One of the most significant environmental forces of displacement in the Indian Sundarbans is the increasing occurrence of cyclones. There has been a noted rise in the strength of cyclonic storms affecting the Sundarbans from 1951 to 2010. Recent studies indicate that this heightened intensity could be linked to a rise in sea surface temperatures (Mahadevia & Vikas, 2010). In a study entitled "Sundarban mangroves: diversity, ecosystem services and climate change impacts" by Sucharit Basu et al. (2016) noted that over the past ten years, the cyclonic storm "Sidr" in 2007 devastated around 30,000 hectares, which is roughly one-fourth of the Sundarban mangrove forests. Additionally, in May 2009, much of the impact of cyclone Aila was mitigated by the mangroves, protecting the city of Kolkata and other nearby urban areas. However, in light of a recent series of severe events, including multiple cyclones in quick succession, there is growing concern that the mangroves' ability to regenerate has been compromised (Neogi et al., 2016).

Natural Disaster

Natural hazards are becoming a significant global issue. In recent decades, disasters have increased in both occurrence and intensity, resulting in extraordinary losses of human lives and property (Dasgupta & Shaw, 2015). It is also thought that natural disasters can influence households over short, medium, and long-term periods (Mazumdar et al., 2014). Low-income families frequently have restricted access to income, clean water, and sanitation facilities. Additionally, many individuals living in poverty reside in poorly built homes, making them more vulnerable to extreme weather events and other direct and indirect consequences of natural disasters (Hajra et al., 2017). Natural disasters lead to a higher availability for traffickers, particularly in regions with fragile institutions and limited capacity. Similar to the Sundarbans region, where institutions are quite weak and natural disasters often occur at their highest frequency, resulting in human losses.

Pandemics

In addition to ecological disasters, pandemics such as COVID-19 have exacerbated the vulnerabilities of Sundarbans inhabitants. Due to decreased income possibilities stemming from COVID-19, the devastation of properties caused by Amphan, and reduced productivity of agricultural land due to climate change, the marginalized communities in the Sundarbans are nearing a state of survival. The cumulative effects of these challenges have compelled "Webinar Report on Impact of Pandemic on Sundarbans" (2021) to state that the economy of the Sundarbans relies significantly on agriculture, fishing, and eco-tourism. All these sectors have been negatively impacted by COVID-19. As migrant workers returned during the lockdown, many found themselves unemployed and unable to sustain themselves through agriculture. With land no longer providing sufficient resources for numerous residents, many island dwellers have turned to the forest to search for food. This change has increased tiger attacks as desperate residents of the Sundarbans assertively enter areas at unsafe times (Webinar Report on Impact of Pandemic on Sundarbans 2021).

Becoming a Child Bride After Displacement

In the wake of climate-related displacement, young women in the Sundarbans face a crisis of transition. Displacement results not only in the loss of land, livelihood, and protection for families. It also disrupts the very development of childhood. Activities once linked with happiness, education, and defining aspects of childhood play are replaced by household duties and early adult duties. In such situations, child marriage becomes a coerced option to a lost childhood, performing not solely as a cultural practice but as a planned existence mechanism. As environmental disruptions such as cyclones and salinity aggravate, the economic and emotional toll on families increases. As reported by Shreya Deb (2022), Climate disasters and the COVID lockdown led to an increase in trafficking and child marriages. A respondent commented in the article that due to the economic hardships faced by families in the Sundarbans islands, many are encouraging their adolescent daughters, aged 14 to 15, to marry. Circumstances contribute to the normalization of early-stage marriage as a reaction to financial strain, reduced food security, and communal expectations. This coping mechanism, though prompt in intent, has far-reaching outcomes, most notably, the limitation of girls' autonomy, education, and sexual freedom. Childhood sexuality, which should be nourished with endorsement, security, and education, becomes altered by early sexual commencement within marital setups, often without the girl's informed choice or willingness. Using the capability approach of Amartya Sen, which emphasizes that we should consider human development not just as theoretical concepts, but as actual abilities within the context of daily life. This approach highlights what every individual can do and become, their 'valuable actions and identities,' in making significant choices among various options; thus, enjoying the freedom to select a life they have valid reasons to appreciate (Walker, 2005). This instance illustrates how displacement in a critical way constrains the substantive freedoms and opportunities of girls. Lack of education, medical care, earning opportunities, and social security, families have very minimal options. Child marriage can be perceived as a means of care or a pathway to have fewer dependents in the household in these limited resource settings. The deprivation of early years' leisure, both in tangible and in sentimental development, further aggravates this concern. Girls are no longer seen as entities with growing capacities, but as economic burdens or exchangeable means. Displacement, thus, reconfigures the social and ethical landscapes of childhood, displacing possibility with insecurity.

Apart from the primary pressure into marriage, displacement-induced risk often reveals climate brides to cycles of exploitation that reflect sexual slavery and sexual exploitation. Many of these girls are displaced far from their settlements, deprived of domestic aid and societal monitoring. In these unstable situations, some are trafficked under the false promise of marriage, only to be trafficked into brothels or compelled into economically based sexual exchange within domestic or seasonal migration cycles. Even within domestic roofs, they may be sexually exploited by husbands or distant relatives, missing any legal option or personal autonomy. Almost no forum talks about this servitude. Due to social reverence surrounding marriage, it stops viewing abuse as an infringement upon one's human rights or an obligation. Climate brides suffer violence, repeated unwanted pregnancies, sexually transmitted infections, and severe psychological trauma. Deprived of mental health support, education, and sources of income, they become trapped in a never-ending cycle of abuse and marginalization. Acceptance of such suffering in displacement-affected regions only offers a deadly silence wherein administrative failure and patriarchal conventions work hand in hand to completely wipe away the very existence of these girls as survivors of sexual slavery masked behind marriage.

Mental Health and Well-being of Child Brides in the Post-Climate Crisis Period

The undergo of climate displacement has mental effects on the psychological and emotional lives of displaced people, with special susceptibility facing young girls forced into early marriages. As cyclone, salinity rise, and livelihood shock intensify, marrying off is more often a means embraced by households as a way of managing as an endurance approach, trying to reduce perceived burden and secure dependents' material needs. All this comes at the price of the effective health and unified well-being of the affected young girls. Post-crisis child brides frequently endure increased emotional trauma as a result of abrupt shifts into adult roles, household work, and reproductive responsibilities. The shortening of education, play, and peer contact breaks down the very constructing blocks of childhood growth and mental well-being. In an article by Rohini Krishnamurthy (2024) reported that the people living in Gosaba, a region of the Sundarbans, are experiencing a concerning surge in mental health issues, largely due to ongoing climate-related disasters, forced displacement, and persistent malnutrition. This rise in psychiatric conditions, such as depression, may put a significant burden on the local healthcare infrastructure, as the community deals with the consequences of regular natural disasters (Krishnamurthy, 2024). Refused autonomy and full informed consent, these girls are brought too rapidly into systems that withhold them support networks, spanning access to healthcare and psychosocial services to protected community space.

Early marriages are a drastic shortening of life options, with underlying freedoms and capabilities significantly limited. Lack of mental health care in the displacement zones and host communities further exacerbates the psychosocial consequences for the girls. They usually express experiences of isolation, anxiety, despair, and fear, compounded by intimate partner violence, poverty, and social safety net collapse. Additionally, the theft of childhood recreation and the perversion of childhood sexuality, under which emotional development is repressed by being advanced too quickly towards adult relations, is the source of long-term mental trauma. Such an interesting development is repressed by being advanced too quickly towards adult relations, which is the source of long-term mental trauma. Such intersecting risks draw attention to the compelling urgency to consider mental health as no afterthought, but rather an integral component of climate resilience and post-disaster healing, most especially among adolescent girls caught between environmental and gender-based crises.

Loss of Leisure and Identity

Climate-induced displacement in the Indian Sundarbans has severely affected the social fabric of childhood, especially for girls, by depriving them of spaces for leisure, unstructured activities, and exploration that are essential to identity structuring (Wearing, 1994). Displacement not only disrupts the physical environment but also deteriorates everyday life's societal and inner infrastructure. Leisure, often overlooked as insignificant, is a critical space where children, notably young girls, navigate their identities, articulate agency, and build emotional resilience (Jun & Kyle, 2012). In many dislocated communities, the crumbling of habitual patterns and safety has altered daily life into an endurance-focused existence, abandoning little room for enjoyment or personal growth. This curtailment of gendered leisure is specifically substantial given the gendered essence of childhood in non-urban and marginalized regions. Girls, who are already obligated with domestic roles from an early age, are now more likely to be restricted to the familial sphere due to intensified insecurity, social surveillance, and the need to aid domestic work after displacement. The deprivation of school, social engagement, and play environments eliminates crucial arenas through which girls undergo joy, freedom, and communal learning, determinants in shaping identity and desires.

Additionally, the deterioration of leisure spaces perpetuates conventional gender norms, where the reduced recognition of girls in public life intensifies choices such as child marriage. In this setting, subjectivity is not only formed by deficit but also by enforced responsibilities and limited opportunities. The loss of childhood leisure thereby aggravates emotional burden and reduces the psychological attention span needed for healthy growth, while forcing girls precipitously into adult roles.

Inadequate Legal Reforms and Social Support

Despite the gravity of the impacts climate-induced displacement has on young girls, including the heightened incidence of child marriage and the deprivation of childhood empowerment, prevalent legal frameworks and communal welfare systems in the Indian Sundarbans persist incoherent, under-enforced, and mismatched with the circumstances on the ground. This systemic gap not only worsens vulnerabilities but also leads to an internalization of rights transgressions under the guise of societal or economic urgency. While statutory frameworks such as the Prohibition of Child Marriage Act (2006), Indian Penal Code (2017), Protection of Children from Sexual Offences Act (2012) exist, they often prove ineffective to deter child marriages in uprooted and marginalized social groups due to insufficient institutional enforcement, and a lack of community level understanding. Families in turmoil are disconnected from reachable legal support systems or protection frameworks. Rather, they are left to manage crisis-induced decision-making without state direction or advocacy.

Societal protective frameworks, including education access, mental healthcare, skill development programs, and child protection welfare provisions, are either lacking or marginally incorporated into the social structure of post-disaster intervention. Humanitarian assistance efforts often prioritize core facilities restoration and survival services distribution, with sparse consideration to gendered psychosocial needs or the consistency of developmental facilities for uprooted adolescents. The shortcoming of improvement and assistance also exposes a disconnect between central policy-making and local socio-cultural settings. Lacking culturally reactive and community-led action, judicial frameworks remain superficial rather than defensive. As a result, many girls persist to fall through the gaps, married off, displaced, and remain unseen within the very systems intended to safeguard them.

The government initiatives aimed at conserving the Sundarbans ecosystem are vital for safeguarding the area against future climatic shocks and their detrimental effects. Alongside global and domestic initiatives, various NGOs have developed in the Sundarbans, focusing on issues like human trafficking, poverty, and the rescue of children. Community-level organizations like the Mukti Foundation focus on enhancing human development. MUKTI is a registered not-for-profit socio-economic development trust that operates in the Sundarbans region and other areas of West Bengal, India, aiming to uplift grassroots communities affected by poverty, social exclusion, and various forms of discrimination. The organization was established by Mr. Sankar Halder, a software engineer who graduated from Jadavpur University. It focuses its efforts on six key areas: Health, Education, Agriculture, Livestock Raising, Economy, and Right to Information (An Insight into the financial rising sun of Sundarbans – Mukti, 2021). While the Sundarbans Social Development Centre (SSDC) aims at empowering women. The Sundarbans Social Development Centre (SSDC) is a non-profit, non-governmental organization located in the Sundarbans delta region of West Bengal, India. Since its establishment as a social development entity in 1986, SSDC has actively worked for the welfare of individuals residing in the disaster-prone islands of Sundarbans, which are often recognized as socio-economically disadvantaged in terms of isolation and poor quality of life (Sundarbans Social Development Centre, 2017). Additionally, the Association for Social and Environmental Development (ASED) concentrates on initiatives for school students.

Eventually, the lack of an attentive and rights-based legislative and social system fosters not just urgent harm but to an extended erosion of confidence, strength across generations. Effective measures must address not only the *presence* of regulations but also their *adaptability*, enforcement, and *access,* especially among oppressed, climate-vulnerable communities like those in the Sundarbans.

CONCLUSION

This research has critically examined the intricate and often ignored relationship between climate-induced displacement and child marriage in the Indian Sundarbans, highlighting that the deterioration of natural surroundings due to climate change is mirrored by a deeper erosion of affective, cognitive, and social well-being among exposed children. Underlying this critical situation lies the loss of early years, sudden, forced, and intensified by intersecting distress of poverty, displacement, and gendered insecurity. A primary consequence of this research is the critical toll of climate displacement on the emotional well-being of climate brides. The encounter of being torn from familiar surroundings, amalgamated with the imposed change into adult responsibilities through marriage, causes prolonged mental distress, depression, and a deep erosion of faith in both family-based and formal structures. These psychological fractures are often unseen but are among the most persistent ramifications of climate change for girls in the Sundarbans. While legal structures and policy measures exist, they often remain detached from the ground-level conditions of uprooted communities. Their performative measures and top-down design fail to reach the emotional and developmental needs of displaced young people. Consequently, the focus must shift from mere statutory safeguards to developing need-based, grassroots-oriented mental health support systems that address mental suffering, facilitate recovery, and uplift resilience in environmentally displaced young girls. Eventually, this research suggests an immediate shift of both scholarly and political debate from legal principle to a focus on psychological well-being and child-sensitive interventions. Without such a change, the Sundarbans will persist in being an area where childhood is not merely forfeited to climate and marriage, but where the internal experiences of children, primarily girls, are portrayed as invisible, unaided, and ignored.

REFERENCES

Afzal, J., & Nishtar, Z. (2023). A substantial study on the history of climate change in South Asia for Sustainable development. *JHSS, 14–14*(1), 102–103. https://jhss-uok.com/index.php/JHSS/article/download/262/130

Aggarwal, S., Francis, K. L., Dashti, S. G., & Patton, G. (2023). Child marriage and the mental health of adolescent girls: A longitudinal cohort study from Uttar Pradesh and Bihar, India. *The Lancet Regional Health. Southeast Asia, 8*. Advance online publication. DOI: 10.1016/j.lansea.2022.100102 PMID: 37384140

Akachi, Y. Goodman, D., & Parker, D. (2009). Global Climate Change and Child Health: A review of pathways, impacts and measures to improve the evidence base. In UNICEF, *Innocenti Discussion Paper* (Report IDP 2009-03). UNICEF Innocenti Research Centre. https://www.unicef-irc.org

Arnall, A., Hilson, C., & McKinnon, C. (2019). Climate displacement and resettlement: The importance of claims-making 'from below.'. *Climate Policy, 19*(6), 665–671. DOI: 10.1080/14693062.2019.1570065

Bandopadhyay, K. (2025). 'Climate change pushing up child labour, trafficking, early marriage in Sundarbans'. The *Times Of India*. 'Climate change pushing up child labour, trafficking, early marriage in Sundarbans' | Kolkata News - The Times of India

Bennett, C., & Friel, S. (2014). Impacts of climate change on inequities in child health. *Children (Basel, Switzerland), 1*(3), 461–473. DOI: 10.3390/children1030461 PMID: 27417491

Bennett, C., & Friel, S. (2014). Impacts of climate change on inequities in child health. *Children (Basel, Switzerland), 1*(3), 461–473. DOI: 10.3390/children1030461 PMID: 27417491

Bhuyan M. (2023). Children of Sundarbans: A Medico-Sociological Study of Their Health Conditions. International Journal of Humanities, Social Science and Management (IJHSSM) Volume 3, Issue 2, Mar.-Apr. 2023, pp 822-825. Bhuyan M. (2023). Children of Sundarbans: A Medico-Sociol... - Google Scholar

Biswas, R. (2024). *Livelihood barriers and coping strategies of Sundarban dwellers* (Vol. 99). Migration and Coping Strategies of Sundarban Dwellers.

Bushra Ali Khan. (2024). Addressing the Legal Void: The Urgent Need for India to Revise Its Laws for Climate Migrants. https://shorturl.at/3sUPI

Currie, J., Deschênes, O., & Princeton University. (2016). Children and climate Change: Introducing the issue. In *The Future of Children* (pp. 3–9). Princeton University. https://oar.princeton.edu/rt4ds/file/14636/Children+and+climate+change+Introducing+the+issue.pdf

DasGupta, R., & Shaw, R. (2015). An indicator based approach to assess coastal communities' resilience against climate related disasters in Indian Sundarbans. *Journal of Coastal Conservation*, *19*(1), 85–101. DOI: 10.1007/s11852-014-0369-1

Deb, S. (2022). Climate disasters, COVID-19 lockdowns, have led to an increase in trafficking and child marriage. Time to Reclaim the Lost Innocence of Sunderbans' Children. https://www.thecitizen.in/index.php/en/NewsDetail/index/13/21704/Time-to-Reclaim-the-Lost-Innocence-of-Sunderbans-Children

Fruttero, A., Halim, D., Broccolini, C., Coelho, B., Gninafon, H., Muller, N., & Gender Global Theme. (2023). Gendered Impacts of Climate Change: Evidence from Weather Shocks. In *Policy Research Working Paper* (Report No. 10442). https://wrd.unwomen.org/sites/default/files/2023-07/WB-gendered-impacts-climate-change.pdf

Ghose, A., & Ali, S. M. (2024). Displaced by Rising Tides: Legal Insights and Policy Imperatives for Climate Refugees in the Sundarbans Region of India. https://doi.org/DOI: 10.60866/CAM.132

Ghosh, U., Bose, S., Bramhachari, R., & STEPS Centre. (2018). Living on the edge: climate change and uncertainty in the Indian Sundarbans. In *STEPS Working Paper 101*. STEPS Centre. https://www.steps-centre.org

Guha, I., & Roy, C. (2016). Climate Change, Migration and Food Security: Evidence from Indian Sundarbans. *International Journal of Theoretical & Applied Sciences*, 45–49. https://www.researchgate.net/publication/309668130

Hajra, R., Szabo, S., Tessler, Z., Ghosh, T., Matthews, Z., & Foufoula-Georgiou, E. (2017). Unravelling the association between the impact of natural hazards and household poverty: Evidence from the Indian Sundarban delta. *Sustainability Science*, *12*(3), 453–464. DOI: 10.1007/s11625-016-0420-2

Hanna, R., & Oliva, P. (n.d.). *Implications of climate change for children in developing countries* (Prashant Bharadwaj, University of California, San Diego, University of California, Santa Barbara, & Harvard University, Eds.). https://files.eric.ed.gov/fulltext/EJ1101592.pdf

Hassan, K., Higham, J., Wooliscroft, B., & Hopkins, D. (2018). Climate change and world heritage: A cross-border analysis of the Sundarbans (Bangladesh–India). *Journal of Policy Research in Tourism, Leisure & Events, 11*(2), 196–219. DOI: 10.1080/19407963.2018.1516073

Jun, J., & Kyle, G. T. (2012). Gender identity, leisure identity, and leisure participation. *Journal of Leisure Research, 44*(3), 353-378. Sci-Hub | Gender Identity, Leisure Identity, and Leisure Participation. Journal of Leisure Research, 44(3), 353–378 | DOI: 10.1080/00222216.2012.11950269

Karmakar, S. (2021). Gender and climate change: The condition of women in the Indian Sundarbans. *Scholars Journal of Arts. Humanities and Social Sciences*, *9*(1), 1–5. DOI: 10.36347/sjahss.2021.v09i01.001

Krishnamurthy, R. (2024). Climate disasters cause a surge in mental health disorders in a Sundarbans block. https://www.downtoearth.org.in/climate-change/climate-disasters-cause-surge-in-mental-health-disorders-in-a-sundarbans-block

Mahadevia, L. K., & Vikas, L. K. (2010). CLIMATE CHANGE – IMPACT ON THE SUNDARBANS: A CASE STUDY. International Scientific Journal Environmental Science http://environment.scientific-journal.com. IMPACT ON THE sundarbans: a case study.pdf

Mahadevia Ghimire, K., & Vikas, M. (2012). Climate change–impact on the Sundarbans, a case study. [PDF] [Climate Change – Impact on the Sundarbans: A case study]. *International Scientific Journal: Environmental Science*, *2*(1), 7–15.

Mazumdar, S., Mazumdar, P. G., Kanjilal, B., & Singh, P. K. (2014). Multiple shocks, coping and welfare consequences: Natural disasters and health shocks in the Indian Sundarbans. *PLoS One*, *9*(8), e105427. DOI: 10.1371/journal.pone.0105427 PMID: 25170846

McLeod, C., Barr, H., Rall, K., & Human Rights Watch. (2021). DOES CLIMATE CHANGE INCREASE THE RISK OF CHILD MARRIAGE? A LOOK AT WHAT WE KNOW—AND WHAT WE DON'T—WITH LESSONS FROM BANGLADESH AND MOZAMBIQUE. In *COLUMBIA JOURNAL OF GENDER AND LAW: Vol. 38.1* (pp. 96–98). https://journals.library.columbia.edu/index.php/cjgl/article/view/4604

Mistri, A. (2019). Is the Migration from the Indian Sundarban an Environmental Migration? Investigating through the Sustainable Livelihood Approach (SLA). In *Asian Profile*. (PDF) Is the Migration from the Indian Sundarban an Environmental Migration? Investigating through the Sustainable Livelihood Approach (SLA)

Neogi, S., Dey, M., Kabir, S. M., Masum, S., Kopprio, G., Yamasaki, S., & Lara, R. (2016). Sundarban mangroves: diversity, ecosystem services and climate change impacts. Asian Journal of Medical and Biological Research ISSN 2411-4472 (Print) 2412-5571. CONICET_Digital_Nro.1c68df3b-9a5e-459d-b1de-152865145bf0_A.pdf

Parsons, J., Edmeades, J., Kes, A., Petroni, S., Sexton, M., & Wodon, Q. (2015). Economic Impacts of Child Marriage: A Review of the Literature. *The Review of Faith & International Affairs*, *13*, 3, 12–22. DOI: 10.1080/15570274.2015.1075757

Pearse, R. (2017). Gender and climate change. In Hulme, M. (Ed.), *WIREs Clim Change*., DOI: 10.1002/wcc.451

Pope, D. H., McMullen, H., Baschieri, A., Philipose, A., Udeh, C., Diallo, J., & McCoy, D. (2022). What is the current evidence for the relationship between the climate and environmental crises and child marriage? A scoping review. *Global Public Health: An International Journal for Research, Policy and Practice*, *18*(1). Advance online publication. DOI: 10.1080/17441692.2022.2095655 PMID: 36403290

Rahman, M., & Asaduzzaman, M. (2010). ECOLOGY OF SUNDARBAN, BANGLADESH. *J. Sci. Foundation*, *8*(1&2), 35–47. http://www.rhinoresourcecenter.com/pdf_files/137/1371726022.pdf

Ghosh, S., & Roy, S. (2022). Climate change, ecological stress and livelihood choices in Indian Sundarban. Climate change and community resilience, 399.

Stephenson, J., Newman, K., & Mayhew, S. (2010). Population dynamics and climate change: What are the links? *Journal of Public Health*, *32*(2), 150–156. DOI: 10.1093/pubmed/fdq038 PMID: 20501867

Sultana, F. & Syracuse University. (2014). Gendering climate Change: geographical insights. In *The Professional Geographer* (Vols. 66–66, Issue 3, pp. 372–381). https://doi.org/DOI: 10.1080/00330124.2013.821730

Uddin, Md. S., Shah, M. A. R., Khanom, S., Nesha, Mst. K. (2015). Climate change impacts on the Sundarbans mangrove ecosystem services and dependent livelihoods in Bangladesh. In the *XIV WORLD FORESTRY CONGRESS*. AJCB-Vol2-No2-Shams-Uddin et al.pdf

Walker, M. (2005). Amartya Sen's capability approach and education. *Educational action research, 13*(1), 103-110. Sci-Hub | Amartya Sen's capability approach and education. Educational Action Research, 13(1), 103–110 | DOI: 10.1080/09650790500200279

Wearing, B. M. (1994). The pain and pleasure of gendered leisure. *World Leisure & Recreation*, *36*(3), 4–10. DOI: 10.1080/10261133.1994.9673919

Webinar Report on Impact of the Pandemic on Sundarbans. (2021). In *Webinar Report* [Report]. https://www.bing.com/search?pglt=299&q=Webinar+Report+on+Impact+of+Pandemic+on+Sundarbans.+(2021).+In+Webinar+Report+%5BReport%5D.&cvid=681922daec1b4f73803bb96197878c0d&gs_lcrp=EgRlZGdlKgYIABBFGDsyBggAEEUYO9IBBzY3OGowajGoAgCwAgA&FORM=ANNTA1&PC=U531

ADDITIONAL READING

Horii, H. (2021). *Child Marriage, Rights and Choice: Rethinking Agency in International Human Rights*. Routledge., DOI: 10.4324/9781003184546

John, M. E. (2021). *Child marriage in an international frame: A feminist review from India*. Routledge India., DOI: 10.4324/9781003158592

Pope, D. H., McMullen, H., Baschieri, A., Philipose, A., Udeh, C., Diallo, J., & McCoy, D. (2023). What is the current evidence for the relationship between the climate and environmental crises and child marriage? A scoping review. *Global Public Health: An International Journal for Research, Policy and Practice*, *18*(1), 2095655. DOI: 10.1080/17441692.2022.2095655 PMID: 36403290

Sen, H. (2019). The Sundarbans: a disaster-prone eco-region. *Berlin: Springer. doi, 10*, 978-3. The Sundarbans: A Disaster-Prone Eco-Region: Increasing Livelihood Security | SpringerLink

Shaw, S. M., Henderson, K. A., & Bialeschki, M. D. (2013). *Leisure, women, and gender*. V. J. Freysinger (Ed.). Venture Publ.LeisureWmnGnderebook-lookinsideOPT-libre.pdf

Stone, K., Blinn, N., & Spencer, R. (2022). Mental health impacts of climate change on women: A scoping review. *Current Environmental Health Reports*, *9*(2), 228–243. DOI: 10.1007/s40572-022-00346-8 PMID: 35267174

Titumir, R. A. M. (2022). Sundarbans and Its Ecosystem Services. Sundarbans and its Ecosystem Services: Traditional Knowledge, Customary Sustainable Use and Community-Based Innovation | SpringerLink

O'Donnell, A., & Wodon, Q. (Eds.). (2015). Climate change adaptation and social resilience in the Sundarbans. London: Routledge. 10.4324_9781315767802_previewpdf.pdf

Mizanur Rahman, M., Rakib Hossain, M., & Nazrul Islam, M. (2021). Climate Change Impact on Sundarbans: Challenges for Mitigation Strategies. Climate Change in Bangladesh: A Cross-Disciplinary Framework, 47-64. Climate Change Impact on Sundarbans: Challenges for Mitigation Strategies | SpringerLink

KEY TERMS AND DEFINITIONS

Climate Brides: The term Climate brides refers to those girls who are vulnerable to the climate crisis and are married off to relieve the burden of their families caused by the climate calamity.

Climate Change: Climate change relates to prolonged changes in global or area-specific climate patterns, particularly those changes that have become more noticeable since the late 20th century due to human-induced processes.

Climate-induced Displacement: Climate-induced displacement refers to the coerced movement of inhabitants from their native due to unfavorable effects of climate change.

Leisure: It refers to the time and space that is available from work and responsibilities, amid which one can relax, involve oneself in activities of their own choice, or find inner satisfaction.

Mental health: It refers to a person's affective, psychological, and relational well–being. It affects how individuals think, feel, act, manage tension, connect to others, and make decisions.

Compilation of References

. Arifin, S. R. M. (2018). Ethical considerations in qualitative study. *International journal of care scholars, 1*(2), 30-33.

. Coccia, M. (2021). Technological innovation. *innovations, 11*, I12.

. Colloquium on the International Protection of Refugees in Central America, Mexico, and Panama. (1984). *Cartagena Declaration on Refugees.*

. Garcia, R., & Calantone, R. (2002). A critical look at technological innovation typology and innovativeness terminology: a literature review. *Journal of Product Innovation Management: An international publication of the product development & management association, 19*(2), 110-132.

. Hasan, N., Rana, R. U., Chowdhury, S., Dola, A. J., & Rony, M. K. K. (2021). Ethical considerations in research. *Journal of Nursing Research, Patient Safety and Practise (JNRPSP) 2799-1210, 1*(01), 1-4.

. Institute on Statelessness and Inclusion [ISI]. (2021). *The statelessness stigma: A global report.* ISI.

. Lienert, C. (2025). Early warning–international collaboration using map-based and user-centered approaches. *International Journal of Cartography*, 1-11.

. Lock, S., Rubin, G. J., Murray, V., Rogers, M. B., Amlôt, R., & Williams, R. (2012). Secondary stressors and extreme events and disasters: a systematic review of primary research from 2010-2011.

. UN General Assembly. (1966a). International Covenant on Civil and Political Rights.

Adger, W. N., Barnett, J., Brown, K., Marshall, N., & O'Brien, K. (2013). Cultural dimensions of climate change impacts and adaptation. *Nature Climate Change, 3*(2), 112–117.

African Union. (2009). The Kampala Convention.

Afzal, J., & Nishtar, Z. (2023). A substantial study on the history of climate change in South Asia for Sustainable development. *JHSS, 14–14*(1), 102–103. https://jhss-uok.com/index.php/JHSS/article/download/262/130

Aggarwal, S., Francis, K. L., Dashti, S. G., & Patton, G. (2023). Child marriage and the mental health of adolescent girls: A longitudinal cohort study from Uttar Pradesh and Bihar, India. *The Lancet Regional Health. Southeast Asia, 8*. Advance online publication. DOI: 10.1016/j.lansea.2022.100102 PMID: 37384140

Akachi, Y. Goodman, D., & Parker, D. (2009). Global Climate Change and Child Health: A review of pathways, impacts and measures to improve the evidence base. In UNICEF, *Innocenti Discussion Paper* (Report IDP 2009-03). UNICEF Innocenti Research Centre. https://www.unicef-irc.org

Albrecht, G. (2022). *Earth emotions: New words for a new world*. Cornell University Press.

Alpass, F., Flett, R., Humphries, S., Massey, C., Morriss, S., & Long, N. (2004). Stress in dairy farming and the adoption of new technology. *International Journal of Stress Management, 11*(3), 270–281. DOI: 10.1037/1072-5245.11.3.270

American College of Emergency Physicians. (n.d.). *Disaster telehealth*. ACEP. https://www.acep.org/patient-care/policy-statements/disaster-telehealth

Anguelovski, I., Shi, L., Chu, E., Gallagher, D., Goh, K., Lamb, Z., & Teicher, H. (2016). Equity impacts of urban land use planning for climate adaptation. *Journal of Planning Education and Research, 36*(3), 333–348. DOI: 10.1177/0739456x16645166

Arnall, A., Hilson, C., & McKinnon, C. (2019). Climate displacement and resettlement: The importance of claims-making 'from below.'. *Climate Policy, 19*(6), 665–671. DOI: 10.1080/14693062.2019.1570065

Askland, H. H., Shannon, B., Chiong, R., Lockart, N., Maguire, A., Rich, J., & Groizard, J. (2022). Beyond migration: A critical review of climate change induced displacement. *Environmental Sociology, 8*(3), 267–278.

Augusterfer, E. F., O'Neal, C. R., Martin, S. W., Sheikh, T. L., & Mollica, R. F. (2020). The role of telemental health, tele-consultation, and tele-supervision in post-disaster and low-resource settings. *Current Psychiatry Reports, 22*(12). Advance online publication. DOI: 10.1007/s11920-020-01209-5 PMID: 33247315

Austin, E. K., Handley, T., Kiem, A. S., Rich, J. L., Lewin, T. J., Askland, H. H., Askarimarnani, S. S., Perkins, D. A., & Kelly, B. J. (2018a). Drought-related stress among farmers: Findings from the Australian Rural Mental Health Study. *The Medical Journal of Australia, 209*(4), 159–165. DOI: 10.5694/mja17.01200 PMID: 30041594

Austin, E. K., Rich, J. L., Kiem, A. S., Handley, T., Perkins, D., & Kelly, B. J. (2020). Concerns about climate change among rural residents in Australia. *Journal of Rural Studies, 75*, 98–109. DOI: 10.1016/j.jrurstud.2020.01.010

Bai, Z., Xu, Z., Xu, X., Qin, X., Hu, W., & Hu, Z. (2020). Association between social capital and depression among older people: Evidence from Anhui Province, China. *BMC Public Health, 20*, 1–11. PMID: 33066764

Bajbouj, M., Ta, T. M. T., Hassan, G., & Hahn, E. (2022). Editorial: The Nine Grand Challenges in Global Mental Health [Editorial]. *Frontiers in Psychiatry, Volume 12 - 2021*. https://doi.org/DOI: 10.3389/fpsyt.2021.822299

Bando, D. H., Teng, C. T., & Volpe, F. M. Masi, E de., Pereirame, L.A., & Braga, A.L. (2017). Suicide and meteorological factors in São Paulo, Brazil, 1996-2011: a time series analysis. *Braz J Psychiatry*, 39:220–7. https://doi.org/DOI: 10.1590/1516-4446-2016-2057

Bandopadhyay, K. (2025). 'Climate change pushing up child labour, trafficking, early marriage in Sundarbans'. The *Times Of India*. 'Climate change pushing up child labour, trafficking, early marriage in Sundarbans' | Kolkata News - The Times of India

Bartram, D. J., Yadegarfar, G., & Baldwin, D. S. (2009). A cross-sectional study of mental health and well-being and their associations in the UK veterinary profession. *Social Psychiatry and Psychiatric Epidemiology, 44*(12), 1075–1085. DOI: 10.1007/s00127-009-0030-8 PMID: 19294320

Batchelor, C. (2002). The 1954 Convention on the Status of Stateless Persons: Past, present and future. *Journal of Refugee Studies, 15*(1), 32–54.

Batchelor, C. (2020). *The legal framework for statelessness: A practical guide*. UNHCR.

Baylis, P., Obradovich, N., Kryvasheyeu, Y., Chen, H., Coviello, L., & Moro, E. (2018). Weather impacts expressed sentiment. *PLoS One, 13*(4), e0195750. DOI: 10.1371/journal.pone.0195750 PMID: 29694424

Bekkar, B., Pacheco, S., Basu, R., & DeNicola, N. (2020). Association of air pollution and heat exposure with preterm birth, low birth weight, and stillbirth in the US: A systematic review. *JAMA Network Open, 3*(6), e208243. DOI: 10.1001/jamanetworkopen.2020.8243 PMID: 32556259

Belton, M. (2019). The invisible children: The mental health of stateless youth. *Journal of International Affairs, 72*(2), 101–116.

Belton, M. (2020). The psychological burden of statelessness on women and children. *Forced Migration Review*, *65*, 30–33.

Benhabib, S. (2021). The right to have rights. In Miller, J. R. (Ed.), *Citizenship and statelessness in the age of climate change* (pp. 45–60). Routledge.

Benjet, C. (2010). Childhood adversities of populations living in low-income countries: Prevalence, characteristics, and mental health consequences. *Current Opinion in Psychiatry*, *23*(4), 356–362. PMID: 20520546

Bennett, C., & Friel, S. (2014). Impacts of climate change on inequities in child health. *Children (Basel, Switzerland)*, *1*(3), 461–473. DOI: 10.3390/children1030461 PMID: 27417491

Bennett, J., & Bennett, R. J. (1982). *Of Time and the Enterprise: North American Family Farm Management in a Context of Resource Marginality*. University of Minnesota Press.

Berry, H. L., Waite, T. D., Dear, K. B., Murray, V., & Shilton, M. (2018). The case for a national strategy on climate, health and well-being. *Public Health Research & Practice*, *28*(3), e2831804.

Betts, A., & Collier, P. (2017). *The new politics of refuge*. Oxford University Press.

Betts, A., Easton, M., & Pincus, L. (2021). The global compact for refugees and the challenges of climate-induced displacement. *International Journal of Refugee Law*, *33*(3), 481–502.

Beyer, H., & Fanning, R. (2021). *Mental health and climate change: A global perspective*. World Health Organization.

Bhatia, S., Naha, L., & Singh, R. (2018). Mental health problems among Rohingya refugees: A systematic review. *Indian Journal of Public Health*, *62*(1), 10–15. PMID: 29512559

Bhullar, N., Davis, M., Kumar, R., Nunn, P., & Rickwood, D. (2022). Climate anxiety does not need a diagnosis of a mental health disorder. *The Lancet. Planetary Health*, *6*(5), e383. DOI: 10.1016/S2542-5196(22)00072-9 PMID: 35550075

Bhuyan M. (2023). Children of Sundarbans: A Medico-Sociological Study of Their Health Conditions. International Journal of Humanities, Social Science and Management (IJHSSM) Volume 3, Issue 2, Mar.-Apr. 2023, pp 822-825. Bhuyan M. (2023). Children of Sundarbans: A Medico-Sociol... - Google Scholar

Biermann, F., & Boas, I. (2010). Preparing for a warmer world: Towards a global governance system of climate-induced migration. *Global Environmental Politics*, *10*(1), 60–88.

Biswas, R. (2024). *Livelihood barriers and coping strategies of Sundarban dwellers* (Vol. 99). Migration and Coping Strategies of Sundarban Dwellers.

Bjelland, I., Dahl, A. A., Haug, T. T., & Neckelmann, D. (2002). The validity of the Hospital Anxiety and Depression Scale. An updated literature review. *Journal of Psychosomatic Research*, *52*(2), 69–77. DOI: 10.1016/s0022-3999(01)00296-3 PMID: 11832252

Bjornestad, A., Cuthbertson, C., & Hendricks, J. (2021). An analysis of suicide risk factors among farmers in the midwestern United States. *International Journal of Environmental Research and Public Health*, *18*(7), 3563. DOI: 10.3390/ijerph18073563 PMID: 33808131

Black, R., Adger, W. N., Arnell, N. W., Dercon, S., Geddes, A., & Thomas, D. S. (2011). The effect of environmental change on human migration. *Global Environmental Change*, *21*, S3–S11.

Black, R., Bennett, S. R. G., Thomas, S. M., & Beddington, J. R. (2011). Migration as adaptation. *Nature*, *478*(7370), 447–449. DOI: 10.1038/478477a PMID: 22012304

Blitz, B. K. (2019). The plight of the world's stateless. *Forced Migration Review*, *62*, 12–15.

Blitz, B. K. (2020). *Statelessness and displacement*. Oxford University Press.

Blitz, B. K., & Sawyer, A. (2021). The psychological impact of statelessness on children. *International Journal of Migration and Border Studies*, *7*(3), 205–222.

Bogic, M., Njoku, A., & Priebe, S. (2015). Long-term mental health of war-refugees: A systematic literature review. *BMC International Health and Human Rights*, *15*(1), 29. DOI: 10.1186/s12914-015-0064-9 PMID: 26510473

Bolton, J. M., Walld, R., Chateau, D., Finlayson, G., & Sareen, J. (2015). Risk of suicide and suicide attempts associated with physical disorders: A population-based, balancing score-matched analysis. *Psychological Medicine*, *45*(3), 495–504. DOI: 10.1017/S0033291714001639 PMID: 25032807

Bolton, P., Bass, J., Betancourt, T., Speelman, L., Onyango, G., Clougherty, K. F., Neugebauer, R., Murray, L., & Verdeli, H. (2007). Interventions for depression symptoms among adolescent survivors of war and displacement in northern Uganda: A randomized controlled trial. *Journal of the American Medical Association, 298*(5), 519–527. PMID: 17666672

Bonilla-Escobar, F., Osorio-Cuellar, G., Pacichana-Quinayaz, S. G., Sanchez-Renteria, G., & Fandino-Losada, A. (2017). Do not forget culture when implementing mental health interventions for violence survivors. *Cienc Saude Coletiva*, 22 (9):3053–9. https://doi.org/DOI: 10.1590/1413-81232017229.12982016

Booth, N. J., & Lloyd, K. (2000). Stress in farmers. *The International Journal of Social Psychiatry, 46*(1), 67–73. DOI: 10.1177/002076400004600108 PMID: 14529080

Bowles, D. C., Butler, C. D., & Morisetti, N. (2015). Climate change, conflict and health. *Journal of the Royal Society of Medicine, 108*(10), 390–395. DOI: 10.1177/0141076815603234 PMID: 26432813

Boykoff, M. T., & Boykoff, J. M. (2007). Climate change and journalistic norms: A case-study of US mass-media coverage. *Geoforum, 38*(6), 1190–1204. DOI: 10.1016/j.geoforum.2007.01.008

Brennan, M., Hennessy, T., & Dillon, E. (2020). Towards a better measurement of the social sustainability of Irish agriculture. *International Journal of Sustainable Development, 23*(3–4), 263–287. DOI: 10.1504/IJSD.2020.115229

Brew, B., Inder, K., Allen, J., Thomas, M., & Kelly, B. (2016). The health and wellbeing of Australian farmers: A longitudinal cohort study. *BMC Public Health, 16*, 988. DOI: 10.1186/s12889-016-3664-y PMID: 27634298

Brun, C., & Fábos, A. (2015). *Refugees and asylum seekers in Europe: Challenges and opportunities for mental health support*. Routledge.

Bryant, L., & Garnham, B. (2015). The fallen hero: Masculinity, shame and farmer suicide in Australia. *Gender, Place and Culture, 22*(1), 67–82. DOI: 10.1080/0966369X.2013.855628

Buguet, A. (2007). Sleep under extreme environments: Effects of heat and cold exposure, altitude, hyperbaric pressure and microgravity in space. *Journal of the Neurological Sciences, 262*(1–2), 145–152. DOI: 10.1016/j.jns.2007.06.040 PMID: 17706676

Burke, M., Hsiang, S. M., & Miguel, E. (2015). Climate and Conflict. *Annual Review of Economics, 7*(1), 577–617. DOI: 10.1146/annurev-economics-080614-115430

Burke, M., Hsiang, S. M., & Miguel, E. (2015). Global non-linear effect of temperature on economic production. *Nature*, *527*(7577), 235–239. DOI: 10.1038/nature15725 PMID: 26503051

Burke, M., Miguel, E., Satyanath, S., Dykema, J., & Lobell, D. B. (2009). Climate and conflict in Africa. *Proceedings of the National Academy of Sciences of the United States of America*, *106*(40), 16035–16040.

Burkett, M. (2011). The climate crisis and the unconstitutionality of statelessness. *The American Journal of International Law*, *105*(3), 517–538.

Burroughs, J. E., & Rindfleisch, A. (2002). Materialism and Well-Being: A Conflicting Values Perspective. *The Journal of Consumer Research*, *29*(3), 348–370. DOI: 10.1086/344429

Burrows, K., Pelupessy, D. C., Khoshnood, K., & Bell, M. L. (2021). Environmental Displacement and Mental Well-Being in Banjarnegara, Indonesia. *Environmental Health Perspectives*, *129*(11), 117002. https://doi.org/doi:10.1289/EHP9391 PMID: 34747632

Burson, C. (2015). *The state of statelessness: A human rights approach to forced migration*. Stanford University Press.

Burton, A., Roberts, R., & Smith, J. (2019). The application of psychological first aid in displacement settings. *Journal of Mental Health (Abingdon, England)*, *28*(4), 405–412. PMID: 30862218

Bushra Ali Khan. (2024). Addressing the Legal Void: The Urgent Need for India to Revise Its Laws for Climate Migrants. https://shorturl.at/3sUPI

Caldwell, J. C. (1990). Cultural and social factors influencing mortality levels in developing countries. *The Annals of the American Academy of Political and Social Science*, *510*(1), 44–59.

Caldwell, T. M., Jorm, A. F., & Dear, K. B. G. (2004). Suicide and mental health in rural, remote and metropolitan areas in Australia. *The Medical Journal of Australia*, *181*(S7), S10–S14. DOI: 10.5694/j.1326-5377.2004.tb06348.x PMID: 15462636

Campbell, C. (2016). The mental health consequences of climate change-related displacement. *International Journal of Environmental Research and Public Health*, *13*(10), 967. PMID: 27690076

Campbell, S., Roberts, R., & Smith, J. (2020). The role of community in building resilience among displaced populations. *Journal of Community Psychology*, *48*(2), 522–535.

Campo-Arias, A., & Herazo, E. (2014). Stigma and mental health in victims of Colombia's internal armed conflict in a situation of forced displacement. *Allman, editor. Estigma Salud Ment En Pers Vict Confl Armado Interno Colomb En Situacion Desplazamiento Forzado*, 43(4):212–7.

Carey, I. M., Anderson, H. R., Atkinson, R. W., Beevers, S. D., Cook, D. G., & Strachan, D. P. (2018). Are noise and air pollution related to the incidence of dementia? A cohort study in London, England. *BMJ Open*, *8*(9), e022404. DOI: 10.1136/bmjopen-2018-022404 PMID: 30206085

Carlisle, K., Kamstra, P., Carlisle, E., McCosker, A., De Cotta, T., Kilpatrick, S., Steiner, A., Kahl, B., & Farmer, J. (2024). A qualitative exploration of online forums to support resilience of rural young people in Australia. *Frontiers in Public Health*, *12*. Advance online publication. DOI: 10.3389/fpubh.2024.1335476 PMID: 38841668

Catani, C., Schauer, E., & Neuner, F. (2008). Beyond individual war trauma: Domestic violence against children in Afghanistan and Sri Lanka. *Journal of Marital and Family Therapy*, *34*(2), 165–176.

Ceptureanu, S. I., & Ceptureanu, E. G. (2019). Community-based healthcare programs sustainability impact on the sustainability of host organizations: A structural equation modeling analysis. *International Journal of Environmental Research and Public Health*, *16*(20), 4035. PMID: 31640288

Charlson, F., Ali, S., Benmarhnia, T., Pearl, M., Massazza, A., & Augustinavicius, J. (2021). Climate Change and Mental Health: A Scoping Review. *International Journal of Environmental Research and Public Health*, *18*(9), 4486. PMID: 33922573

Charlson, F., van Ommeren, M., & Silove, D. (2019). Mental health and psychosocial support in humanitarian emergencies. *The Lancet. Psychiatry*, *6*(8), 653–654.

Chen, S., Oliva, P., & Zhang, P. (2022). *Air Pollution and Mental Health: Evidence from China*. National Bureau of Economic Research (NBER Working Papers). Report No.: w24686. http://www.nber.org

Chen, S., Oliva, P., & Zhang, P. (2024, May). Air pollution and mental health: evidence from China. In *AEA Papers and Proceedings* (Vol. 114, pp. 423-428). 2014 Broadway, Suite 305, Nashville, TN 37203: American Economic Association.

Chersich, M. F., Pham, M. D., Areal, A., Haghighi, M. M., Manyuchi, A., Swift, C. P., & Luchters, S. (2020). Associations between high temperatures in pregnancy and risk of preterm birth, low birth weight, and stillbirths: A meta-analysis. *BMJ (Clinical Research Ed.)*, *371*, m3811. DOI: 10.1136/bmj.m3811 PMID: 33148618

Chindarkar, N. (2019). Livelihoods and psychological well-being among climate migrants. *Environmental Research Letters, 14*(12), 125001.

Clayton, S., & Manning, C. M. (2018). *Psychology and climate change: Human perception, impacts, and responses*. Academic Press.

Coffey, Y., Bhullar, N., Durkin, J., Islam, M. S., & Usher, K. (2021). Understanding eco-anxiety: A systematic scoping review of current literature and identified knowledge gaps. *The Journal of Climate Change and Health, 3*, 100047. DOI: 10.1016/j.joclim.2021.100047

Connell, J. (2016). Climate change-induced displacement in the Pacific Islands: A review of the literature. *The Geographical Journal, 182*(2), 133–143.

Corrigan, P. (2004). How stigma interferes with mental health care. *The American Psychologist, 59*(7), 614. PMID: 15491256

Cruwys, T., Macleod, E., Heffernan, T., Walker, I., Stanley, S. K., Kurz, T., Greenwood, L., Evans, O., & Calear, A. L. (2023). Social group connections support mental health following wildfire. *Social Psychiatry and Psychiatric Epidemiology, 59*(6), 957–967. DOI: 10.1007/s00127-023-02519-8 PMID: 37428193

Cuijpers, P., Heim, E., Ramia, J. A., Burchert, S., & Carswell, K.. (2023). Correction: Effects of a WHO-guided digital health intervention for depression in Syrian refugees in Lebanon: A randomized controlled trial. *PLoS Medicine, 20*(4), e1004231. DOI: 10.1371/journal.pmed.1004231 PMID: 37058690

Cunsolo, A., & Ellis, N. R. (2018). Ecological grief as a new mental health challenge. *The Lancet. Planetary Health, 2*(12), e515–e516. PMID: 30526935

Currie, J., Deschênes, O., & Princeton University. (2016). Children and climate Change: Introducing the issue. In *The Future of Children* (pp. 3–9). Princeton University. https://oar.princeton.edu/rt4ds/file/14636/Children+and+climate+change+Introducing+the+issue.pdf

Daghagh Yazd, S., Wheeler, S. A., & Zuo, A. (2019). Key risk factors affecting farmers' mental health: A systematic review. *International Journal of Environmental Research and Public Health, 16*(23), 4849. DOI: 10.3390/ijerph16234849 PMID: 31810320

Das, A. (2011). Farmers' suicide in India: Implications for public mental health. *The International Journal of Social Psychiatry, 57*(1), 21–29. DOI: 10.1177/0020764009103645 PMID: 21252353

DasGupta, R., & Shaw, R. (2015). An indicator based approach to assess coastal communities' resilience against climate related disasters in Indian Sundarbans. *Journal of Coastal Conservation*, *19*(1), 85–101. DOI: 10.1007/s11852-014-0369-1

Das, S., & Chaturvedi, V. (2019). Colonialism, displacement, and the transmission of intergenerational trauma. *Journal of Postcolonial Studies*, *21*(3), 256–271.

Davis-Brown, K., & Salamon, S. (1987). Farm families in crisis: An application of stress theory to farm family research. *Family Relations*, *36*(4), 368. DOI: 10.2307/584485

De Hert, M., Correll, C. U., Bobes, J., Cetkovich-Bakmas, M., Cohen, D., & Asai, I. (2011). Physical illness in patients with severe mental disorders. I. Prevalence, impact of medications and disparities in health care. *World Psychiatry; Official Journal of the World Psychiatric Association (WPA)*, *10*(1), 52–77. DOI: 10.1002/j.2051-5545.2011.tb00014.x PMID: 21379357

Deb, S. (2022). Climate disasters, COVID-19 lockdowns, have led to an increase in trafficking and child marriage. Time to Reclaim the Lost Innocence of Sunderbans' Children. https://www.thecitizen.in/index.php/en/NewsDetail/index/13/21704/Time-to-Reclaim-the-Lost-Innocence-of-Sunderbans-Children

Deegan, A., & Dunne, S. (2022). An investigation into the relationship between social support, stress, and psychological well-being in farmers. *Journal of Community Psychology*, *50*(7), 3054–3069. DOI: 10.1002/jcop.22814 PMID: 35132638

Di Giorgi, E., Michielin, P., & Michielin, D. (2020). Perception of climate change, loss of social capital and mental health in two groups of migrants from African countries. *Annali dell'Istituto Superiore di Sanita*, *56*(2), 150–156. PMID: 32567564

Dickson, K., Ko, S. Y. J., Nguyen, C., Minchenko, D., & Bangpan, M. (2024). Mental health and psychosocial support programmes for displaced populations in low-and middle-income countries (LMICs): A systematic review of process, perspectives and experiences. *Cambridge Prisms: Global Mental Health*, *11*, e62. PMID: 38774885

Dijkerman, S., Hossain, J., Persson, M., Konika, R., & Paul, D. (2025). Applying an intersectional climate justice lens to understand climate crisis impacts on sexual and reproductive health and rights and identify local solutions: Qualitative findings from Khulna, Bangladesh. *Women's Health*, *21*. Advance online publication. DOI: 10.1177/17455057251339283 PMID: 40418759

Ding, N., Berry, H. L., & Bennett, C. M. (2016). The importance of humidity in the relationship between heat and population mental health: Evidence from Australia. *PLoS One*, *11*(10), e0164190. DOI: 10.1371/journal.pone.0164190 PMID: 27727320

Dittmar, H., Bond, R., Hurst, M., & Kasser, T. (2014). The relationship between materialism and personal well-being: A meta-analysis. - PsycNET. *Journal of Personality and Social Psychology*, *107*(5), 879–924. DOI: 10.1037/a0037409 PMID: 25347131

Dixon, P. G., Sinyor, M., Schaffer, A., Levitt, A., Haney, C. R., & Ellis, K. N. (2014). Association of weekly suicide rates with temperature anomalies in two different climate types. *International Journal of Environmental Research and Public Health*, *11*(11), 11627–11644. DOI: 10.3390/ijerph111111627 PMID: 25402561

Docherty, B., & Giannini, T. (2009). Confronting a rising tide: A proposal for a convention on climate change displacement. *Georgetown Environmental Law Review*, *21*, 621–682.

Dodge, R., Daly, A., Huyton, J., & Sanders, L. (2012). The challenge of defining wellbeing. *International Journal of Wellbeing*, *2*(3), 222–235. DOI: 10.5502/ijw.v2i3.4

Dumke, L. (2024). *Mental health care for refugees in Germany: Needs and barriers*. Universitätsbibliothek Bielefeld.

Ebi, K. L., & Burton, I. (2008). Identifying practical adaptation options: An approach to address climate change-related health risks. *Environmental Science & Policy*, *11*(4), 359–369.

Ebi, K. L., & Hess, J. J. (2020). Health risks due to climate change: inequity in causes and consequences: study examines health risks due to climate change. *Health Affairs*, *39*(12), 2056–2062. PMID: 33284705

Ejiohuo, O., Onyeaka, H., Unegbu, K. C., Chikezie, O. G., Odeyemi, O. A., Lawal, A., & Odeyemi, O. A. (2024). Nourishing the Mind: How Food Security Influences Mental Wellbeing. *Nutrients*, *16*(4). Advance online publication. DOI: 10.3390/nu16040501 PMID: 38398825

Elgar, F. J., Pickett, W., Pförtner, T. K., Gariépy, G., Gordon, D., & Georgiades, K. (2021). Relative food insecurity, mental health and wellbeing in 160 countries. *Social Science & Medicine*, *268*, 113556. DOI: 10.1016/j.socscimed.2020.113556 PMID: 33293171

Ellegala, D. B., Simpson, L., Mayegga, E., Nuwas, E., Samo, H., Naman, N., & Nicholas, J. S. (2014). Neurosurgical capacity building in the developing world through focused training. *Journal of Neurosurgery*, *121*(6), 1526–1532. PMID: 25216067

Ellis, D. (2022). *Social Identification and Mental Health*. https://www.news-medical.net/health/Social-Identification-and-Mental-Health.aspx

Eloul, L., Quosh, C., Ajlani, R., Avetisyan, N., Barakat, M., Barakat, L., Ikram, M. W., Shammas, L., & Diekkamp, V. (2013). Inter-agency coordination of mental health and psychosocial support for refugees and people displaced in Syria. *Intervention (Amstelveen, Netherlands), 11*(3), 340–348.

EU. E. C. (2022). *Fostering responsive mental health systems in the Syrian refugee crisis*. https://cordis.europa.eu/project/id/733337

Ezeonu, N. A., Hertelendy, A. J., Adu, M. K., Kung, J. Y., Itanyi, I. U., Da Luz Dias, R., Agyapong, B., Hertelendy, P., Ohanyido, F., Agyapong, V. I. O., & Eboreime, E. (2024). Mobile apps to support mental health response in natural disasters: Scoping review. *Journal of Medical Internet Research, 26*, e49929. DOI: 10.2196/49929 PMID: 38520699

Factor, R., Mahalel, D., & Yair, G. (2007). The social accident: A theoretical model and a research agenda for studying the influence of social and cultural characteristics on motor vehicle accidents. *Accident; Analysis and Prevention, 39*(5), 914–921. PMID: 17291438

Fagan, B. (2000). *The Little Ice Age: How climate made history 1300-1850*. Basic Books.

FAO (Food and Agriculture Organization). (2021). *The state of food security and nutrition in the world 2021*. FAO, IFAD, UNICEF, WFP, and WHO. https://doi.org/DOI: 10.4060/cb4474en

Farbotko, C., & Lazrus, H. (2012). The psychological effects of climate change-induced displacement. *International Journal of Human Rights, 16*(3), 489–505.

Farny, S., & Dentoni, D. (2025). Social identity and place-based dynamics in community resilience building for natural disasters: An integrative framework. *Ecology and Society, 30*(2).

Fekadu, D., Alem, A., & Hägglöf, B. (2006). The prevalence of mental health problems in Ethiopian child laborers. *Journal of Child Psychology and Psychiatry, and Allied Disciplines, 47*(9), 954–959.

Fennell, K. M., Jarrett, C. E., Kettler, L. J., Dollman, J., & Turnbull, D. A. (2016). "Watching the bank balance build up then blow away and the rain clouds do the same": A thematic analysis of South Australian farmers' sources of stress during drought. *Journal of Rural Studies, 46*, 102–110. DOI: 10.1016/j.jrurstud.2016.05.005

Ferrero, G., Setty, K., Rickert, B., George, S., Rinehold, A., DeFrance, J., & Bartram, J. (2019). Capacity building and training approaches for water safety plans: A comprehensive literature review. *International Journal of Hygiene and Environmental Health, 222*(4), 615–627. PMID: 30744955

Ferris, E. (2019). Climate change, migration and the Global Compact for Migration. *Forced Migration Review, 62*, 22–25.

Filler, T., Benipal, P. K., Torabi, N., & Minhas, R. S. (2021). A chair at the table: A scoping review of the participation of refugees in community-based participatory research in healthcare. *Globalization and Health, 17*(1), 103. DOI: 10.1186/s12992-021-00756-7 PMID: 34488810

Fitzpatrick, S. (2005). The 1961 Convention on the Reduction of Statelessness. *International Journal of Refugee Law, 17*(2), 401–428.

Fong, K. C., Hart, J. E., & James, P. (2018). A review of epidemiologic studies on greenness and health: Updated literature through 2017. *Current Environmental Health Reports, 5*, 77–87. PMID: 29392643

Ford, J. D., & King, D. (2015). Coverage and framing of climate change adaptation in the media: A review of influential North American newspapers during 1993–2013. *Environmental Science & Policy, 48*, 137–146. DOI: 10.1016/j.envsci.2014.12.003

Forray, A. I., Oltean, O., Hanft-Robert, S., Madzamba, R., Liem, A., Schouten, B., Anthonissen, C., Swartz, L., Chereche , R. M., Higgen, S., Hall, B. J., & Mösko, M. (2024). Uncovering multi-level mental healthcare barriers for migrants: A qualitative analysis across China, Germany, Netherlands, Romania, and South Africa. *BMC Public Health, 24*(1), 1593. DOI: 10.1186/s12889-024-19046-z PMID: 38877460

Frankish, H., Boyce, N., & Horton, R. (2018). Mental health for all: A global goal. *Lancet, 392*(10157), 1493–1494. DOI: 10.1016/S0140-6736(18)32271-2 PMID: 30314858

Fraser, C. E., Smith, K. B., Judd, F., Humphreys, J. S., Fragar, L. J., & Henderson, A. (2005). Farming and mental health problems and mental illness. *The International Journal of Social Psychiatry, 51*(4), 340–349. DOI: 10.1177/0020764005060844 PMID: 16400909

Fricchione, G. L., Borba, C. P., Alem, A., Shibre, T., Carney, J. R., & Henderson, D. C. (2012). Capacity building in global mental health: Professional training. *Harvard Review of Psychiatry, 20*(1), 47–57. PMID: 22335182

Fritze, J. G., Schnyder, U., & Stiess, G. (2008). The psychological effects of natural disasters: A review of the literature. *Environmental Health Perspectives, 116*(6), 705–712.

Fruttero, A., Halim, D., Broccolini, C., Coelho, B., Gninafon, H., Muller, N., & Gender Global Theme. (2023). Gendered Impacts of Climate Change: Evidence from Weather Shocks. In *Policy Research Working Paper* (Report No. 10442). https://wrd.unwomen.org/sites/default/files/2023-07/WB-gendered-impacts-climate-change.pdf

Furey, E. M., O'Hora, D., McNamara, J., Kinsella, S., & Noone, C. (2016). The roles of financial threat, social support, work stress, and mental distress in dairy farmers' expectations of injury. *Frontiers in Public Health, 4*, 126. DOI: 10.3389/fpubh.2016.00126 PMID: 27446893

Gaber, N., Silva, A., Lewis-Patrick, M., Kutil, E., Taylor, D., & Bouier, R. (2021). Water insecurity and psychosocial distress: Case study of the Detroit water shutoffs. *J Public Health, 43*(4), 839–845. DOI: 10.1093/pubmed/fdaa157 PMID: 32930795

Galán, J. (2021). CREWS: a CAMELS-based early warning system of systemic risk in the banking sector. *Banco de Espana Occasional Paper*, (2132).

Gao, J., Cheng, Q., Duan, J., Xu, Z., Bai, L., & Zhang, Y. (2019). Ambient temperature, sunlight duration, and suicide: A systematic review and meta-analysis. *The Science of the Total Environment, 646*, 1021–1029. DOI: 10.1016/j.scitotenv.2018.07.098 PMID: 30235587

García, E., & Doocy, S. (2020). National responses to climate change displacement. *The Lancet. Planetary Health, 4*(11), e510–e511. PMID: 33159877

Garud, R., Nayyar, P. R., Shapira, Z., & Shapira, Z. B. (Eds.). (1997). *Technological innovation: Oversights and foresights*. Cambridge University Press.

Gemenne, F. (2015). *The state of climate change displacement: Legal and policy challenges*. Routledge.

Gemenne, F., & Blocher, J. (2017). Climate change and statelessness: A new dimension of human rights protection. *Journal of Human Rights Practice, 9*(3), 323–342.

Ghose, A., & Ali, S. M. (2024). Displaced by Rising Tides: Legal Insights and Policy Imperatives for Climate Refugees in the Sundarbans Region of India. https://doi.org/DOI: 10.60866/CAM.132

Ghosh, S., & Roy, S. (2022). Climate change, ecological stress and livelihood choices in Indian Sundarban. Climate change and community resilience, 399.

Ghosh, U., Bose, S., Bramhachari, R., & STEPS Centre. (2018). Living on the edge: climate change and uncertainty in the Indian Sundarbans. In *STEPS Working Paper 101*. STEPS Centre. https://www.steps-centre.org

Ghosh, R., Dubey, M. J., Chatterjee, S., & Dubey, S. (2020). Impact of COVID-19 on children: Special focus on the psychosocial aspect. *Minerva Pediatrica*, *72*(3), 226–235. PMID: 32613821

Gibson, K., Haslam, N., & Kaplan, I. (2019). Distressing encounters in the context of climate change: Idioms of distress, determinants, and responses to distress in Tuvalu. *Transcultural Psychiatry*, *56*(4), 667–696. PMID: 31067153

Gifford, R., & Gifford, J. (2016). Psychology and climate change: The human dimension. In *Handbook of Climate Change and Human Behavior* (pp. 3–18). Springer.

Gill, A., & Singh, L. (2006). Farmers' suicides and response of public policy: Evidence, diagnosis and alternatives from Punjab. *Economic and Political Weekly*, *41*(26), 2762–2768.

Gillespie, A. M., Singh, P., & Fernandez, M. (2021). Gender and climate change: Bridging data gaps to strengthen adaptation policies. *Climate Policy*, *21*(7), 945–961. DOI: 10.1080/14693062.2021.1953728

Giosan, L., Clift, P. D., Macklin, M. G., Durcan, J. A., Adger, W. N., & Ellis, N. R. (2012). Holocene Indian monsoon variability and the collapse of the Indus Valley civilization. *Nature*, *484*(7392), 79–82.

Gislason, M. K., Kennedy, A. M., & Witham, S. M. (2021). The interplay between social and ecological determinants of mental health for children and youth in the climate crisis. *International Journal of Environmental Research and Public Health*, *18*(9), 4573. PMID: 33925907

Gleeson, C., Frost, R., Sherwood, L., Shevlin, M., Hyland, P., & Halpin, R. (2020). Post-migration factors and mental health outcomes in asylum-seeking and refugee populations: A systematic review. *European Journal of Psychotraumatology*, *11*(1), 1793567. DOI: 10.1080/20008198.2020.1793567 PMID: 33312450

Global Burden of Diseases. (2019). Mental Disorders Collaborators. Global, regional, and national burden of 12 mental disorders in 204 countries and territories, 1990–2019: A systematic analysis for the Global Burden of Disease Study 2019. *The Lancet. Psychiatry*, *9*(2), 137–150.

Godden, N. J., Farrant, B. M., Yallup Farrant, J., Heyink, E., Carot Collins, E., Burgemeister, B., & Cooper, T. (2021). Climate change, activism, and supporting the mental health of children and young people: Perspectives from Western Australia. *Journal of Paediatrics and Child Health, 57*(11), 1759–1764. PMID: 34792244

Goldschmidt, W. (1978). *As you sow: Three studies in the social consequences of agribusiness. Monclair: Allanheld.* Osmun and Company.

Goodkind, S., Shaffer, J., & Leff, M. (2019). The psychosocial impact of statelessness: An intersectional approach. *Journal of Human Rights and Social Work, 4*(3), 187–196.

Goodwin-Gill, G. S., & McAdam, J. (2007). *The refugee in international law.* Oxford University Press.

Government of Canada. (2021). *Climate change and gender equality framework.* Retrieved from https://www.canada.ca/en/services/environment/climate.html

Government of Germany. (2019). *Gender equality in climate policy.* Retrieved from https://www.bmu.de/en

Government of India. (2008). *National action plan on climate change (NAPCC).* Retrieved from https://www.india.gov.in/spotlight/national-action-plan-climate-change

Government of India. (2017). *Mental Healthcare Act 2017.* Ministry of Law and Justice. https://www.indiacode.nic.in/handle/123456789/2249

Government of Kenya. (2010). *National climate change response strategy.* Retrieved from https://www.environment.go.ke

Government of Mexico. (2012). *National strategy on climate change.* Retrieved from https://www.gob.mx/semarnat

Government of South Africa. (2013). *National climate change adaptation strategy.* Retrieved from https://www.environment.gov.za

Government of Sweden. (2017). *Climate policy action plan.* Retrieved from https://www.government.se

Government of the Philippines. (2009). *Climate change act (Republic Act No. 9729).* Retrieved from https://www.officialgazette.gov.ph/2009/10/23/republic-act-no-9729/

Green, A. R., Betancourt, J. R., & Carrillo, J. E. (2002). Integrating social factors into cross-cultural medical education. *Academic Medicine, 77*(3), 193–197. PMID: 11891153

Green, G. P. (1985). Large-scale farming and the quality of life in rural communities: Further specification of the Goldschmidt hypothesis. *Rural Sociology*, *50*(2), 262–274.

Grover, D., Singh, J. M., Kaur, A., & Kumar, S. (2017). *State Agricultural Profile-Punjab* [Technical report]. Punajb Agricultural University. https://rgdoi.net/10.13140/RG.2.2.29375.87203

Guha, I., & Roy, C. (2016). Climate Change, Migration and Food Security: Evidence from Indian Sundarbans. *International Journal of Theoretical & Applied Sciences*, 45–49. https://www.researchgate.net/publication/309668130

Guillien, A., Laurent, L., Soumagne, T., Puyraveau, M., Laplante, J.-J., Andujar, P., Annesi-Maesano, I., Roche, N., Degano, B., & Dalphin, J.-C. (2017). Anxiety and depression among dairy farmers: The impact of COPD. *International Journal of Chronic Obstructive Pulmonary Disease*, *13*, 1–9. DOI: 10.2147/COPD.S143883 PMID: 29296078

Gulati, A., Roy, R., & Hussain, S. (2021). Performance of Agriculture in Punjab. In Gulati, A., Roy, R., & Saini, S. (Eds.), *Revitalizing Indian Agriculture and Boosting Farmer Incomes* (pp. 77–112). Springer Nature., DOI: 10.1007/978-981-15-9335-2_4

Gupta, S., & Murray, R. M. (1992). The Relationship of Environmental Temperature to the Incidence and Outcome of Schizophrenia. *The British Journal of Psychiatry*, *160*(6), 788–792. DOI: 10.1192/bjp.160.6.788 PMID: 1617362

Habib, S., Khan, S. A., & Ahmad, N. (2021). The mental health of Rohingya refugees in Bangladesh: A systematic review. *BMC Psychiatry*, *21*(1), 1–15. PMID: 33388029

Haines, A. (2017). Health co-benefits of climate action. *The Lancet. Planetary Health*, *1*(1), e4–e5. PMID: 29851591

Haines, A., & Ebi, K. L. (2019). The imperative for climate action to protect health. *The New England Journal of Medicine*, *380*(3), 263–273. DOI: 10.1056/NEJMra1807873 PMID: 30650330

Hajra, R., Szabo, S., Tessler, Z., Ghosh, T., Matthews, Z., & Foufoula-Georgiou, E. (2017). Unravelling the association between the impact of natural hazards and household poverty: Evidence from the Indian Sundarban delta. *Sustainability Science*, *12*(3), 453–464. DOI: 10.1007/s11625-016-0420-2

Han, G., Schoolman, E. D., Arbuckle, J. G., & Morton, L. W. (2022). Weather, values, capacity and concern: Toward a social-cognitive model of Specialty crop farmers' perceptions of climate change risk. *Environment and Behavior*, *54*(2), 327–362. DOI: 10.1177/00139165211026607

Hanlon, C., Whitley, R., Wondimagegn, D., Alem, A., & Prince, M. (2009). Postnatal mental distress in relation to the sociocultural practices of childbirth: An exploratory qualitative study from Ethiopia. *Social Science & Medicine*, *69*(8), 1211–1219. PMID: 19709793

Hanna, R., & Oliva, P. (n.d.). *Implications of climate change for children in developing countries* (Prashant Bharadwaj, University of California, San Diego, University of California, Santa Barbara, & Harvard University, Eds.). https://files.eric.ed.gov/fulltext/EJ1101592.pdf

Hansen, B. G., Bugge, C. T., & Skibrek, P. K. (2020). Automatic milking systems and farmer wellbeing–exploring the effects of automation and digitalization in dairy farming. *Journal of Rural Studies*, *80*, 469–480. DOI: 10.1016/j.jrurstud.2020.10.028

Haque, M. D. R., & Rubya, S. (2023). An overview of chatbot-based mobile mental health apps: Insights from app description and user reviews. *JMIR mHealth and uHealth*, *11*, e44838. DOI: 10.2196/44838 PMID: 37213181

Haque, M. R., & Rubya, S. (2022). *"For an app supposed to make its users feel better, it sure is a joke":* An analysis of user reviews of mobile mental health applications. *arXiv*. https://arxiv.org/abs/2209.07796

Haque, S., & Etzold, B. (2022). Climate-induced displacement and psychological distress in coastal communities: A case study of the Maldives. *International Journal of Environmental Research and Public Health*, *19*(17), 10822. PMID: 36078536

Hassan, K., Higham, J., Wooliscroft, B., & Hopkins, D. (2018). Climate change and world heritage: A cross-border analysis of the Sundarbans (Bangladesh–India). *Journal of Policy Research in Tourism, Leisure & Events*, *11*(2), 196–219. DOI: 10.1080/19407963.2018.1516073

Hathaway, J. C. (2005). *The rights of refugees under international law*. Cambridge University Press.

Haug, G. H., Günther, D., & Peterson, L. C. (2003). Climate and the collapse of the Maya civilization. *Science*, *299*(5613), 1731–1735. PMID: 12637744

Hayes, K., & Poland, B. (2018). Addressing Mental Health in a Changing Climate: Incorporating Mental Health Indicators into Climate Change and Health Vulnerability and Adaptation Assessments. *International Journal of Environmental Research and Public Health*, *15*(9), 1806. DOI: 10.3390/ijerph15091806 PMID: 30131478

Henderson, S. B., McLean, K. E., Lee, M. J., & Kosatsky, T. (2022). Analysis of community deaths during the catastrophic 2021 heat dome. *Environmental Epidemiology (Philadelphia, Pa.)*, *6*(1), e189. DOI: 10.1097/EE9.0000000000000189 PMID: 35169667

Heo, W., Lee, J. M., & Park, N. (2020). Financial-related psychological factors affect life satisfaction of farmers. *Journal of Rural Studies*, *80*, 185–194. DOI: 10.1016/j.jrurstud.2020.08.053

Hobfoll, S. E. (1988). *The Ecology of Stress*. Taylor & Francis.

Hobfoll, S. E. (2001). The influence of culture, community, and the nested-self in the stress process: Advancing Conservation of Resources theory. *Applied Psychology*, *50*(3), 337–370. DOI: 10.1111/1464-0597.00062

Holliday, K., & Kraler, A. (2022). A legal framework for climate change-related displacement. *International Journal of Refugee Law*, *34*(1), 1–25.

Honadle, G. H., & Hannah, J. P. (1982). Management performance for rural development: Packaged training or capacity building. *Public Administration and Development*, *2*(4), 295–307.

Horn, R., O'May, F., Esliker, R., Gwaikolo, W., Woensdregt, L., Ruttenberg, L., & Ager, A. (2019). The myth of the 1-day training: The effectiveness of psychosocial support capacity-building during the Ebola outbreak in West Africa. *Global Mental Health (Cambridge, England)*, *6*, e5. PMID: 31143466

Hounsome, B., Edwards, R. T., Hounsome, N., & Edwards-Jones, G. (2012). Psychological morbidity of farmers and non-farming population: Results from a UK survey. *Community Mental Health Journal*, *48*(4), 503–510. DOI: 10.1007/s10597-011-9415-8 PMID: 21556782

Hsiang, S. M., & Burke, M. (2014). Climate, conflict, and social stability: What does the evidence say? *Climatic Change*, *123*(1), 39–55. DOI: 10.1007/s10584-013-0868-3

Hsiang, S. M., & Jina, A. S. (2014). The effect of climate change on international migration. *Climatic Change*, *123*(2), 173–184.

Hugo, G. (2020). Migration and psychological well-being: The role of economic security. *Journal of International Migration and Integration*, *21*(4), 1145–1165.

Hu, J., Wen, Y., Duan, Y., Yan, S., Liao, Y., & Pan, H. (2020). The impact of extreme heat and heat waves on emergency ambulance dispatches due to external cause in Shenzhen, China. *Environ Pollut*, *261*, 114156. DOI: 10.1016/j.envpol.2020.114156 PMID: 32092626

Hurley, E. A., Dalglish, S. L., & Sacks, E. (2022). Supporting young people with climate anxiety: Mitigation, adaptation, and resilience. *The Lancet. Planetary Health*, *6*(3), e190. DOI: 10.1016/S2542-5196(22)00015-8 PMID: 35278385

Hwong, A. R., Wang, M., Khan, H., Chagwedera, D. N., Grzenda, A., Doty, B., & Compton, W. M. (2022). Climate change and mental health research methods, gaps, and priorities: A scoping review. *The Lancet. Planetary Health*, *6*(3), e281–e291. PMID: 35278392

Ibragimov, K., Palma, M., Keane, G., Ousley, J., Crowe, M., Carreño, C., Casas, G., Mills, C., Llosa, A., Marques, A., Tijerino, A., Droste, A., Echeverri, C., Fredj, N., Sleit, R., & Sultan, S. F. (2022). Shifting to tele-mental health in humanitarian and crisis settings: An evaluation of Médecins Sans Frontières experience during the COVID-19 pandemic. *Conflict and Health*, *16*(1). Advance online publication. DOI: 10.1186/s13031-022-00437-1 PMID: 35164807

Ibrahim, A., Abdalla, S. M., Jafer, M., Abdelgadir, J., & De Vries, N. (2019). Child labor and health: A systematic literature review of the impacts of child labor on child's health in low-and middle-income countries. *Journal of public health*, *41*(1), 18–26. PMID: 29409061

IFRC. (2018). *A guide to psychological first aid.* International Federation of Red Cross and Red Crescent Societies.

IIED. I. I. f. E. a. D. (2022). *People fleeing impacts of climate change suffering physical and mental health crises.* https://www.iied.org/people-fleeing-impacts-climate-change-suffering-physical-mental-health-crises

Intergovernmental Panel on Climate Change. (2014). [*Summary for Policymakers.* https://www.ipcc.ch]. *Climatic Change*, *2014*, •••.

Internal Displacement Monitoring Centre. (2023). *Displacement data.* Retrieved from https://www.internal-displacement.org/database/displacement-data/

Israel, B. A., Checkoway, B., Schulz, A., & Zimmerman, M. (1994). Health education and community empowerment: Conceptualizing and measuring perceptions of individual, organizational, and community control. *Health Education Quarterly*, *21*(2), 149–170. DOI: 10.1177/109019819402100203 PMID: 8021145

Jacobs, J. A., Duggan, K., Erwin, P., Smith, C., Borawski, E., Compton, J., & Brownson, R. C. (2014). Capacity building for evidence-based decision making in local health departments: Scaling up an effective training approach. *Implementation Science : IS*, *9*, 1–11. PMID: 25253081

Jafry, T. (2016). Making the case for gender-sensitive climate policy – lessons from South Asia/IGP. *International Journal of Climate Change Strategies and Management*, *8*(4), 559–577. DOI: 10.1108/ijccsm-04-2015-0049

Jagga, R. (2022, April 24). *Depressed over low wheat yield, farmer dies by suicide; 14th such death in Punjab in April.* https://indianexpress.com/article/chandigarh/depressed-over-low-wheat-yield-farmer-dies-by-suicide-14th-such-death-in-punjab-in-april-7884497/

James, P., Iyer, A., & Webb, T. L. (2019). The impact of post-migration stressors on refugees' emotional distress and health: A longitudinal analysis. *European Journal of Social Psychology*, *49*(7), 1359–1367. DOI: 10.1002/ejsp.2589

Jarman, D., Naimi, T., Pickard, S., Daley, W., & De, A. (2007). *Binge drinking and occupation, North Dakota, 2004–2005. 4*(4). https://stacks.cdc.gov/view/cdc/4124

Jaskulowski, K. (2020). Statelessness and vulnerability: The case of the elderly. *Journal of Refugee Studies*, *33*(3), 511–529.

Jennings, N., Fecht, D., & De Matteis, S. (2020). Mapping the co-benefits of climate change action to issues of public concern in the UK: A narrative review. *The Lancet. Planetary Health*, *4*(9), e424–e433. PMID: 32918888

Jiménez, F. N., Brazier, J. F., Davoodi, N. M., Florence, L. C., Thomas, K. S., & Gadbois, E. A. (2021). A Technology Training Program to Alleviate Social Isolation and Loneliness Among Homebound Older Adults: A Community Case Study [Community Case Study]. *Frontiers in Public Health, Volume 9 - 2021.* https://doi.org/DOI: 10.3389/fpubh.2021.750609

Jones, M. S., Reed, D. B., & Hunt, M. L. (2018). Suicide: An unrecognized epidemic among farmers. *Workplace Health & Safety*, *66*(9), 464. DOI: 10.1177/2165079918784055 PMID: 30015597

Jones, S., & Faas, D. (2017). The role of social networks in the psychological well-being of displaced populations. *The Journal of Social Psychology*, *157*(4), 450–467.

Joseph, S., & Castan, M. (2013). *The International Covenant on Civil and Political Rights*. Oxford University Press.

Jun, J., & Kyle, G. T. (2012). Gender identity, leisure identity, and leisure participation. *Journal of Leisure Research, 44*(3), 353-378. Sci-Hub | Gender Identity, Leisure Identity, and Leisure Participation. Journal of Leisure Research, 44(3), 353–378 | DOI: 10.1080/00222216.2012.11950269

Kahsay, G. A., Kassie, W. A., Medhin, H., & Hansen, L. G. (2022). Are religious farmers more risk taking? Empirical evidence from Ethiopia. *Agricultural Economics*, *53*(4), 617–632. DOI: 10.1111/agec.12697

Kaijser, A., & Kronsell, A. (2013). Climate change through the lens of intersectionality. *Environmental Politics*, *23*(3), 417–433. DOI: 10.1080/09644016.2013.835203

Kallioniemi, M. K., Simola, A., Kaseva, J., & Kymäläinen, H.-R. (2016). Stress and burnout among Finnish dairy farmers. *Journal of Agromedicine*, *21*(3), 259–268. DOI: 10.1080/1059924X.2016.1178611 PMID: 27081893

Kandlur, R., Sardana, S., & Richardson-Vejlgaard, R. (2022). The Agrarian distress: Factors explaining the will to live among rural and distressed family farmers. *Psychiatry Research Communications*, *2*(1), 100019. DOI: 10.1016/j.psycom.2021.100019

Karmakar, S. (2021). Gender and climate change: The condition of women in the Indian Sundarbans. *Scholars Journal of Arts. Humanities and Social Sciences*, *9*(1), 1–5. DOI: 10.36347/sjahss.2021.v09i01.001

Karyotaki, E., Sijbrandij, M., Purgato, M., Acarturk, C., Lakin, D., Bailey, D., Peckham, E., Uygun, E., Tedeschi, F., Wancata, J., Augustinavicius, J., Carswell, K., Välimäki, M., van Ommeren, M., Koesters, M., Popa, M., Leku, M. R., Anttila, M., Churchill, R., & Barbui, C. (2023). Self-Help Plus for refugees and asylum seekers: An individual participant data meta-analysis. *BMJ Mental Health*, *26*(1). Advance online publication. DOI: 10.1136/bmjment-2023-300672 PMID: 37524517

Kearney, G. D., Rafferty, A. P., Hendricks, L. R., Allen, D. L., & Tutor-Marcom, R. (2014a). A cross-sectional study of stressors among farmers in Eastern North Carolina. *North Carolina Medical Journal*, *75*(6), 384–392. DOI: 10.18043/ncm.75.6.384 PMID: 25402689

Kelman, I. (2021). *Island vulnerability and climate-induced displacement*. Routledge.

Kelman, I., Ayeb-Karlsson, S., Rose-Clarke, K., Prost, A., Ronneberg, E., & Wheeler, N. (2021). A review of mental health and wellbeing under climate change in small Island developing states (SIDS). *Environ Res Lett ERL Web Site*, *16*(3), 033007. DOI: 10.1088/1748-9326/abe57d PMID: 34149865

Khanal, S., Bolte, G., & Boeckmann, M. (2024). Sector silos in climate action: A missed opportunity to prioritize health equity. *PLOS Climate*, *3*(2), e0000349. DOI: 10.1371/journal.pclm.0000349

Khanal, S., Shrestha, R., & Boeckmann, M. (2024). Examining health equity in Nepal's climate change and health policies through the lens of environmental justice: Insights from a content analysis. *Global Health Action, 17*(1). Advance online publication. DOI: 10.1080/16549716.2024.2432069 PMID: 39611321

Kiliç, C., Kaya, E., Karadağ, Ö., & Üner, S. (2024). Barriers To Accessing Mental Health Services Among Syrian Refugees: A Mixed-Method Study. [Suriye Kökenli Sığınmacıların Ruh Sağlığı Hizmetlerine Ulaşmasını Engelleyen Etmenler: Karma Yöntemli Bir Çalışma.]. *Türk Psikiyatri Dergisi = Turkish Journal of Psychiatry, 35*(2), 87–94. DOI: 10.5080/u27044 PMID: 38842150

Kirmayer, L. J., Brass, G. M., & Bélanger, K. (2011). Cultural issues in the mental health of refugees and asylum seekers. *Lancet, 378*(9799), 1664–1675.

Knight, J., & Gunatilaka, R. (2012). Aspirations, adaptation and subjective well-being of rural–urban migrants in China. In *Adaptation, Poverty and Development* (Clark, D.A., pp. 91–110). Palgrave Macmillan UK. https://doi.org/DOI: 10.1057/9781137002778_4

Kothari, U. (2020). The psychological effects of climate-induced displacement on women and families. *International Journal of Gender, 5*(1), 78–92.

Koubi, V., Spilker, G., Schultze, P. E., & Böhmelt, T. (2016). The role of climate change in forced migration. *Climatic Change, 139*(3), 391–402.

Krishnamurthy, R. (2024). Climate disasters cause a surge in mental health disorders in a Sundarbans block. https://www.downtoearth.org.in/climate-change/climate-disasters-cause-surge-in-mental-health-disorders-in-a-sundarbans-block

LaMonica, H. M., Davenport, T. A., Ottavio, A., Rowe, S. C., Cross, S. P., Iorfino, F., Jackson, T. A., Easton, M. A., Melsness, J., & Hickie, I. B. (2021). Optimising the integration of technology-enabled solutions to enhance primary mental health care: A service mapping study. *BMC Health Services Research, 21*(1), 68. DOI: 10.1186/s12913-021-06069-0 PMID: 33451328

Larrieta, J., Esponda, G. M., Gandhi, Y., Simpson, N., Biriotti, M., Kydd, A., Eaton, J., & Ryan, G. K. (2022). Supporting community-based mental health initiatives: Insights from a multi-country programme and recommendations for funders. *BMJ Global Health, 7*(5), e008906. PMID: 35537763

Lau, L. S., & Rodgers, G. (2021). Cultural Competence in Refugee Service Settings: A Scoping Review. *Health Equity, 5*(1), 124–134. DOI: 10.1089/heq.2020.0094 PMID: 33778315

Lawrence, J., Berry, H., & Williams, S. (2022). Integrating mental health into national climate change adaptation plans. *The Journal of Climate and Health*, *3*(4), 1–10.

Le Masson, V., Norton, A., & Wilkinson, E. (2019). *Gender and resilience: From theory to practice*. Overseas Development Institute (ODI). https://www.odi.org/en/publications/gender-and-resilience-from-theory-to-practice/

Le Monde. (2024, November 22). Climat: L'offensive de l'Arabie Saoudite, de l'Iran et du Vatican sur l'égalité entre les genres à la COP29. *Le Monde*. Retrieved from https://www.lemonde.fr/planete/article/2024/11/22/climat-l-offensive-de-l-arabie-saoudite-de-l-iran-et-du-vatican-sur-l-egalite-entre-les-genres-a-la-cop29_6408741_3244.html

Leap, B., & Thompson, D. (2018). Social solidarity, collective identity, resilient communities: Two case studies from the rural US and Uruguay. *Social Sciences*, *7*(12), 250.

Lee, S., Lee, H., Myung, W., Kim, E. J., & Kim, H. (2018). Mental disease-related emergency admissions attributable to hot temperatures. *The Science of the Total Environment*, *616*, 688–694. DOI: 10.1016/j.scitotenv.2017.10.260 PMID: 29126638

Leonardsen, A.-C. L., Hardeland, C., Helgesen, A. K., & Grøndahl, V. A. (2020). Patient experiences with technology enabled care across healthcare settings- a systematic review. *BMC Health Services Research*, *20*(1), 779. DOI: 10.1186/s12913-020-05633-4 PMID: 32838784

Leslie, H. H., Sun, Z., & Kruk, M. E. (2017). Association between infrastructure and observed quality of care in 4 healthcare services: A cross-sectional study of 4,300 facilities in 8 countries. *PLoS Medicine*, *14*(12), e1002464. PMID: 29232377

Lindskov, S. (2021). The rights of elderly stateless persons: A human rights perspective. *Journal of Human Rights and Social Work*, *6*(1), 60–75.

Linkowski, P., Martin, F., & De Maertelaer, V. (1992). Effect of some climatic factors on violent and non-violent suicides in Belgium. *Journal of Affective Disorders*, *25*(3), 161–166. DOI: 10.1016/0165-0327(92)90001-m PMID: 1527270

Liu, J., Varghese, B., Hansen, A., Xiang, J., Zhang, Y., & Dear, K. (2021). Is there an association between hot weather and poor mental health outcomes? A systematic review and meta-analysis. *Environment International*, •••, 153. PMID: 33799230

Logstein, B. (2016a). Farm-related concerns and mental health status among Norwegian farmers. *Journal of Agromedicine*, *21*(4), 316–326. DOI: 10.1080/1059924X.2016.1211055 PMID: 27420178

Logstein, B. (2016b). Predictors of mental complaints among Norwegian male farmers. *Occupational Medicine*, *66*(4), 332–337. DOI: 10.1093/occmed/kqw019 PMID: 26944593

Lopes Cardozo, M. T., Sijbrandij, M., & Thung, M. (2019). The psychological well-being of stateless people. *International Journal of Migration and Border Studies*, *5*(4), 312–330.

Lowder, S. K., Sánchez, M. V., & Bertini, R. (2021). Which farms feed the world and has farmland become more concentrated? *World Development*, *142*, 105455. DOI: 10.1016/j.worlddev.2021.105455

Lowder, S. K., Skoet, J., & Raney, T. (2016). The Number, size, and distribution of farms, smallholder afrms, and family farms worldwide. *World Development*, *87*, 16–29. DOI: 10.1016/j.worlddev.2015.10.041

Lunner Kolstrup, C., Kallioniemi, M., Lundqvist, P., Kymäläinen, H.-R., Stallones, L., & Brumby, S. (2013). International perspectives on psychosocial working conditions, mental health, and stress of dairy farm operators. *Journal of Agromedicine*, *18*(3), 244–255. DOI: 10.1080/1059924X.2013.796903 PMID: 23844791

Maciuba, S. A., Westneat, S. C., & Reed, D. B. (2013). Active coping, personal satisfaction, and attachment to land in older African-American farmers. *Issues in Mental Health Nursing*, *34*(5), 335–343. DOI: 10.3109/01612840.2012.753560 PMID: 23663020

Madhav, N., Oppenheim, B., Gallivan, M., Mulembakani, P., Rubin, E., & Wolfe, N. (2017). Pandemics: Risks, impacts, and mitigation. In Jamison, D. T., Gelband, H., Horton, S., Jha, P., Laxminarayan, R., Mock, C. N., & Nugent, R. (Eds.), *Disease control priorities: Improving health and reducing poverty* (3rd ed., Vol. 9, pp. 315–345). The World Bank., DOI: 10.1596/978-1-4648-0527-1_ch17

Mahadevia Ghimire, K., & Vikas, M. (2012). Climate change–impact on the Sundarbans, a case study. [PDF] [Climate Change – Impact on the Sundarbans: A case study]. *International Scientific Journal: Environmental Science*, *2*(1), 7–15.

Mahadevia, L. K., & Vikas, L. K. (2010). CLIMATE CHANGE – IMPACT ON THE SUNDARBANS: A CASE STUDY. International Scientific Journal Environmental Science http://environment.scientific-journal.com. IMPACT ON THE sundarbans: a case study.pdf

Manly, J., & Kim, S. (2023). The generational transmission of trauma in stateless families. *Journal of Intergenerational Trauma*, *2*(1), 12–25.

Manly, J., & van Waas, L. (2022). Statelessness and the psychological burden of a lost identity. *Journal of Human Rights and Social Work, 7*(2), 110–125.

Mazumdar, S., Mazumdar, P. G., Kanjilal, B., & Singh, P. K. (2014). Multiple shocks, coping and welfare consequences: Natural disasters and health shocks in the Indian Sundarbans. *PLoS One, 9*(8), e105427. DOI: 10.1371/journal.pone.0105427 PMID: 25170846

McAdam, J. (2012). Climate change and displacement. In *The Oxford Handbook of International Refugee Law* (pp. 581–598). Oxford University Press.

McAdam, J. (2020). *Climate change, statelessness and human rights*. Oxford University Press.

McAdam, J. (2022). The rights of persons with disabilities in situations of statelessness. *Forced Migration Review, 71*, 45–48.

McCubbin, H. I., & Patterson, J. M. (1983). The family stress process. *Marriage & Family Review, 6*(1–2), 7–37. DOI: 10.1300/J002v06n01_02

McGregor, M., Willock, J., & Deary, I. (1995). Farmer stress. *Journal of Farm Management, 9*, 57.

McHale, S., Pearsons, A., Neubeck, L., & Hanson, C. L. (2020). Green health partnerships in Scotland; pathways for social prescribing and physical activity referral. *International Journal of Environmental Research and Public Health, 17*(18), 6832. PMID: 32962081

McIntosh, R. J. (1998). *The ancient West African states: A millennium of change*. University of California Press.

McLeman, R. A., & Gemenne, F. (2018). Climate-induced migration: A new frontier in international relations. In C. B. C. (Ed.), Handbook of international relations (pp. 249–264). Sage Publications.

McLeod, C., Barr, H., Rall, K., & Human Rights Watch. (2021). DOES CLIMATE CHANGE INCREASE THE RISK OF CHILD MARRIAGE? A LOOK AT WHAT WE KNOW—AND WHAT WE DON'T—WITH LESSONS FROM BANGLADESH AND MOZAMBIQUE. In *COLUMBIA JOURNAL OF GENDER AND LAW: Vol. 38.1* (pp. 96–98). https://journals.library.columbia.edu/index.php/cjgl/article/view/4604

McManus, S., Gunnell, D., Cooper, C., Bebbington, P. E., Howard, L. M., & Brugha, T. (2019). Prevalence of non-suicidal self-harm and service contact in England, 2000–14: Repeated cross-sectional surveys of the general population. *The Lancet. Psychiatry, 6*(7), 573–581. DOI: 10.1016/S2215-0366(19)30188-9 PMID: 31175059

McMichael, A. J., Friel, S., & Butler, C. D. (2021). Climate change and human health: A new challenge for the health sector. *Globalization and Health, 17*(1), 1–10. PMID: 33397377

McMichael, C., Schwerdtle, P. N., & Ayeb-Karlsson, S. (2023). Waiting for the wave, but missing the tide: Case studies of climate-related (im) mobility and health. *Journal of Migration and Health, 7*, 100147. PMID: 36619800

Meyer, K., & Lobao, L. (2003). Economic hardship, religion and mental health during the midwestern farm crisis. *Journal of Rural Studies, 19*(2), 139–155. DOI: 10.1016/S0743-0167(02)00069-4

Middleton, J., Cunsolo, A., Jones-Bitton, A., Wright, C. J., & Harper, S. L. (2020). Indigenous mental health in a changing climate: A systematic scoping review of the global literature. *Environmental Research Letters, 15*(5), 053001.

Middleton, J., Cunsolo, A., Pollock, N., Jones-Bitton, A., Wood, M., & Shiwak, I. (2021). Temperature and place associations with Inuit mental health in the context of climate change. *Environmental Research*, •••, 198. PMID: 33857460

Miller, K. E., & Rasmussen, A. (2020). Mental health and psychosocial support for refugees and asylum seekers. In M. R. T. (Ed.), Refugee mental health (pp. 185–204). Routledge.

Miller, K. E., & Rasmussen, A. (2017). Mental health in displaced populations. *World Psychiatry; Official Journal of the World Psychiatric Association (WPA), 16*(2), 120–129.

Miller, K. E., & Rasmussen, A. (2022). Peer support programs for displaced populations. *Journal of Mental Health (Abingdon, England), 31*(2), 195–205.

Minayo, M. C. D. S., Souza, E. R. D., Silva, M. M. A. D., & Assis, S. G. D. (2018). Institucionalização do tema da violência no SUS: Avanços e desafios. *Ciencia & Saude Coletiva, 23*, 2007–2016. PMID: 29972507

Miner, A. S., Kuhn, E., Hoffman, J. E., Owen, J. E., & Ruzek, J. I. (2016). Feasibility, acceptability, and potential efficacy of the PTSD Coach app: A pilot randomized controlled trial with community trauma survivors. *Psychological Trauma: Theory, Research, Practice, and Policy, 8*(3), 384–392. DOI: 10.1037/tra0000126 PMID: 27046668

Ministry of Health & Family Welfare. (2021). *Kiran Helpline implementation report*. Government of India. Retrieved July 4, 2025, from https://www.mohfw.gov.in/?q=/press-info/8322

Ministry of Health & Family Welfare. (2022). *National Suicide Prevention Strategy*. Government of India. https://www.mohfw.gov.in/?q=/press-info/8322

Ministry of Health & Family Welfare. (2023). *Tele-MANAS quarterly progress report*. Government of India. Retrieved July 4, 2025, from https://www.mohfw.gov.in/?q=/press-info/7787

Mistri, A. (2019). Is the Migration from the Indian Sundarban an Environmental Migration? Investigating through the Sustainable Livelihood Approach (SLA). In *Asian Profile*. (PDF) Is the Migration from the Indian Sundarban an Environmental Migration? Investigating through the Sustainable Livelihood Approach (SLA)

Moldes, O., & Ku, L. (2020). Materialistic cues make us miserable: A meta-analysis of the experimental evidence for the effects of materialism on individual and societal well-being. *Psychology and Marketing*, *37*(10), 1396–1419. DOI: 10.1002/mar.21387

Molnar, J. J. (1985). Determinants of subjective well-being among farm operators: Characteristics of the individual and the firm. *Rural Sociology*, *50*(1), 141–162.

Monaghan, C., Bizumic, B., Williams, T., & Sellbom, M. (2020). Two-dimensional Machiavellianism: Conceptualization, theory, and measurement of the views and tactics dimensions. *Psychological Assessment*, *32*, 277–293. DOI: 10.1037/pas0000784 PMID: 31750680

Montaud, J. (2020). Enhancing food security through climate-smart agriculture in Mali. *Journal of Environmental Policy and Planning*, *22*(4), 567–583. DOI: 10.1080/1523908X.2020.1750679

Montgomery, D. R., Biklé, A., Archuleta, R., Brown, P., & Jordan, J. (2022). Soil health and nutrient density: Preliminary comparison of regenerative and conventional farming. *PeerJ*, *10*, e12848. PMID: 35127297

Müller, L., Rapoport, O., & Rahe, M. (2024). Mindfulness meditation improves mental health in flood survivors and disaster volunteers: A randomized Wait-List controlled trial. *Mindfulness*. Advance online publication. DOI: 10.1007/s12671-024-02467-7

Muntaner, C., Solar, O., Vanroelen, C., Martínez, J. M., Vergara, M., & Santana, V. EMCONET network. (2010). Unemployment, informal work, precarious employment, child labor, slavery, and health inequalities: Pathways and mechanisms. *International Journal of Health Services*, *40*(2), 281–295. PMID: 20440971

Mzoughi, N. (2014). Do organic farmers feel happier than conventional ones? An exploratory analysis. *Ecological Economics*, *103*, 38–43. DOI: 10.1016/j.ecolecon.2014.04.015

Namboodiri, V., George, S., & Singh, S. P. (2019). The Mental Healthcare Act 2017 of India: A challenge and an opportunity. *Asian Journal of Psychiatry*, *44*, 25–28. DOI: 10.1016/j.ajp.2019.07.016 PMID: 31302439

Naslund, J. A., Aschbrenner, K. A., Araya, R., Marsch, L. A., Unützer, J., Patel, V., & Bartels, S. J. (2017). Digital technology for treating and preventing mental disorders in low-income and middle-income countries: A narrative review of the literature. *The Lancet. Psychiatry*, *4*(6), 486–500. DOI: 10.1016/s2215-0366(17)30096-2 PMID: 28433615

National Institute of Mental Health. (n.d.). *Digital global mental health program.* NIMH. Retrieved July 4, 2025, from https://www.nimh.nih.gov/about/organization/cgmhr/digital-global-mental-health-program

Neogi, S., Dey, M., Kabir, S. M., Masum, S., Kopprio, G., Yamasaki, S., & Lara, R. (2016). Sundarban mangroves: diversity, ecosystem services and climate change impacts. Asian Journal of Medical and Biological Research ISSN 2411-4472 (Print) 2412-5571. CONICET_Digital_Nro.1c68df3b-9a5e-459d-b1de-152865145bf0_A.pdf

Nishikiori, N., Ohnishi, M., & Nakashima, A. (2017). The role of education in building resilience among displaced populations. *International Journal of Educational Development*, *53*, 1–10.

Nwazor, J. C. (2012). Capacity building, entrepreneurship and sustainable development. *Journal of Emerging Trends in Educational Research and Policy Studies*, *3*(1), 51–54.

Odabasi, S., & Hartarska, V. (2021). Farmer suicides: Effects of socio-economic, climate, and mental health factors. *The Journal of Mental Health Policy and Economics*, *24*(2), 61–71. PMID: 34151778

Ofori, D. O., Bandauko, E., Kutor, S. K., Odoi, A., Asare, A. B., Akyea, T., & Arku, G. (2023). A Systematic Review of International and Internal Climate-Induced Migration in Africa. *Sustainability*, *15*(22), 16105.

Oppenheim, C. E., Axelrod, K., Menyongai, J., Chukwuezi, B., Tam, A., Henderson, D. C., & Borba, C. P. C. (2019). The HEAAL Project: Applying Community-Based Participatory Research (CBPR) Methodology in a Health and Mental Health Needs Assessment With an African Immigrant and Refugee Faith Community in Lowell, Massachusetts. *Journal of Public Health Management and Practice*, *25*(1), E1–e6. DOI: 10.1097/phh.0000000000000707 PMID: 30507809

Osofsky, H. J., Stevens, D. M., & Osofsky, J. D. (2010). The psychological effects of disasters on children and adolescents. *Child and Adolescent Psychiatric Clinics of North America*, *19*(2), 191–207.

Otorkpa, C., Otorkpa, O. J., Olaniyan, O. E., & Adebola, O. A. (2024). Mental health issues of children and young people displaced by conflict: A scoping review. *PLOS Mental Health*, *1*(6), e0000076.

Owen, J. E., Kuhn, E., Jaworski, B. K., McGee-Vincent, P., & Juhasz, K. J. (2018). VA mobile apps for PTSD and related problems: Public health resources for veterans and those who care for them. *JMIR mHealth and uHealth*, *6*(7), e156. DOI: 10.2196/mhealth.9388 PMID: 30021707

Palmer, D., & Ward, K. (2007). 'Lost': Listening to the voices and mental health needs of forced migrants in London. *Medicine, Conflict, and Survival*, *23*(3), 198–212. DOI: 10.1080/13623690701417345 PMID: 17822063

Pandey, R., Gupta, S., Upadhyay, A., Gupta, R. P., Shukla, M., Mishra, R. C., & Kumari, V. (2020). Childhood maltreatment and its mental health consequences among Indian adolescents with a history of child work. *The Australian and New Zealand Journal of Psychiatry*, *54*(5), 496–508. PMID: 32156147

Panichkriangkrai, W., Detkong, T., Kruahong, N., Thamarangsi, T., & Tangcharoensathien, V. (2022). Mental health preparedness and response during the COVID-19 pandemic: From global to national implementation. *Global Mental Health (Cambridge, England)*, *9*, 111–114. PMID: 36606233

Paraschakis, A., Michopoulos, I., Christodoulou, C., Koutsaftis, F., & Douzenis, A. (2016). Psychiatric medication intake in suicide victims: Gender disparities and implications for suicide prevention. *Journal of Forensic Sciences*, *61*(6), 1660–1663. DOI: 10.1111/1556-4029.13195 PMID: 27643812

Pargament, K. I. (2011). *The psychology of religion and coping: Theory, research, practice*. Guilford Press.

Pargament, K. I., Koenig, H. G., & Perez, L. M. (2000). The many methods of religious coping: Development and initial validation of the RCOPE. *Journal of Clinical Psychology, 56*(4), 519–543. DOI: 10.1002/(sici)1097-4679(200004)56:4<519:aid-jclp6>3.0.co;2-1 PMID: 10775045

Parry, S., McCarthy, S. R., & Clark, J. (2022). Young people's engagement with climate change issues through digital media–a content analysis. *Child and Adolescent Mental Health, 27*(1), 30–38. DOI: 10.1111/camh.12532 PMID: 34904349

Parsons, J., Edmeades, J., Kes, A., Petroni, S., Sexton, M., & Wodon, Q. (2015). Economic Impacts of Child Marriage: A Review of the Literature. *The Review of Faith & International Affairs, 13*, 3, 12–22. DOI: 10.1080/15570274.2015.1075757

Patel, V., Chowdhary, N., Rahman, A., & Verdeli, H. (2011). Improving access to psychological treatments: Lessons from developing countries. *Behaviour Research and Therapy, 49*(9), 523–528. PMID: 21788012

Patel, V., Khan, S., & Singh, R. (2023). The mental health of displaced children: A systematic review. *Journal of Refugee Studies, 36*(1), 1–18.

Patel, V., Saxena, S., Lund, C., Thornicroft, G., Baingana, F., & Bolton, P. (2018). The Lancet Commission on global mental health and sustainable development. *Lancet, 392*(10157), 1553–1598. PMID: 30314863

Pearlin, L. I., Lieberman, M. A., Menaghan, E. G., & Mullan, J. T. (1981). The stress process. *Journal of Health and Social Behavior, 22*(4), 337–356. PMID: 7320473

Pearse, R. (2017). Gender and climate change. In Hulme, M. (Ed.), *WIREs Clim Change.*, DOI: 10.1002/wcc.451

Peconga, E. K., & Hogh Thogersen, M. (2020). Post-traumatic stress disorder, depression, and anxiety in adult Syrian refugees: What do we know? *Scandinavian Journal of Public Health, 48*(7), 677–687. DOI: 10.1177/1403494819882137 PMID: 31814514

Peel, D., Berry, H. L., & Schirmer, J. (2016). Farm exit intention and wellbeing: A study of Australian farmers. *Journal of Rural Studies, 47*, 41–51. DOI: 10.1016/j.jrurstud.2016.07.006

Peluso, N. L., & Watts, M. (2001). *Violent environments.* Cornell University Press.

Perceval, M., Ross, V., Kõlves, K., Reddy, P., & Leo, D. (2018). *Social factors and Australian farmer suicide: A qualitative study. 18*(1), 1367. https://doi.org/DOI: 10.1186/s12889-018-6287-7

Perkins, P. (2018). Climate justice, gender, and intersectionality. In *Climate justice: A global perspective* (pp. 349-358). https://doi.org/DOI: 10.4324/9781315537689-26

Phang, J. K., Kwan, Y. H., Yoon, S., Goh, H., Yee, W. Q., Tan, C. S., & Low, L. L. (2023). Digital intergenerational program to reduce loneliness and social isolation among older adults: Realist review. *JMIR Aging*, *6*(1), e39848. PMID: 36598801

Piguet, É. (2013). From "primitive migration" to "climate refugees": The curious fate of the natural environment in migration studies. *Annals of the Association of American Geographers*, *103*(1), 148–162.

Pihkala, P. (2022). Toward a Taxonomy of Climate Emotions. *Front Clim*, 3(3:738154).

Pihkala, P. (2020). Climate anxiety. *The Lancet. Planetary Health*, *4*(11), e507–e508.

Polanco-Roman, L., Gomez, J., Miranda, R., & Jeglic, E. (2016). Stress-related symptoms and suicidal ideation: The roles of rumination and depressive symptoms vary by gender. *Cognitive Therapy and Research*, *40*(5), 606–616. DOI: 10.1007/s10608-016-9782-0 PMID: 27695146

Pope, D. H., McMullen, H., Baschieri, A., Philipose, A., Udeh, C., Diallo, J., & McCoy, D. (2022). What is the current evidence for the relationship between the climate and environmental crises and child marriage? A scoping review. *Global Public Health: An International Journal for Research, Policy and Practice*, *18*(1). Advance online publication. DOI: 10.1080/17441692.2022.2095655 PMID: 36403290

Posklinsky, I., & Nouman, H. (2025). Community Building in Crisis: Interventions With Displaced Communities in Temporary Housing Shelters. *Journal of Community Psychology*, *53*(1), e23170. DOI: 10.1002/jcop.23170 PMID: 39711305

Prajapati, B., Dunne, M., & Armstrong, R. (2010, July 16). *Sample size estimation and statistical power analyses*. https://www.semanticscholar.org/paper/Sample-size-estimation-and-statistical-power-Prajapati-Dunne/8b940fa7c816608266e71832e021f4cd036ded76

Pritchard, A., Richardson, M., Sheffield, D., & McEwan, K. (2020). The relationship between nature connectedness and eudaimonic well-being: A meta-analysis. *Journal of Happiness Studies*, *21*(3), 1145–1167. DOI: 10.1007/s10902-019-00118-6

Purc-Stephenson, R., Doctor, J., & Keehn, J. E. (2023). Understanding the factors contributing to farmer suicide: A meta-synthesis of qualitative research. *Rural and Remote Health*, *23*(3), 8189. DOI: 10.22605/RRH8189 PMID: 37633833

Putnam, R. D. (2000). *Bowling alone: The collapse and revival of American community*. Simon and schuster.

Rahman, M., & Asaduzzaman, M. (2010). ECOLOGY OF SUNDARBAN, BANGLADESH. *J. Sci. Foundation*, *8*(1&2), 35–47. http://www.rhinoresourcecenter.com/pdf_files/137/1371726022.pdf

Rahman, S., & Alam, K. (2020). Women's leadership in community-based adaptation to climate change: A case study from Sub-Saharan Africa. *Climate and Development*, *12*(5), 462–473. DOI: 10.1080/17565529.2020.1754246

Ramaswamy, S., & Seshadri, S. (2020). Children on the brink: Risks for child protection, sexual abuse, and related mental health problems in the COVID-19 pandemic. *Indian Journal of Psychiatry*, *62*(Suppl 3), S404–S413. PMID: 33227060

Reifels, L. (2018). *Reducing Mental Health Risks in Emergencies: MHPSS through a Disaster Risk Reduction Lens*. MHPSS Network Webinar: MHPSS & DRR.

Reis, A. C., Saheb, R., Moyo, T., Smith, C., & Sperandei, S. (2022). The impact of mental health literacy training programs on the mental health literacy of university students: A systematic review. *Prevention Science*, *23*(4), 648–662. PMID: 34272642

Repper, J., & Carter, T. (2011). A review of the literature on peer support in mental health services. *Journal of Mental Health (Abingdon, England)*, *20*(4), 392–411. PMID: 21770786

Reuben, H. (1949). *Families under stress: Adjustment to the crises of war separation and reunion*. Harper.

Reuters. (2024, December 2). Comment: Women are custodians of nature and key to improving the integrity of carbon markets. *Reuters*. Retrieved from https://www.reuters.com/sustainability/land-use-biodiversity/comment-women-are-custodians-nature-key-improving-integrity-carbon-markets-2024-12-02/

Rhodes, J., Chan, C., Paxson, C., Rouse, C. E., Waters, M., & Fussell, E. (2010). The impact of hurricane Katrina on the mental and physical health of low-income parents in New Orleans. *The American Journal of Orthopsychiatry*, *80*(2), 237–247. DOI: 10.1111/j.1939-0025.2010.01027.x PMID: 20553517

Riethmuller, M. L., Dzidic, P. L., McEvoy, P. M., & Newnham, E. A. (2023). Change, connection and community: A qualitative exploration of farmers' mental health. *Journal of Rural Studies*, *97*, 591–600. DOI: 10.1016/j.jrurstud.2023.01.018

Riley, E., Smith, J., & Patel, K. (2017). The mental health of the Rohingya: A literature review. *Forced Migration Review*, *56*, 45–48.

Riza, E., Kalkman, S., Coritsidis, A., Koubardas, S., Vassiliu, S., Lazarou, D., Karnaki, P., Zota, D., Kantzanou, M., & Psaltopoulou, T. (2020). Community-based healthcare for migrants and refugees: a scoping literature review of best practices. Healthcare, Rosenlund, M., Kinnunen, U.-M., & Saranto, K. (2023). The use of digital health services among patients and citizens living at home: scoping review. *Journal of Medical Internet Research*, *25*, e44711.

Rizkalla, N., Mallat, N. K., Arafa, R., Adi, S., Soudi, L., & Segal, S. P. (2020). "Children are not children anymore; they are a lost generation": Adverse physical and mental health consequences on Syrian refugee children. *International Journal of Environmental Research and Public Health*, *17*(22), 8378. PMID: 33198333

Roberts, L., Evans, A., & Jones, S. (2021). The therapeutic role of group counseling for displaced populations. *Journal of Clinical Psychology*, *77*(3), 601–615.

Roitman, Y., & Yeshua-Katz, D. (2021). WhatsApp group as a shared resource for coping with political violence: The case of mothers living in an ongoing conflict area. *Mobile Media & Communication*, *10*(1), 3–20. DOI: 10.1177/20501579211021455

Röös, E., Fischer, K., Tidåker, P., & Nordström Källström, H. (2019). How well is farmers' social situation captured by sustainability assessment tools? A Swedish case study. *International Journal of Sustainable Development and World Ecology*, *26*(3), 268–281. DOI: 10.1080/13504509.2018.1560371

Rosenblatt, P. C., & Anderson, R. M. (1982). Interaction in farm families: Tension and stress 1. In *The Family In Rural Society*. Routledge.

Roy, A. (1999). *The greater common good*. Penguin.

Roy, P., Duplessis-Brochu, É., & Tremblay, G. (2019). Responses to adversity faced by farming men: A gender-transformative analysis. *International Journal of Child, Youth & Family Studies : IJCYFS*, *10*(1), 49–69. DOI: 10.18357/ijcyfs101201918806

Roy, P., Tremblay, G., Oliffe, J. L., Jbilou, J., & Robertson, S. (2013). Male farmers with mental health disorders: A scoping review. *The Australian Journal of Rural Health*, *21*(1), 3–7. DOI: 10.1111/ajr.12008 PMID: 23384130

Rubel, A. J., & Garro, L. C. (1992). Social and cultural factors in the successful control of tuberculosis. *Public Health Reports*, *107*(6), 626. PMID: 1454974

Rudolphi, J. M., Berg, R. L., & Parsaik, A. (2020). Depression, anxiety and stress among young farmers and ranchers: A pilot study. *Community Mental Health Journal*, *56*(1), 126–134. DOI: 10.1007/s10597-019-00480-y PMID: 31583619

Rural Mental Health Overview—Rural Health Information Hub. (n.d.). Retrieved March 24, 2023, from https://www.ruralhealthinfo.org/topics/mental-health

Ryan, B., Evans, A., & Jones, S. (2021). The role of community-based support in fostering resilience among stateless populations. *Journal of Human Rights and Social Work*, *6*(4), 312–325.

Saini, S., Siraj, H., & Khatri, P. (2022, June 30). *A Fresh Spike in Farmer Suicides in Punjab* |. https://www.theindiaforum.in/article/spike-farmer-suicides-punjab

Sanne, B. (2004). Farmers are at risk for anxiety and depression: The Hordaland health study. *Occupational Medicine*, *54*(2), 92–100. DOI: 10.1093/occmed/kqh007 PMID: 15020727

Sanne, B., Mykletun, A., Dahl, A. A., Moen, B. E., & Tell, G. S. (2003). Occupational differences in levels of anxiety and depression: The Hordaland Health Study. *Journal of Occupational and Environmental Medicine*, *45*(6), 628–638. DOI: 10.1097/01.jom.0000069239.06498.2f PMID: 12802216

Santos, E. G. de O., Queiroz, P. R., Nunes, A. D. da S., Vedana, K. G. G., & Barbosa, I. R. (2021). Factors associated with suicidal behavior in farmers: A systematic review. *International Journal of Environmental Research and Public Health*, *18*(12), 6522. DOI: 10.3390/ijerph18126522 PMID: 34204344

Sawyer, A., & Blitz, B. K. (2022). Economic exploitation and the mental health of stateless populations. *International Journal of Migration, Health and Social Care*, *18*(2), 121–135.

Schatz, B. R., & Berlin, R. B.Jr. (2011). *Healthcare infrastructure: Health systems for individuals and populations*. Springer Science & Business Media.

Schilling, J., Haras, D., & O'Brien, C. (2018). The long-term psychological effects of climate-induced displacement. *Climatic Change*, *149*(3), 369–382. PMID: 31258223

Schweitzer, R., Evans, A., & Jones, S. (2018). Community resilience in the face of forced migration. *Journal of Community Psychology*, *46*(6), 701–715.

Schwerdtle, P., Bowen, K., & McMichael, C. (2018). The health impacts of climate-related migration. *BMC Medicine*, *16*, 1–7. PMID: 29301536

Scudder, T. (2012). *The future of large dams: Dealing with social, environmental, institutional and political costs*. Routledge.

Seager, M., Sullivan, L., & barry, J. (2014). Gender-related schemas and suicidality: Validation of the male and female gender scripts questionnaire. *New Male Studies*, *3*(3), 32–54. https://discovery.ucl.ac.uk/id/eprint/1472580/9/Gender_related_pdf.pdf

Shonkoff, S., Morello-Frosch, R., Pastor, M., & Sadd, J. (2011). The climate gap: Environmental health and equity implications of climate change and mitigation policies in California—a review of the literature. *Climatic Change, 109*(S1), 485–503. DOI: 10.1007/s10584-011-0310-7

Shri Guru Granth Sahib. (2000Singh, G., Trans.). World Sikh Center Inc.

Siddiq, H., Elhaija, A., & Wells, K. (2023). An Integrative Review of Community-Based Mental Health Interventions Among Resettled Refugees from Muslim-Majority Countries. *Community Mental Health Journal, 59*(1), 160–174. DOI: 10.1007/s10597-022-00994-y PMID: 35751790

Silove, D., Ventevogel, P., & Rees, S. (2017). The contemporary refugee crisis: An overview of mental health challenges. *World Psychiatry; Official Journal of the World Psychiatric Association (WPA), 16*(2), 130–139. PMID: 28498581

Singh, I., Kumar, S., & Panchal, J. (2023). Employee Training and Development Enhancing Employee Performance-A Study. *ResearchGate. Retrieved October, 14*, 2024.

Siriwardhana, C., & Stewart, R. (2013). Forced migration and mental health: Prolonged internal displacement, return migration and resilience. *International Health, 5*(1), 19–23. DOI: 10.1093/inthealth/ihs014 PMID: 24029841

Smith, L. C., Kosec, K., & Frankenberger, T. (2021). *Food security and gender: A review of key issues.* International Food Policy Research Institute (IFPRI). https://doi.org/DOI: 10.2499/9780896292001

Snowden, B., Lahiri, S., Dutton, R., & Morton, L. (2023). Achieving and sustaining change through capacity building train-the-trainer health initiatives in low-and middle-income countries: A systematic review. *The Journal of Continuing Education in the Health Professions, 43*(2), 96–103. PMID: 36215155

Sorensen, C., Murray, V., Lemery, J., & Balbus, J. (2018). Climate change and women's health: Impacts and policy directions. *PLoS Medicine, 15*(7), e2002603. DOI: 10.1371/journal.pmed.2002603 PMID: 29990343

Staples, L. G., Dear, B. F., Nielssen, O., & Titov, N. (2025). Characteristics and Treatment Outcomes for Patients of a Digital Psychology Service in Regional and Remote Parts of Australia. *The Australian Journal of Rural Health, 33*(2), e70032. DOI: 10.1111/ajr.70032 PMID: 40110934

Steiner, A., Farmer, J., Kamstra, P., Carlisle, K., McCosker, A., & Kilpatrick, S. (2023). Online mental health forums and rural resilience: Mixed methods study and logic model. *JMIR Mental Health, 10*, e47459. DOI: 10.2196/47459 PMID: 37379080

Stephenson, J., Newman, K., & Mayhew, S. (2010). Population dynamics and climate change: What are the links? *Journal of Public Health, 32*(2), 150–156. DOI: 10.1093/pubmed/fdq038 PMID: 20501867

Stewart-Brown, S. (1998). Emotional wellbeing and its relation to health. *BMJ (Clinical Research Ed.), 317*(7173), 1608–1609. DOI: 10.1136/bmj.317.7173.1608 PMID: 9848897

Sturrock, S., & Hodes, M. (2016). Child labour in low-and middle-income countries and its consequences for mental health: A systematic literature review of epidemiologic studies. *European Child & Adolescent Psychiatry, 25*, 1273–1286. PMID: 27217156

Sue, D. W., Arredondo, P., & McDavis, R. J. (1992). Multicultural counseling competencies and standards: A call to the profession. *Journal of Counseling and Development, 70*(4), 477–486.

Sukiasyan, S. (2024). The Mental Health of Refugees and Forcibly Displaced People: A Narrative Review. *Consortium Psychiatricum, 5*(4), 78–92. DOI: 10.17816/cp15552 PMID: 39980620

Sultana, F. & Syracuse University. (2014). Gendering climate Change: geographical insights. In *The Professional Geographer* (Vols. 66–66, Issue 3, pp. 372–381). https://doi.org/DOI: 10.1080/00330124.2013.821730

Syme, S. L., Hyman, M. M., & Enterline, P. E. (1964). Some social and cultural factors associated with the occurrence of coronary heart disease. *Journal of Chronic Diseases, 17*(3), 277–289. PMID: 5878595

Szreter, S., & Woolcock, M. (2004). Health by association? Social capital, social theory, and the political economy of public health. *International Journal of Epidemiology, 33*(4), 650–667. PMID: 15282219

Tahernejad, A., Sohrabizadeh, S., & Mashhadi, A. (2024). Exploring factors affecting psychological resilience of farmers living in drought-affected regions in Iran: A qualitative study. *Frontiers in Psychology, 15*, 1418361. DOI: 10.3389/fpsyg.2024.1418361 PMID: 39286558

Taki, F., Witte, M., Jaijee, I. S., Jessel, S., Mauer, E., Lurie, J., & Kaur, G. (2021). Suicide in rural Punjab, India: Implications for ensuring farmer rights. *Journal of Global Health Reports, 5*, e2021037. DOI: 10.29392/001c.22236

Tan, E. Y., Wee, R. R., Ern, S. Y., Heng, K. J., Chin, J. W., Tong, E. M., & Liu, J. C. (2020). Tracking private WhatsApp discourse about COVID-19: A longitudinal infodemiology study in Singapore. *medRxiv*. https://doi.org/DOI: 10.1101/2020.09.29.20203646

Theodosopoulos, L., Fradelos, E. C., Panagiotou, A., Dreliozi, A., & Tzavella, F. (2024). Delivering Culturally Competent Care to Migrants by Healthcare Personnel: A Crucial Aspect of Delivering Culturally Sensitive Care. *Social Sciences*, *13*(10), 530. https://www.mdpi.com/2076-0760/13/10/530

Think Global Health. (2025). *Mental health after India's glacial floods*. https://www.thinkglobalhealth.org/article/mental-health-after-indias-glacial-floods

Thompson, R., Hagen, B. N. M., Lumley, M. N., Winder, C. B., Gohar, B., & Jones-Bitton, A. (2023). "An incredible amount of stress before you even put a shovel in the ground": A mixed methods analysis of farming stressors in Canada. *Sustainability*, *15*(8), 6336. DOI: 10.3390/su15086336

Thornton, P. H., Ribeiro-Soriano, D., & Urbano, D. (2011). Socio-cultural factors and entrepreneurial activity: An overview. *International Small Business Journal*, *29*(2), 105–118.

Tirado, M. C., Hunnes, D., Cohen, M. J., & Lartey, A. (2019). Climate change and nutrition in Africa. *Annals of Nutrition & Metabolism*, *75*(2), 137–145. DOI: 10.1159/000503669

Titcomb, G., Uelmen, J., Janko, M., & Nunn, C. (2024). Infectious disease responses to human climate change adaptations. *Global Change Biology*, *30*(8), e17433. PMID: 39171421

Tol, W. A., Barbui, C., Galappatti, A., Silove, D., Betancourt, T. S., Souza, R., Golaz, A., & Van Ommeren, M. (2011). Mental health and psychosocial support in humanitarian settings: Linking practice and research. *Lancet*, *378*(9802), 1581–1591. PMID: 22008428

Tol, W. A., Hall, B. J., & van Ommeren, M. (2020). Mental health and psychosocial support in humanitarian settings: A review. *Current Opinion in Psychiatry*, *33*(4), 362–370.

Torske, M. O., Hilt, B., Glasscock, D., Lundqvist, P., & Krokstad, S. (2016). Anxiety and depression symptoms among farmers: The HUNT study, Norway. *Journal of Agromedicine*, *21*(1), 24–33. DOI: 10.1080/1059924X.2015.1106375 PMID: 26488439

Trautmann, S., Rehm, J., & Wittchen, H. U. (2016). The economic costs of mental disorders: Do our societies react appropriately to the burden of mental disorders? *EMBO Reports*, *17*(9), 1245–1260. PMID: 27491723

Troup, J., Fuhr, D. C., Woodward, A., Sondorp, E., & Roberts, B. (2021). Barriers and facilitators for scaling up mental health and psychosocial support interventions in low- and middle-income countries for populations affected by humanitarian crises: A systematic review. *International Journal of Mental Health Systems*, *15*(1), 5. DOI: 10.1186/s13033-020-00431-1 PMID: 33413526

Truchot, D., & Andela, M. (2018). Burnout and hopelessness among farmers: The farmers stressors Inventory. *Social Psychiatry and Psychiatric Epidemiology*, *53*(8), 859–867. DOI: 10.1007/s00127-018-1528-8 PMID: 29725701

Tschakert, P., Ellis, N., Anderson, C., Kelly, A., Obeng, J., & Porter, R. (2019). Climate change and loss, grief, and trauma. *Climatic Change*, *152*(1), 75–89.

Uddin, Md. S., Shah, M. A. R., Khanom, S., Nesha, Mst. K. (2015). Climate change impacts on the Sundarbans mangrove ecosystem services and dependent livelihoods in Bangladesh. In the *XIV WORLD FORESTRY CONGRESS*. AJCB-Vol2-No2-Shams-Uddin et al.pdf

UN General Assembly. (1966b). *International Covenant on Economic*. Social and Cultural Rights.

UN. U. N. (2023). *Review of mental health and well-being policies and practices in United Nations system organizations*. https://docs.un.org/en/JIU/REP/2023/4

UNFPA (United Nations Population Fund). (2019). *The impact of climate change on maternal and newborn health*. UNFPA Reports. https://www.unfpa.org/publications/climate-change-and-maternal-health

Ungar, M. (2006). Resilience across Cultures. *British Journal of Social Work*, *38*(2), 218–235. DOI: 10.1093/bjsw/bcl343

UNHCR. (2014). *A global action plan to end statelessness*. United Nations High Commissioner for Refugees.

UNHCR. (2018). *Global report 2018: Climate change and displacement*. United Nations High Commissioner for Refugees.

UNHCR. (2022). *Climate change and disaster displacement*. United Nations High Commissioner for Refugees.

United Nations Foundation. (n.d.). *Five facts about gender equality and climate change*. Retrieved from https://unfoundation.org/blog/post/five-facts-about-gender-equality-and-climate-change/

United Nations. (2023). *What is climate change?* United Nations. Retrieved July 1, 2025, from https://www.un.org/en/climatechange/what-is-climate-change

United Nations. (n.d.). *Women & climate change*. Retrieved from https://www.un.org/en/climatechange/science/climate-issues/women

Uscher-Pines, L. (2009). The mental health needs of climate-displaced populations. *Journal of Climate and Health*, *2*(3), 1–10.

USCRI. (2024). U. S. [*Mental Health Awareness Month: Barriers and Access to Mental Health Care*. https://refugees.org/mental-health-awareness-month-barriers-and-access-to-mental-health-care/]. *CR (East Lansing, Mich.)*, *I*, •••.

Utterback, J. M. (1971). The process of technological innovation within the firm. *Academy of Management Journal*, *14*(1), 75–88.

Van Daalen, K. R., Jung, L., Dhatt, R., & Phelan, A. L. (2020). Climate change and gender-based health disparities. *The Lancet. Planetary Health*, *4*(2), e44–e45. DOI: 10.1016/S2542-5196(20)30001-4 PMID: 32112742

van Ommeren, M., Hameed, W., & Qureshi, M. (2018). The role of mobile mental health clinics in humanitarian settings. *World Psychiatry; Official Journal of the World Psychiatric Association (WPA)*, *17*(1), 58–60.

van Sint Fiet, A., de la Rie, S., van der Aa, N., Bloemen, E., & Wind, T. (2022). The relevance of social capital and sense of coherence for mental health of refugees. *SSM - Population Health*, *20*, 101267. PMID: 36281249

van Waas, L. (2017). The psychological effects of statelessness: A review of the literature. *International Journal of Human Rights*, *21*(5), 650–666.

van Waas, L., & Khanna, A. (2021). Economic insecurity and the mental health of stateless populations. *Forced Migration Review*, *68*, 28–31.

Vergunst, F., & Berry, H. L. (2021). Climate Change and Children's Mental Health: A Developmental Perspective. *Clinical Psychological Science*, •••, 21677026211040788. PMID: 35846172

Versey, H. S., Murad, S., McPhee, I., & Schwarz, W. (2023). A Bridging-Community (ABC) Project: A Community Building Social Participation Intervention Among Resettled Refugees in Boston. *Journal of International Migration and Integration*, *24*(1), 441–465. DOI: 10.1007/s12134-021-00908-z

Villalonga-Olives, E., Wind, T., Armand, A., Yirefu, M., Smith, R., & Aldrich, D. P. (2022). Social-capital-based mental health interventions for refugees: A systematic review. *Villalonga-Olives, E., Wind, TR, Armand, A., Yirefu, M., Smith, R., and Aldrich, DP (2022). Social capital based mental health interventions for refugees: A systematic review.* [Record #43 is using a reference type undefined in this output style.]. *Social Science & Medicine*.

Vince, G. (2022). How scientists are coping with 'ecological grief'. *The Observer*. https://www.theguardian.com

Vu, H. T., Liu, Y., & Tran, D. V. (2019). Nationalizing a global phenomenon: A study of how the press in 45 countries and territories portrays climate change. *Global Environmental Change*, *58*, 101942. DOI: 10.1016/j.gloenvcha.2019.101942 PMID: 32288343

Waheed, S., & Waheed, H. (2022). Impact of climate finance on gender equity for sustainable global development: Can aid for climate action also aid gender equity? *Brazilian Journal of Science*, *1*(7), 82–94. DOI: 10.14295/bjs.v1i7.145

Walker, M. (2005). Amartya Sen's capability approach and education. *Educational action research, 13*(1), 103-110. Sci-Hub | Amartya Sen's capability approach and education. Educational Action Research, 13(1), 103–110 | DOI: 10.1080/09650790500200279

Watts, N., Adger, W. N., & O'Brien, K. (2015). The role of public health in addressing the psychological effects of climate change. *Lancet*, *386*(10008), 2101–2102. PMID: 26713340

Wearing, B. M. (1994). The pain and pleasure of gendered leisure. *World Leisure & Recreation*, *36*(3), 4–10. DOI: 10.1080/10261133.1994.9673919

Weber, J. G. (2011). *Individual and Family Stress and Crises* (1st ed.). SAGE., https://us.sagepub.com/en-us/nam/individual-and-family-stress-and-crises/book228977

Webinar Report on Impact of the Pandemic on Sundarbans. (2021). In *Webinar Report* [Report]. https://www.bing.com/search?pglt=299&q=Webinar+Report+on+Impact+of+Pandemic+on+Sundarbans.+(2021).+In+Webinar+Report+%5BReport%5D.&cvid=681922daec1b4f73803bb96197878c0d&gs_lcrp=EgRlZGdlKgYIABBFGDsyBggAEEUYO9IBBzY3OGowajGoAgCwAgA&FORM=ANNTA1&PC=U531

Weigel, D. J., & Weigel, R. R. (1990). Family satisfaction in two-generation farm families: The role of stress and resources. *Family Relations*, *39*(4), 449–455. DOI: 10.2307/585227

Weintrobe, S. (2021). *Psychological roots of the climate crisis: neoliberal exceptionalism and the culture of uncare*. Bloomsbury Publishing.

Weiss, H., & Bradley, R. S. (2001). What drives human migration? A new approach. *Climatic Change*, *49*(1-2), 1–18.

Werz, C., & Hoffman, M. (2018). Economic opportunities and resilience for climate migrants. *Journal of International Development*, *30*(2), 221–235.

White, S. (2014). The Little Ice Age in Europe. *The Journal of Interdisciplinary History*, *45*(1), 101–121.

WHO. (2011). *Psychological first aid: A guide for fieldworkers*. World Health Organization.

WHO. (2018). *Mental health and psychosocial support in humanitarian settings*. World Health Organization.

WHO. (2021). *Mental disorders*. https://www.who.int

WHO. (2021). *Mental health and forced displacement*. https://www.who.int/news-room/fact-sheets/detail/mental-health-and-forced-displacement

WHO. (2021). *Mental health: strengthening our response*. https://www.who.int

WHO. (World Health Organization). (2021). *Gender, climate change and health*. WHO Reports. https://www.who.int/publications/i/item/9789240026389

Wojewódzka-Wiewiórska, A., Kłoczko-Gajewska, A., & Sulewski, P. (2020). Between the social and economic dimensions of sustainability in rural areas—In search of farmers' quality of life. *Sustainability*, *12*(1), 1. Advance online publication. DOI: 10.3390/su12010148

Women, U. N. (2022). *Gender-responsive climate action: A call for gender-disaggregated data*. UN Women Publications. https://www.unwomen.org/en/digital-library/publications/2022/07/gender-responsive-climate-action

Wong, S. (2016). Can climate finance contribute to gender equity in developing countries? *Journal of International Development*, *28*(3), 428–444. DOI: 10.1002/jid.3212

World Health Organization (WHO). (2005). *Promoting mental health: concepts, emerging evidence, practice: a report of the World Health Organization*. Department of Mental Health and Substance Abuse in collaboration with the Victorian Health Promotion Foundation and the University of Melbourne. World Health Organization.

World Health Organization. (2004). *The World Health Organization quality of life (WHOQOL)—BREF* (No. WHO/HIS/HSI Rev.2012.02). World Health Organization. https://apps.who.int/iris/handle/10665/77773

World Health Organization. (2021). *Global strategy on digital health 2020–2025.* https://www.who.int/publications/i/item/9789240020924

World Health Organization. (2022). *The top 10 causes of death.* Retrieved from https://www.who.int/news-room/fact-sheets/detail/the-top-10-causes-of-death

World Health Organization. (2022). WHO digital mental health intervention effective in reducing depression among Syrian refugees in Lebanon. https://www.who.int/news-room/feature-stories/detail/who-digital-mental-health-intervention-effective-in-reducing-depression-among-syrian-refugees-in-lebanon?

World Health Organization. (2023). Climate change and non-communicable diseases: Connections. World Health Organization. https://www.who.int/news/item/02-11-2023-climate-change-and-noncommunicable-diseases-connections

World Meteorological Organization. (2024, December 30). Climate change impacts grip globe in 2024. *WMO.* Retrieved from https://wmo.int/media/news/climate-change-impacts-grip-globe-2024

Zeira, A. (2022). Mental Health Challenges Related to Neoliberal Capitalism in the United States. *Community Mental Health Journal, 58*(2), 205–212. DOI: 10.1007/s10597-021-00840-7 PMID: 34032963

Zetter, R. (2017). *The international refugee regime: A human rights perspective.* Oxford University Press.

Zigmond, A. S., & Snaith, R. P. (1983). The hospital anxiety and depression scale. *Acta Psychiatrica Scandinavica, 67*(6), 361–370. DOI: 10.1111/j.1600-0447.1983.tb09716.x PMID: 6880820

About the Contributors

Jeffrey Kurebwa is a Senior Lecturer and Chairperson in the Department of Peace and Governance at Bindura University of Science Education, Bindura, Zimbabwe. He holds a PhD in Public Administration from Nelson Mandela University, South Africa; a Master of Public Administration, and a Bachelor of Science Honours in Administration Degree from the University of Zimbabwe. Jeffrey Kurebwa teaches undergraduate, master, and Doctoral level courses in Gender, Peace and Security, Local Governance, Research Methods, and Public Policy and Management. Jeffrey has over 90 publications including books, book chapters, and peer-reviewed journal articles. Jeffrey is editing a book on Mental Health Challenges in Academia: Stressors Faced by Students and Faculty to be published by IGI-GLOBAL Publishers (2026). Kurebwa has edited three books on Managing the Health Risks of Climate Change (2025), Understanding Gender in the African Context (2020), and Participation of Young People in Governance Processes in Africa (2019) published by IGI-GLOBAL Publishers. Dr. Kurebwa has mentored and guided a cohort of undergraduate and graduate students in their research, fostering professional development. He has examined dissertations and theses for Africa University, Women's University in Africa, Zimbabwe Open University, Bindura University of Science Education, Africa Leadership Management Academy (ALMA), (Zimbabwe), Durban University of Technology, University of Kwazulu Natal, University of the Free State, and Cape Peninsula University of Technology (South Africa), and University of Namibia. Jeffrey is a part-time Lecturer at ALMA an affiliate of the National University of Science and Technology (NUST) Graduate Business School (Zimbabwe) where he teaches Management Theory and Practice, and Strategic Management. He was a part-time Lecturer for the Peace Building and Post Conflict Recovery course for the PhD programme in the College of Business, Peace, Leadership and Governance at Africa University, Zimbabwe (2021). Dr Kurebwa is a Programme Peer Reviewer for the Zimbabwe Council for Higher Education (ZIMCHE). Kurebwa has peer-reviewed programmes for Women's

University in Africa, Zimbabwe Open University, University of Zimbabwe, Arrupe Jesuit University, Great Zimbabwe University, and Midlands State University. Dr Kurebwa has done consultancy work for UN Women, the National AIDS Council (NAC), and the National Peace and Reconciliation Commission (NPRC). Jeffrey is a Research Fellow with the Institute for Research and Policy Integration in Africa (IRPIA). He completed the Climate Governance, Diplomacy and Negotiations Leadership Program facilitated by The African Group of Negotiators Experts Support (AGNES) held from 06 May 2024 to 12 July 2024. Kurebwa is an Advisory Board member for Shamva Agricultural College. He has designed and implemented an innovative curriculum for Monitoring, Evaluation Accountability and Learning (MEAL), Projects Monitoring and Evaluation; Project Planning and Management; and Gender and Transformative Studies fostering a forward-thinking learning environment.

***

Bimal Anjum is working as a Faculty of Commerce and Management since 1997, presently with DAV College Chandigarh. Supervised 14 Ph.D Research Schools, 15 M.Phil, more than 200 PG Research projects. More than 90 research papers in journals of repute.

Ayushi Bhatt is currently pursuing an MBA at Graphic Era Deemed to be University, Dehradun with a focus on Specialization in Marketing. Passionate about the world of business and management, and actively involved in academic projects, practical training, and industry exposure to build a well-rounded skill set. With strong communication abilities and an analytical approach, Bhatt wants to build a successful career in the corporate sector, contributing to innovative and result-driven solutions.

Debapriya Ganguly is an Assistant Professor of English in the Department of Humanities and Social Sciences at Graphic Era University, Dehradun, India. She earned her doctorate from the Indian Institute of Technology, Dhanbad, India. Ganguly's research interests span diverse topics, including masculinity and men's studies, Indian cinema, gerontology, and health humanities. She has previously published in South Asian Popular Culture, Intersections (Gender and Sexuality in Asia and the Pacific), Humanities and Social Sciences Communications, and Vernon Press.

Nishant Goyal is currently pursuing an MBA at Graphic Era Deemed to be University, Dehradun with a focus on specialization in Marketing. Passionate about the world of business and management, and actively involved in academic projects, practical training, and industry exposure to build a well-rounded skill set. With strong communication abilities and an analytical approach, Goyal wants to build a successful career in the corporate sector, contributing to innovative and result-driven solutions.

Mubashir Gull is working as an Assistant Professor in the Department of Applied Psychology, GITAM School of Humanities and Social Sciences. Prior to this, he worked as an Assistant Professor at the Department of Psychology, Akal University, Bathinda, Punjab. He did his M. Phil and Ph.D. from the Aligarh Muslim University, Aligarh, and has been awarded a National Fellowship from the Indian Council of Social Science Research (ICSSR). Related to his topic of expertise, he has successfully presented and published papers in National and International Journals. He also holds a PG Diploma in Rehabilitation Psychology and has been registered as a Rehabilitation Psychologist by the Rehabilitation Council of India (A50125). His research interests are Psychometrics, Research Methodology, Psychological Assessment, and Rehabilitation Psychology.

Navneet Kaur completed her B.A. (Honours) in Psychology and a Postgraduate Certificate in Autism and Behavioural Science. She is currently enrolled in the Postgraduate Certificate program in Mental Health and Substance Use at Conestoga College, Doon Valley Drive, Canada. She is also working part-time as an ABA Instructor Therapist at A Balanced Approach, Waterloo.

Mumba Thaddeus Mahela holds a PhD in Public Administration from the University of Namibia. He is the Associate Dean for the School of Military Science at the University of Namibia.

|Aishwarya Majumdar - Contributing Author|Aishwarya, a distinguished law scholar, is currently pursuing her PhD in law from Galgotias University. She holds both an LL.B and an LL.M from Sharda University. Additionally, she has contributed extensively to academic literature, having published numerous papers and chapters in various books.

Chandni Rani completed her B.A. (Hons.) and M.A. in Psychology from Panjab University, Chandigarh, and earned her Ph.D. from Punjabi University, Patiala. She has contributed to psychological research through various national and international publications and presentations. She has served as an Assistant Professor at Govt. Mohindra College, Patiala, Akal University, Bathinda, and is currently working at Amity University Punjab, Mohali.

Apoorva Roy is a distinguished academician with teaching and research experience of around eight years. She is currently working as an Assistant Professor of Law at the Galgotias University, India. Law concerning Environment and Climate Change, Intellectual Property Rights and Related aspects of International Law are some of the areas of her forte.

Sonali Sejwal is an MBA candidate in Sustainability Management at the TERI School of Advanced Studies, New Delhi. She earned a bachelor's degree in Psychology (Hons.) from O.P. Jindal Global University. Her undergraduate thesis, titled "Understanding the Psychosocial Impact of Climate Change and Its Correlates among College-going Young Indian Adults," explores the intersection of climate and mental health. Her current focus lies in exploring how technological innovation can be harnessed to generate positive outcomes for both the environment and society, guided by the frameworks of Environmental, Social, and Governance (ESG).

Vaidehi Uniyal is pursuing a PhD (English) in the Department of Social Science, Graphic Era Deemed to be University, Dehradun, Uttarakhand. Her areas of interest encompass feminism, gender studies, mythology, and literary analysis, with a focus on exploring cultural narratives. Currently, she is working on research projects that explore the interconnection of sexual slavery and religious institutions, sustainable feminine health practices, and gender-related policy reform within the context of the global south

Index

C

D

E

F

G

H

I

L

M

P

R

S

V

W

www.ingramcontent.com/pod-product-compliance
Lightning Source LLC
LaVergne TN
LVHW060003031026
849161LV00025B/1340
9798337332819